NASA AND THE AMERICAN SOUTH

NASA AND
THE AMERICAN SOUTH

EDITED BY
Brian C. Odom
and Stephen P. Waring

UNIVERSITY OF FLORIDA PRESS

Gainesville

Published by the University of Florida Press, 2024

29 28 27 26 25 24   6 5 4 3 2 1

Library of Congress Cataloging-in-Publication Data
Names: Odom, Brian C., editor. | Waring, Stephen P., editor.
Title: NASA and the American South / edited by Brian C. Odom and Stephen P. Waring.
Other titles: NASA & the American South
Description: 1st. | Gainesville : University of Florida Press, 2023. | Includes bibliographical references
and index. | Summary: "This volume examines NASA's strong ties to the American South, explor-
ing how the space program and the region have influenced each other since NASA's founding in
1958"—Provided by publisher.
Identifiers: LCCN 2023032345 (print) | LCCN 2023032346 (ebook) | ISBN 9781683404019 (cloth) |
ISBN 9781683404095 (pdf) | ISBN 9781683404255 (epub)
Subjects: LCSH: United States. National Aeronautics and Space Administration—History. | John F.
Kennedy Space Center—History. | Southern States—History. | Southern States—Social life and
customs. | Florida—History. | BISAC: TECHNOLOGY & ENGINEERING / Aeronautics & Astro-
nautics | HISTORY / United States / State & Local / South (AL, AR, FL, GA, KY, LA, MS, NC, SC,
TN, VA, WV)
Classification: LCC TL521.312 .N3618 2023 (print) | LCC TL521.312 (ebook) | DDC 629.40973—
dc23/eng/20230808
LC record available at https://lccn.loc.gov/2023032345
LC ebook record available at https://lccn.loc.gov/2023032346

University of Florida Press
2046 NE Waldo Road
UF PRESS   Suite 2100
Gainesville, FL 32609
UNIVERSITY
OF FLORIDA   http://upress.ufl.edu

CONTENTS

FIGURES

FOREWORD

Douglas Brinkley

John Kennedy and Lyndon Johnson didn't start out as space enthusiasts. As career politicians, their business was promoting practical solutions to earth-bound challenges. But the men who would come to be known as America's space presidents were quick to recognize the opportunities of an American Space Age. While others dreamed of the stars, they imagined the changes that investments in space technology could deliver at warp speed, driving America toward a more modern, forward-thinking, merit-based, and prosperous future. In particular, the space industry could be used to remake the American South into everything it was not at the dawn of the 1960s. When JFK toured Cape Canaveral on November 16, 1963, he was stunned to see what a gargantuan, well-funded, centralized federal project could achieve in breakneck time.

In the eyes of the rest of the nation, the southern states had stagnated in every way. Economically, agriculture was by far their leading industry, dominated by a system not far from the plantation model, with subsistence share-croppers working the fields of large landowners. Other southern industries, including textiles and steelmaking (centered in Alabama), changed very little in at least half a century. This lack of economic progress stymied social progress, locking the South into a cycle of prejudice, segregation, and racial abuse that was unlikely to end without major, systemic changes across the region.

This is the sociopolitical moment Brian Odom and Stephen Waring began exploring in their 2019 award-winning book *NASA and the Long Civil Rights Movement*, which illuminated the crucial relationship between space-race spending and improved conditions for African Americans in the South. This second volume continues tracing the subtleties of that connection while expanding the discussion to NASA's effects across the breadth of southern life and culture—and the resentments that sometimes followed.

When Alan Shepard and John Glenn undertook NASA's earliest manned flights in 1961 and 1962, respectively, space research and preparation were generally mid-Atlantic endeavors. Although Shepard and Glenn launched from Cape Canaveral Air Force Station on Florida's Atlantic coast, their training had taken place at facilities spread across Ohio, Virginia, and Pennsylvania and Mission Control was in Maryland. But in the wake of Kennedy's challenge to put a man on the moon by the end of the decade and the vast expansion of NASA's budget that followed, nearly every state in the union vied to become home to the agency's core facilities and the research and manufacturing operations that supported them.

For some time, Massachusetts was considered a front-runner for NASA's launch site; supporters claimed that the commonwealth's many outstanding universities would provide a ready pool of scientists for the space program. Boosters of this plan assumed they'd get support from the state's most famous native son, President Kennedy, but JFK boldly reversed their logic, envisioning instead that the space program could be the draw that attracted scientists elsewhere, including to the South.

Along with those transplants, space facilities in the South could also spur the growth of homegrown scientific research institutions. Alabama Polytechnic, for example, was ready for new challenges. In 1960, it changed its name to Auburn University to reflect a shift in emphasis from agricultural sciences and engineering to a broader array of high-tech subjects, including aerospace. A smattering of similarly respected colleges across the South were also primed for long-overdue transformation, spurred by the push and pull of NASA requirements. NASA's influence even reached into some public schools located near its facilities.

Cape Canaveral ultimately won the competition to be NASA's permanent launch site, both due to Washington's ambitions to kickstart southern advancement and to Florida's sunny, launch-friendly weather. The latter was commonly believed to be NASA's principal reason for focusing on the South, although facilities for engineering, manufacturing, assembly, and mission control could really have been sited anywhere. But Vice President Lyndon Johnson, in his role as the chair of President Kennedy's Space Committee, was as determined as his boss to use NASA's budget to advantage the rest of the deep South. He steered major expansions of rocket development in Alabama, rocket construction in Mississippi, and spacecraft assembly in Louisiana. Johnson's home state of Texas was chosen as the site for Mission Control, completing what was known as the "Space Crescent" across states that line the Gulf of Mexico, and nearby Missouri was chosen for manufacture of the lunar module. The transformation of the crescent states into the center of the US aerospace industry happened almost overnight, and NASA announced plans to hire 200,000 people across its southern sites.

For Kennedy and then Johnson, it was critically important that the recruitment process benefit a broad range of Americans, especially African Americans. As *NASA and the American South* describes, however, that goal proved difficult to achieve in a region where mechanisms of exclusion were both complex and deeply entrenched. Many in the Kennedy and Johnson administrations were disappointed at the relatively small number of African Americans the program eventually recruited, but to observers in the South in the mid-1960s, even those modest successes were startling.

Modern scholars, including Odom and Waring, are revealing how the federal government laid the groundwork and effectively paid for the social changes it was demanding through NASA investment. That strategy had its contemporary critics, who complained that the "moondoggle" was sending inordinate pork to southern states and scorned southern abilities whenever there was a setback or worse, a tragedy. But in the hands of President Johnson—a master of the backroom deal—that cash infusion did its work, among other things by helping soften southern resistance to civil rights legislation in Congress.

As *NASA and the American South* ably illuminates, the new tech colonies the federal government seeded across the South quickly worked as intended, bolstering local communities in spheres from education and public engineering to civics and tourism. New winds of prosperity and sophistication not only lifted those communities but also forced the rest of the country to reappraise their southern neighbors. In one unforeseen effect, several of the facilities in the Space Crescent quickly became tourist attractions. Space City in Cocoa Beach, Florida; Rocket City in Huntsville, Alabama; and the Space Center in Houston all drew unprecedented numbers of middle Americans, who arrived expecting a ragged rural landscape and boll weevils but found something entirely new instead. True, within the region there were some who resented the implication that the South needed to become more like the North or the West and risked losing its culture and history in the pursuit of prosperity. Those critics may have thought NASA's influence was changing what was best about the South, but in truth it was only changing the attributes that were holding it back. Besides working to land Apollo astronauts on the moon, NASA engineers and computer technicians were laying the foundation for solar panels, carbon monoxide detectors, satellite television, GPS, and microchips.

For nearly six decades, few Americans have recognized the sweeping domestic vision that drove Kennedy's and Johnson's aerospace plans for the South—plans not unlike those that transformed the entire country during

World War II: massive construction, a wave of hiring and upskilling, people on the move, and advanced research, all on an accelerated schedule. Without a hot war to spur change, Kennedy used the peaceful exploration of space to help win the Cold War against the Soviet Union in a noncombative way. NASA's massive budget gave Washington leverage to address entrenched problems in economics, race relations, law, and culture. In this book, a wide range of writers manage to paint a uniformly vivid picture of not only the overarching Kennedy-Johnson scheme but also the triumphs and the frustrations that resulted as the New South began to emerge.

Introduction

BRIAN C. ODOM

On April 3, 1963, D. Brainerd Holmes, NASA's director of manned space-flight, spoke to the Yale University chapter of Sigma Xi. During his speech, Holmes recounted how American history was a story of progress—the story of a country built on a vast network of internal improvements that included canals, railroads, bridges, and a nationwide grid of electrical power. Holmes noted that over the previous centuries, the country had "developed a system of industry, agriculture, and distribution" that provided the American people with all the "necessities and conveniences" and placed the country in an "unprecedented position of wealth and power in the world." Holmes argued that with the closing of the American frontier, only the exploration of space could provide a fresh challenge and unlock a new "great national purpose."[1] Speaking to the privileged few in New Haven, Connecticut, Holmes's topic was the theme of technological advancement and an ideal of progress capable of continually reshaping American society.

Nowhere would Holmes's vision be tested more than in the southern United States. The transition from the benighted South to the Sunbelt took decades of economic investment in improving infrastructure, military build-up, research and development, and the Cold War space program. At its peak in 1965, NASA funding constituted an overall investment of $25 billion (in 1960s dollars), or just over 4 percent of the federal budget. The decision to construct most of NASA's major new facilities in the South—including those in Alabama, Florida, Louisiana, Mississippi, and Texas—represented a massive investment worth more than $2.5 billion in fiscal years 1962 and 1963 alone. Beyond the initial outlay, the presence of those expansive, federally funded technology development centers exerted an outsized influence on the society, economics, and politics of the region, as they continue to do. Ultimately, these investments created ripple effects in the communities, towns, and cities in the space crescent that continue to reverberate today.

Many contemporary observers in the age of Apollo noted the impact of aerospace funding on regional economies. Early on, scholars struggled to convey the potential long-term impacts of the aerospace industry on local communities. In 1964, Massachusetts Institute of Technology graduate student Thomas Morring conducted a study of the impact of space spending on Huntsville, Alabama. Morring recognized the obvious benefits, but he questioned whether the city's overreliance on federal funding for aerospace jobs and lack of economic diversification was sustainable.[2] This question of economic sustainability has long plagued the communities surrounding Kennedy Space Center, Marshall Space Flight Center, Stennis Space Center in particular. The federal government's initial investments certainly transformed local economies, and an influx of supporting aerospace infrastructure followed that initial funding. But how could local elites work to solidify those gains through economic diversification? Clearly, the issue was not limited to white elites, as both Black and white working-class residents in these locations increasingly gravitated to new jobs in the aerospace sector and support services for that sector. Overreliance on federal funding for the aerospace industry remains foundational to economic planning in these communities.

Another important subject has been the region's influence on the development of aerospace technologies and how the presence of that technology development has impacted the region. In 1968, historian Loyd S. Swenson Jr. argued that the South deserved much of the credit for developing the technology for Project Apollo. Swenson pointed to the predominance of southerners in NASA leadership positions, including NASA administrator James Webb, spokesperson Julian Scheer, and a host of others trained at Langley Research Center in Virginia. Like many others, Swenson hoped that the "reaction engines for space" might compel the region to "leave behind reactionary thought."[3] Historian Bruce Schulman has pointed out that many southern leaders, for whom the memory of New Deal investment was fresh, understood aerospace investment as a "conduit to a new 'New South' of science and technology-based enterprise."[4] Likewise, historian Walter McDougall argues that many, including President Lyndon Johnson, James Webb, and several members of Congress, viewed the "political distributivism not as palm-greasing but as the economic and intellectual component of the Second Reconstruction" of the South. McDougall insists that although Department of Defense funding dwarfed the government's overall investment in aerospace, the "post-1961 NASA construction went beyond the pork barrel into the realm of social planning. Technological infusion was to call to life a New South, and the space program thus addressed several large items on the national agenda all at once."[5]

Other contemporary observers argued that the benefits of the investment were not equitably distributed. A 1967 assessment by New York University graduate student Robert A. Myers pointed out that the continually decreasing Black population in the city had been "completely bypassed by Huntsville's rapid economic growth" and had "not participated equally with the White community in reaping the benefits" of the city's expanding economy.[6] The more recent historiography of the civil rights movement reflects the economic impact of federal funding in the South and the transition of the southern economy from one dominated by agriculture to one characterized by increasing economic diversity. Gavin Wright has explored the role of economics in the aftermath of the Civil Rights Act of 1964, arguing that the legislation impelled southern business to "take actions serving its own best economic interests."[7] Bobby Wilson contends that the new "entrepreneurial regimes" in the South showed a greater willingness than the planter elites to "sacrifice old racial beliefs and practices to maintain and attract new investments." For Wilson, it was this willingness that provided advocates of the civil rights movement with important leverage in their battle against continued segregation.[8] The connection to civil rights activism in the southern aerospace communities was just one of many points of impact.

While our previous volume, *NASA and the Long Civil Rights Movement*, explored the intersection of race, class, and gender in the space program, *NASA and the American South* adds additional layers of context and previously overlooked perspectives, building on and moving beyond those significant themes.[9] Aside from the struggle for race and gender equality in aerospace, what other areas of society, politics, and economics in the South were impacted by federal investment in the space race? This volume offers additional context about the influence of aerospace on the region. The aim of this volume is to scrutinize the economic, social, and political impact of aerospace investment on the South since 1958, with a particular focus on the critical early decade plus of this process, roughly from 1960 to 1975. In the following pages, authors address overarching frameworks, assessments of aerospace tourism in the region, and investigations of the influence of aerospace on regional art and architecture, antecedent technologies, and Black institutions of higher education. We have attempted to be as inclusive as possible in terms of both contexts and perspectives. A principal objective was to engage scholars from diverse backgrounds and at various levels—from those in the early years of their careers to advanced scholars. As with *NASA and the Long Civil Rights Movement*, we believe this diversity of perspectives and expanded context is critical to any deeper understanding of aerospace's impact on the region.

The essays that make up this volume build on a narrow historiography by shifting a previously compartmentalized approach that views Project Apollo in isolation from the larger national transformation occurring simultaneously to one that repositions the immense investment in the space program in the context of regional issues. We hope that by doing so, this volume will spark a more expansive discourse on the South's relationship to the space program and on other perspectives, including postcolonial and transnational views of aerospace developments from the beginning to the present day. But the reverse is also true. We hope that examining the history of aerospace from these new perspectives will impact our understanding of space exploration and scientific discovery. If science and technology are in fact socially constructed (and I argue that they are), how can these new perspectives inform national and international space programs—not to mention the increasing role played by nonstate actors?

The organization follows the approach of *NASA and the Long Civil Rights Movement*, moving from general frameworks to specific case studies in each theme. The essays in part 1 present an expansive view of NASA's southern context and the challenging regional issues, from politics to race, associated with NASA's presence in the South. Each chapter positions the space agency in a broader context that encompasses southern politics, economics, and society. These essays address issues such as regional bias, Black education, and political wrangling on a national and local level. In the opening essay, Roger Launius provides an overview of NASA through a southern lens, arguing that Project Apollo has cast a long shadow over expectations about the American exploration of space, a shadow that has yet to be overcome. Launius explores several aspects of the Apollo story through extended discussions of its place in the context of LBJ's Great Society effort, Project Apollo, and America's original sin of slavery and race relations, nostalgia for Apollo, and the Huntsville school of space history.

Going deeper in the political landscape, Andrew Dunar assesses the intersection of space and politics as he investigates the outsized role of Lyndon B. Johnson in the development and progress of the space program in the United States. As a member of the United States Congress, as vice president, and as president from November 1963 to January 1969, Johnson was in key positions to influence the American space program in its formative years. His outsized influence on the early years of the space age is reflected in the sites selected for government aerospace facilities and echoed Johnson's philosophical outlook that technology could reshape society. His thumbprint would make Project Apollo reflective of the New Deal. As Dunar points out, Johnson leveraged funding for the space program to further other national goals, including his

Great Society programs, modernization, and particularly the advancement of civil rights and the amelioration of poverty. Dunar's view of Johnson's role from the southern perspective offers new layers to the already extensive historiography of Johnson and the space program.[10]

Although most southern politicians fought it, federal action related to equal employment opportunity gradually opened new doors of economic opportunity for Black southerners. My essay explores the evolution of one institution of Black education in Huntsville, Alabama, in the context of both federal investment in aerospace and the new legal framework established by challenges to Jim Crow segregation. No one person embodied that transition at Alabama A&M University more than its president, Dr. Richard D. Morrison, who in a March 1963 speech proclaimed that the rise of the Space Age and the rapid development of the aerospace industry in the South provided Black students with an opportunity to enter new careers on equal terms with whites. I also examine how Morrison developed an integrated educational program at the university while at the same time maintaining the moderate position on civil rights in order to placate state funding agencies. Viewing the aerospace investment from this perspective enables us to understand the delicate balancing act administrators at Historically Black Colleges and Universities who sought access to economic opportunities for their students faced.

General perceptions of the region often become a lens through which understanding of activities and accidents are filtered. Stephen Waring examines several aspects of the aftermath of the *Challenger* disaster, noting how stereotypes about the South played out in investigations and subsequent interpretations of the accident and reinterrogating the events and perspectives that emerged after the accident. Waring introduces key leaders from NASA's Marshall Space Flight Center to see if they engineered with a regional accent. Waring studies the events and dialogue that emerged during the Rogers Commission's investigation, looking for regional rhetoric and bias. However, post-accident interpretations of managerial miscommunication and malfeasance had undertones of negative stereotypes about white Southerners that deflected attention from weaknesses in neoliberal national space policy and the political economy of the space program. Taken together, the essays in part 1 call previous assumptions in question while offering fresh perspectives and new investigatory frameworks for future scholars to consider.

The essays in part 2 shift from the macro to the micro with investigations of how these transformations impacted southerners at a local level. NASA's Kennedy Space Center holds a unique place in American memory as a pillar of technology amid a landscape of lagoons, mosquitos, and alligators. But as

Rachael Kirschenmann argues, the center was not constructed out of a barren wilderness. Before the arrival of NASA in the early 1960s, communities on North Merritt Island were home to residents who fished on the creek adjacent to the Vehicle Assembly Building, went to school on land that is now a wildlife refuge, and picked oranges where the Visitor Complex now stands. Kirschenmann focuses on the citrus workers and growers who were displaced by eminent domain and urban expansion to make room for the space center and the technical workforce it attracted. A comparison of the agricultural technology citrus workers used with the technology of Project Apollo reveals a long history of adaptation punctuated by a leap of innovation in a national project. Kirschenmann argues that memories documented in oral histories reflect the primacy of the local and the lasting importance of preexisting industries for the area.

Regional investments in Cold War aerospace were not unique to the American context. Arslan Jumaniyazov draws an important parallel to the Soviet program, comparing the Russian investment at the Baikonur Cosmodrome in Soviet-controlled Kazakhstan to the investment at Kennedy Space Center at Cape Canaveral and Cocoa Beach. Jumaniyazov argues that political rivalry and aerospace competition propelled the United States and the Soviet Union to invest heavily in their respective space programs. By shifting the perspective from one South to another, Jumaniyazov underscores how local communities received and impacted these national programs at the two major launch centers. Studying the history from the bottom up also reveals powerful contrasts. Yet each "federal" government developed their launch centers in relatively poor and peripheral regions: the American south and the Soviet east. Taken together, these essays from Kirschenmann and Jumaniyazov offer a comparative framework that draws close parallels between two forms of colonization.

Turning back to the space crescent, Stuart Simms surveys the development of what would become Stennis Space Center in Mississippi in the early 1960s. In the beginning, many residents viewed the coming of the testing facilities as a blessing. It quickly became a curse. From their establishment, these facilities played a central role in the development of the Saturn V launch vehicle and the transformation of the local region's economy. With the 1969 moon landing and the end of expansive 7/28/2023 budgets, discussions of closure became more prevalent and stirred conflicting opinions among residents. Simms argues that while many residents remained willing to sacrifice their homes and land for the American space program, in the end, some no longer saw the value in view of what was lost.

Investment in the American science and technology infrastructure were always part of larger presidential political, economic, and social programs. Max Campbell compares two massive government programs in the 1960s: President John F. Kennedy's challenge to conquer the moon through Project Apollo and the efforts of his administration to alleviate the plight of Appalachia, one of the country's poorest regions. Several recent studies, for example those by Matthew Tribbe and Neil Maher, have highlighted the paradox of this era: the country was divided between a grand effort in science and engineering and a social justice program.[11] Yet the Apollo-Appalachia dynamic reveals just how carefully and equitably the American political system and public weighed these issues, albeit with significantly different problems and results.

How did the families moving to these regions adapt to the strange new environments? Using oral history to explore the collective memory of the early years in Houston, Jennifer Ross-Nazzal explores the social ramifications of NASA's decision to locate what would become Johnson Space Center in Houston in the fall of 1961. Over the course of a decade, thousands of employees working on the lunar program moved to the region, reshaping the area's economy from one that relied on fishing, shrimping, and agriculture to one that depended on aerospace. Writing from the perspective of white women who settled in the new NASA-dominated suburbs, Ross-Nazzal shows how participants saw the transformation of the Clear Lake area and residents of the surrounding communities. Project Apollo and the space program permeated every aspect of this unique community, from church sermons to community events.

The economic benefits of aerospace in the South went beyond industry and government. The birth of aerospace also intersected with the explosion of automobile tourism. In the booming postwar economy, many families in the United States found themselves with disposable income and increasing leisure time. Boosters across the region attempted to leverage the presence of the aerospace industry to cash in on the trend. Kari Edwards examines the efforts of Carl McIntire, a bombastic fundamentalist radio preacher, to turn a large tract of land in Cape Canaveral, Florida, into a Christian community called Gateway to the Stars. McIntire's vision was for a fundamentalist Bible conference center with a large hotel, condominiums, restaurants, a Bible college, and everything else like-minded Christians might desire. But the highlight of the complex would be its view of the Kennedy Space Center's launch area. Edwards positions McIntire's efforts around the Gateway to the Stars community in the context of the Cold War battle with the "Godless" Soviet Union.

Edwards argues that it was a unique and relatively undocumented example of the often-precarious marriage between religion and scientific progress that characterized the Cold War–era American religious landscape.

Continuing the theme of aerospace tourism, Emily A. Margolis focuses on Walter V. Linde and the struggle to create the Alabama Space and Rocket Center (now the U.S. Space & Rocket Center) in Huntsville, Alabama. Margolis argues that Linde recognized the potential of space-age tourism to transform the local economy and spent the next twelve years advocating for a space museum and enlisting the most prominent Alabamians and Alabama institutions in his cause. Linde rallied diverse stakeholders who saw in this project a means to further their individual and institutional agendas, which often had little to do with spaceflight. Margolis locates the origins of the Alabama Space and Rocket Center at the intersection of popular culture, postwar heritage travel, political necessity, the civil rights struggle, and southern industrial expansion.

Failure was a common theme in most aerospace tourism endeavors. In his essay, Drew Adan examines an ill-fated attempt by a small business group in Huntsville to develop Space City USA, a space theme park. Adan traces the history of this failed effort from initial fund-raising and planning in 1959 to bankruptcy and liquidation in 1967. Through a study of archival resources and documents of private collectors in the region, Adan places the Space City story in the context of Huntsville's rise to prominence and the growing pains of a nascent space-related tourism industry. Unlike the eventual success of the U.S. Space & Rocket Center, Space City demonstrates the limitations of the aerospace investment to sustain even the most promising entrepreneurial efforts.

The essays in part 4 shift the focus from economic concerns to the regional impact of aerospace on the broader culture, particularly art and architecture. Jeffrey S. Nesbit examines how architectural modernity was represented in the design of NASA administrative office buildings. Nesbit focuses on the contributions of Charles Luckman, the architect behind both Madison Square Garden and the Kennedy Space Center Headquarters in Florida, both of which were completed in 1965. Going beyond the known connections, Nesbit offers answers to the questions of where and how construction of the space complex combined a modern aesthetic with rational choice. Nesbit illustrates how the culture of architectural modernization from American cities such as New York, Houston, and Chicago became implicitly connected to the production of the space complex in the remote South through form and increased projections of rationality.

Caroline Swope examines how the aerospace investment in Huntsville, Alabama, was reflected in the local architectural landscape. As the US Army post Redstone Arsenal and NASA's Marshall Space Flight Center continued to grow, the architectural styles that defined the city changed significantly. Huntsvillians demanded modern buildings for their space-age city, and many saw the construction of the International Style Madison County Courthouse and the Neo-Formalist First Baptist Church as pivotal in forming Huntsville's new identity. First Baptist Church, with its space-age-themed massive mosaic and rocket-shaped bell tower, grew exponentially in size along with Huntsville; its membership more than doubled from 1953 to 1963, when it reached 2,626 members. Swope's essay explores the development of these two buildings in terms of their local significance and any wider part they played in the development of post–World War II architecture.

Katarzyna Balug explores the influence of the space program on more abstract art forms. In the 1968 catalog for the Vera Simons exhibit at the Contemporary Arts Museum Houston (CAMH), curator James Harithas wrote that Simons was an artist who had the scientific knowledge to use space-age technology for space-age forms. Emerging in response to the technologies of the 1960s space program, these forms were often embodied in inflatable sculptures and structures. Balug examines the role that institutions such as CAMH played in networking artists and architects who created inflatable works, serving but as gravitational forces whose orbit facilitated conversation and contributed to the sudden and expansive output of inflatable projects in the late 1960s and early 1970s. Closing out the volume, Stephen Waring offers potential new directions for building on these histories by considering the regional context.

This volume critically examines the numerous impacts of aerospace on the American South. It also points to the amount of work yet to be done, not only in understanding the scope and scale of this aerospace investment on the South but beyond to other regions of the United States and around the world. This has not been a one-way street. Regional factors have also constrained and reshaped the direction aerospace has taken over the previous six plus decades. We hope by examining even this limited number of issues, this volume will stimulate further research in many more important questions.

Notes

1 *Aeronautics and Astronautics, 1963: Chronology on Science, Technology, and Policy* (Washington DC: NASA, 1964), 125.

2 Thomas Franklin Morring, "The Impact of Space Age Spending on the Economy of Huntsville, Alabama" (MS thesis, Massachusetts Institute of Technology, 1964).

3 Loyd S. Swenson Jr., "The Fertile Crescent: The South's Role in the National Space Program," *Southwestern Historical Quarterly* 71, no. 3 (1968): 377–392.

4 Bruce Schulman, *From Cotton Belt to Sunbelt: Federal Policy, Economic Development, and the Transformation of the South*, 1938–1980 (Durham, NC: Duke University Press, 1994), 148.

5 Walter A. McDougall, *The Heavens and the Earth: A Political History of the Space Age* (New York: Basic Books, 1985), 376.

6 Robert A. Myers, "Planning for Impact: A Case Study on the Impact of the Space Program on Huntsville, Alabama" (MA thesis, New York University, 1967), 91.

7 Gavin Wright, *Sharing the Prize: The Economics of the Civil Rights Revolution in the American South* (Cambridge, MA: Harvard University Press, 2013).

8 Bobby Wilson, *American Johannesburg: Industrial Transformation in Birmingham* (New York: Rowman and Littlefield, 2000), 226.

9 Brian C. Odom and Stephen P. Waring, eds., *NASA and the Long Civil Rights Movement* (Gainesville: University Press of Florida, 2019).

10 A notable example of this literature is Robert Dallek's essay "Johnson, Project Apollo, and the Politics of Space Program Planning," in *Spaceflight and the Myth of Presidential Leadership*, ed. Roger Launius and Howard McCurdy (Urbana: University of Illinois Press, 1997).

11 Matthew Tribbe, *No Requiem for the Space Age: The Apollo Moon Landings and American Culture* (Oxford: Oxford University Press, 2014); Neil Maher, *Apollo in the Age of Aquarius* (Cambridge, MA: Harvard University Press, 2017).

Part 1

Politics and Prejudice
in a Southern Context

1

Constructing Apollo

The American South and the Moon Landings

ROGER D. LAUNIUS

The Apollo moon landing program represented the ultimate in the nation's space exploration agenda and a high point in national pride when it succeeded in 1969. The American South experienced this pride more than most other regions because of the larger number of NASA human spaceflight centers there and the number of people in the region working on the program. The memory of Apollo has evoked especially powerful connotations for most Americans as a triumphalist, exceptionalist history of the nation overcoming adversity in the face of the Soviet Union's rivalry.[1] This essay explores several of aspects of the Apollo story through extended discussions of its place in the context of LBJ's Great Society effort, America's original sin of slavery and race relations, and Project Apollo and the Huntsville school of space history.[2]

Apollo and the Great Society

In the summer of 1961, the new NASA administrator, James E. Webb, discussed with E. F. Buryan, president of Motec Industries, Inc. in North Carolina, "the urgent necessity for a strong technological underpinning for any regional economic system that has survival qualities. Indeed, the presence of basic research and the kinds of people who do basic research is of urgent importance for the long run and should be effectively worked out along with the technological and industrial competence." Webb viewed NASA as critical to that economic development, especially in regions of the United States that were left behind by the post–World War II economic boom. "We are going to spend 30 to 35 billion dollars pushing the most advanced science and technol-

Figure 1.1. No image bespeaks the juxtaposition of NASA in the South better than this one from 1967 showing a NASA rocket test stand in the background with a mule and plow trenching in the area in the foreground. Courtesy of NASA.

ogy," he wrote, and then "endeavoring in every way possible to feed-back what we learn into the total national economy."[3]

Around the same time, Webb addressed a memorandum to NASA officials stating his position on the role of science and technology in advancing the nation's welfare: "One of the most important aspects of the space program is the possibility of the feed-back of valuable, new technological ideas and know-how for use in the American economy."[4] There was nothing new about this idea. Webb had been demanding efforts in this direction since 1954, when he cofounded the Frontiers of Science Foundation. Webb stated at the time that "the future lay in the area of scientific knowledge and its development." He used the foundation to press for the development of curricula for K-12 and university students in science and mathematics; aggressive research programs that would attract government grants and contracts and industry dollars to underdeveloped regions, especially in the South; and the relocation of science and technology organizations to the area.[5]

These early perceptions of what NASA would accomplish in the 1960s suggest that it was not just about winning a space race with the Soviet Union or demonstrating technological verisimilitude in a game of geopolitical one-upmanship or any of the myriad other reasons given for an opening of the national treasury to expend 5 percent a year on NASA during the middle part of the decade. It was also about advancing the general economic well-being of the nation as a whole, especially in regions where there seemed to be lag such as the Deep South.

This approach became a centerpiece of Project Apollo during the presidency of Lyndon Baines Johnson, whose career had been built on the assumption that federal monies well spent on infrastructure, social programs, and defense could serve the national well-being, especially in the less affluent South. Apollo was tailor made for this use. In 1963, when critics began to talk about "the Moondoggle," Johnson opined, "The space program is expensive, but it can be justified as a solid investment which will give ample returns in security, prestige, knowledge, and material benefits."[6]

Once Johnson had been elected in his own right in November 1964, he embarked on a broad agenda to use the power of the federal government to accomplish goals that would extend the efforts of the New Deal. The civil rights crusade was coming to fruition and the quest for egalitarianism across American society combined with that social movement to make possible the creation of a new Great Society. As president, Lyndon B. Johnson remarked in a speech at the University of Michigan in May 1964:

> For better or for worse, your generation has been appointed by history to deal with those problems and to lead America toward a new age. You have the chance never before afforded to any people in any age. You can help build a society where the demands of morality, and the needs of the spirit, can be realized in the life of the Nation. So, will you join in the battle to give every citizen the full equality which God enjoins and the law requires, whatever his belief, or race, or the color of his skin?[7]

Johnson rolled out a broad domestic agenda aimed at creating a more equitable society that achieved early success but was sidetracked by the Vietnam conflict and civil unrest in the latter part of the decade.

Project Apollo, like the other endeavors that were part of the Great Society, represented LBJ's vision of the positive liberal state in the 1960s. Through it, the president sought to use state power to advance the public welfare. Without seeking to do so, Johnson offered an important perspective on a debate that has raged over the proper place of state power since the beginning of the republic. As only one example of how this has played out over time, the Whig Party of the early nineteenth century sought an activist government that would accomplish important tasks for the benefit of all. Historian Daniel

Walker Howe eloquently called the Whigs the champions of the positive liberal state. He wrote:

> This ideal implied the belief that the state should actively seek "to promote the general welfare, raise the level of opportunity for all men, and aid all individuals to develop their full potentialities." The Democrats, by contrast, believed in a "negative liberal state," which left men free to pursue their own definition of happiness. A great advantage of this distinction between the parties is that it implies a connection between the economic and moral aspects of Whiggery. In both cases, the Whigs believed in asserting active control. They wanted "improvements," both economic and moral, and they did not believe in leaving others alone.[8]

Like the Whigs, the Democrats of the 1960s believed in activist government, and Apollo represented one of its major accomplishments (fig. 1.2). The War on Poverty, the Peace Corps, support for civil rights, the Great Society programs, and a host of other initiatives are also examples.[9]

In this context, James Webb emphasized NASA's efforts to increase expenditures in the American South. More important, he carefully tracked how NASA's investments in the South affected the region's economy and took opportunities to apprise Johnson of these gains. In a 1965 report to the president, for example, he pointed out that in the previous year 94 percent of NASA's "procurement dollars" had gone to 20,000 private US industrial companies. Of those expenditures, $331 million had been spent in 120 cities in 22 states with high unemployment rates, especially in the South, and nearly 1 million people had worked directly or indirectly on NASA-related business. NASA tracked down to the congressional district where expenditures on Project Apollo were made to show each member of Congress how their constituents were affected by NASA spending.[10]

Both Johnson and Webb understood that much more than pork barrel spending would result from Project Apollo. They promoted the economic and political gains NASA spending brought to Florida, Alabama, Mississippi, Louisiana, Tennessee, Texas, Virginia, and other states around the country.[11] Webb also advertised the longer-term national advances NASA's efforts might produce. For example, he told LBJ in 1964 that

> NASA has something to offer law enforcement in terms of data processing and communication systems; to the construction industry through NASA developed materials; to pollution control through the development of an outlook whereby the Earth's air and water are beginning

Figure 1.2. Former president Lyndon B. Johnson and vice president Spiro Agnew are among the spectators at the launch of Apollo 11, which lifted off from Pad 39A at Kennedy Space Center at 9:32 am EDT on July 16, 1969. Johnson wrote in his presidential memoir that "space was the platform from which the social revolution of the 1960s was launched. We broke out of far more than the atmosphere with our space program. . . . If we could send a man to the Moon, we knew we should be able to send a poor boy to school and to provide decent medical care for the aged. In hundreds of other forms the space program had an impact on our lives." Courtesy of NASA.

to be viewed as finite resources operating as closed systems; to transportation of people in and out of the inner city through research on short-haul aircraft; to improvement of economic opportunities for all citizens by stimulating business through new inventions and transfers of space technology to industry; and to a richer life by development of techniques making possible cheaper, lighter, and more reliable television sets and other electronic items for use in the home.[12]

LBJ was convinced by these arguments, In 1965, he announced that NASA's work not only increased "our knowledge of technology" but also led "to a better life for all." In his 1972 memoir, he wrote: "Space was the platform from which the social revolution of the 1960s was launched. We broke out of far more than the atmosphere with our space program. . . . If we could send a man to the Moon, we knew we should be able to send a poor boy to school and to provide decent medical care for the aged. In hundreds of other forms the space program had an impact on our lives."[13]

A belief in the primacy of NASA in transforming the South might have been overblown, but it certainly helped advance the economic interests of the region and strengthened LBJ's political hold there, especially when racists emerged to undermine the Democratic agenda. While Johnson denied that he had had any part in selecting which southern companies would receive Project Apollo funds, such declarations were suspect and politicos knew better. Senator George Smathers (D-Florida) said he confronted Johnson about the establishment of the Manned Spacecraft Center in Houston. "Johnson tried to act like he didn't know," Smathers recalled, but it never "made sense to have a big operation at Cape Canaveral and another big operation in Texas." Clearly, this was about spreading NASA dollars around the American South. Indeed, with many key southern members of Congress running critical committees, the South profited from the expansion of NASA's spending far beyond what had been envisioned at the beginning of the accelerated moon landing program in 1961.[14] For Johnson, NASA's largesse in the South represented a critical element of the positive liberal state that would assist less developed parts of the country.

Webb also sought to help less advantaged regions, especially cities, not only with dollars but also with NASA know-how. At a NASA-sponsored conference on space and urban problems held in Oakland, California, in March 1963, participants considered two important questions: "(1) Can a national program of space exploration be applicable to the daily tasks of people who live and work in our central cities? (2) How may new knowledge, developing in these days of scientific and technological revolution, be used to seek answers to the critical issues of expanding urban populations?"[15] Those are important questions, but the answers at the time were less than impressive. With participants like Webb, southern members of Congress, presidential science advisor Jerome B. Weisner, and a host of academics and industry leaders in attendance, one might have expected more concrete results. Instead, participants offered broad assertions of possibilities but not much in the way of concrete ideas.

Some participants criticized the vision that space program technology could have wider applications that would benefit society. As one person put it, "Science and technology have done to the city what they have done to any part of human endeavor they have touched. They have freed us more and more from our environment; they have given us more opportunity to manipulate it."[16] The upshot was that while science and technology could help resolve some of the challenges faced in the inner cities of the United States through the development of cleaner-burning fuels, more efficient automobiles, better mass transit systems, and more sophisticated city management, NASA's expertise could only solve problems that were inherently technological. The truly dicey problems of race, class, economic disparities, and the like defied a technological fix.[17]

Nevertheless, NASA's monetary investment and its ideas had significant ramifications in the South. The impact was profound around the sites of NASA activity—Kennedy Space Center in Florida, Marshall Space Flight Center in Alabama, Langley Research Center and Wallops Flight Facility in Virginia, Goddard Space Flight Center in Maryland, Michoud Assembly Facility in Louisiana, Mississippi Test Facility, and the Manned Spacecraft Center in Texas. NASA has measured its impact at these sites in several ways:

- An adaptation of a macroeconomic production function model estimates impacts of technological change attributed to R&D spending.
- A microeconomic model evaluates the returns to specific technologies using benefit-cost ratios.
- An examination of data provides evidence of the direct transfer of technology from federal space R&D programs to the private sector.

For example, in 1971, the Midwest Research Center showed large economic returns to the region— a 7:1 ratio of long-term economic benefits to expenditures. Other studies showed an even greater positive return on investment.[18] Despite criticisms, all these studies pointed to a significant economic boost to the regions in which NASA facilities were located (fig. 1.3).

NASA emphasized the divergence between high technology activities and economic advancement in the South. In each of the areas where NASA facilities and contractors were located, the large influx of well-educated, well-compensated individuals made unique differences in the areas where they lived and worked. For example, one 1968 study found that

NASA and NASA-contractor personnel have contributed to . . . upgrading . . . the environment in each community in a variety of ways, from running for and being elected to local political offices to providing pressure through neighborhood and community organizations and volunteer, charitable, and religious groups. Furthermore . . . in many cases, a substantial portion of the teaching staff in local grade and high schools was made up of wives of engineers and scientists on NASA

Figure 1.3. Wernher von Braun represented as a "southerner" in this cartoon illustration from ca. 1970. Courtesy of NASA.

projects. These women are generally well educated, often from a more cosmopolitan environment than that found in many of the NASA locations in the South, and thus able to bring to school children a broader experience and a greater appreciation for education than they would have otherwise.

Researchers also found that "it appeared . . . that there was something distinct about the infusion into a community of federal funds for research and development, as opposed to federal funds for other uses."[19]

Other studies supported this finding. In 1971, a NASA-supported study concluded: "The $25 billion, in 1958 dollars, spent on civilian space R&D during the 1959–1969 period has returned $52 billion through 1970 and will continue to produce pay-off through 1987, at which time the total pay-off will have been $181 billion. . . . The discounted rate of return for this investment will have been 33 percent."[20] Such findings, even when discounted somewhat to adjust for an overzealous estimated return rate, demonstrated that NASA's economic investment in the American South helped to fundamentally transform the region from an agricultural economy into an economy with a significantly larger mix of industry and high technology activities.

Apollo and America's Original Sin

Without question, the great American original sin is the tragedy of slavery and race relations. The space program had a role in coming to grips with this issue during Project Apollo's effort to reach the moon. At some level, NASA was born into this original sin, and the space program struggled with race during Apollo. In the aftermath of the Soviet launch of Sputnik 1, Democratic strategist George E. Reedy addressed a memorandum to Senator Lyndon Johnson about the opportunity Sputnik presented to make political hay at the expense of the Republicans. He was disappointed in his fellow Democratic Party members for their failure to advocate for racial desegregation—mostly because of southern Democrat segregationists, but he felt that championing the Sputnik crisis was the next best option. Reedy summarized his position: "The simple fact is that we can no longer consider the Russians to be behind us in technology. It took them four years to catch up to our atomic bomb and nine months to catch up to our hydrogen bomb. Now we are trying to catch up to their satellite."[21] Johnson listened. The birth of NASA arose directly from LBJ's agitation for action to respond to the Soviets in space.

The race issue also directly affected the space program during the race to the moon in the 1960s. Throughout the decade, fewer than 3 percent of the NASA professional workforce were African Americans or other people of color. There were virtually none at NASA facilities in the South. For example, African Americans constituted 18 percent of Huntsville's population throughout much of the 1960s, but less than 1 percent of Marshall's workforce. After the bombing of the 16th Street Baptist Church in Birmingham in 1963 and the riots that followed, the Kennedy administration made addressing this disparity a moral imperative. Attorney General Robert M. Kennedy berated Webb at a meeting on the matter. "Mr. Webb," he said, "I just raised a question of whether you can do this job and run a Center and administer its $3.9 billion worth of contracts and make sure that Negroes and nonwhites have jobs. . . . I am trying to ask some questions. I don't think I am able to get the answers, to tell you the truth."[22]

This infuriated Webb. He pushed for more action on the issue from all of his center directors, especially Wernher von Braun in Huntsville, Alabama. At Webb's insistence representatives from NASA and other federal entities in Huntsville met on June 18, 1963, and decided to conduct surveys of housing and federal employment practices, to assist Alabama A&M College and Tuskegee Institute in placing African American engineers and other professionals at NASA, to push contractors to ensure equal employment opportunities, and to increase African American employment in all arenas.[23] Marshall

Space Flight Center established an Affirmative Action Program, hiring Dr. Frank R. Albert to coordinate. Albert hired Charlie Smoot as a professional staffing recruiter, "possibly the first Negro recruiter in government service."[24]

With the passage of the Civil Rights Act in 1964 and the Voting Rights Act in 1965, the Johnson administration pressed every federal agency and government contractors and other entities to adhere to this new law. As a loyal Johnson supporter, James Webb led the charge in NASA to press for more minority hiring, especially of engineers and scientists. These efforts were only moderately successful.

Because the key NASA facilities were located in the southern United States where racial segregation was still in place, it was difficult to recruit mathematicians, engineers, scientists, and technicians of color throughout the decade. Julius Montgomery, Clyde Foster, and a few other African American engineers that came to Kennedy Space Center in Florida and Marshall Space Flight Center in Alabama in the mid-1960s were notable exceptions. Arriving for his first day at work, Montgomery had to deal with NASA coworkers who were members of the Ku Klux Klan and refused to acknowledge him or shake his hand. Whether his white supremacist colleagues liked it or not, Montgomery represented the vanguard of a growing movement toward diversity at NASA. Some African American women served important roles in Project Apollo. Among the most famous was Katherine Johnson, whose career began at Langley Research Center in 1953. After the facility was absorbed into NASA in 1958, Johnson was recruited to work on rocket trajectories for Project Mercury. She continued to make significant contributions to the space agency's work until her retirement in 1986.[25]

This poor showing on minority employment was even more evident in the astronaut corps. Many astronauts were career military officers and fighter pilots from the South who had been educated at military academies or state universities. None were African American. It might have been otherwise. In 1963, Air Force pilot Captain Edward Dwight Jr. became the first astronaut trainee of color for a military program, a position that catapulted him to instant fame. He was featured in news magazines around the world and on the cover of a variety of publications aimed at a Black readership in the United States, including *Ebony* and *Jet*. However, the severe discrimination he faced from many of his fellow trainees and from government officials eventually prompted him to resign from the Air Force. After that, the Air Force selected Major Robert Lawrence to join its military astronaut program, but he died soon after in a training accident. The military astronaut program was subsequently canceled. It would be another eleven years before NASA recruited its

Figure 1.4. Construction on the Vertical (now Vehicle) Assembly Building at NASA's Kennedy Space Center began on August 2, 1963, as part of NASA's massive effort to send astronauts to the moon for the Apollo program. Most of this new construction took place in the American South. This photo from November 9, 1970, shows a ground-level view of Apollo 14 leaving the VAB for Launch Complex-39A atop a huge crawler-transporter. Courtesy of NASA.

first African American astronaut, Guion "Guy" Bluford, who flew on space shuttle *Challenger* on August 30, 1983. Astronauts from other ethnic minorities would follow, but none went to the moon.[26]

NASA compiled a somewhat better record of fighting off segregationist politicians who wanted to use astronauts or space facilities as backdrops for their machinations. Webb let it be known that no one from NASA would participate in any event at which the organizers were segregationists. Senator John Stennis (D-Mississippi) was an especially difficult case. As an ardent NASA supporter—NASA's test facility in Mississippi was named for him—James Webb naturally wanted to appease him whenever possible. But because he was a staunch segregationist, the space agency needed to avoid public appearances with him. When Webb was asked to speak in Jackson, Mississippi, his staff had initially agreed but then he had to back out at the last minute. He called Stennis to gracefully bow out, telling him why and making clear that Stennis could say whatever he wanted about the reason. Stennis chose not to admit that it was because Webb deemed him a racist. Later, when Alabama governor George Wallace tried to use NASA to boost his public persona for a presidential run, Webb and Wernher von Braun conspired to "big foot" the affair with a rocket test that conflicted with the event. Inviting the media that would have otherwise been covering Wallace's speech, they fired the mighty F-1 rocket engine—five of which would power the Saturn V's first stage to the Moon—at one of Marshall's test stands. Thereafter, Webb and von Braun criticized Wallace's segregationist ideas as "backward" for Alabama. Von Braun pointedly commented that the future would "belong to those who can shed the shackle of the past."[27]

Finally, at the time of the Apollo 11 launch, Rev. Ralph Abernathy, the successor to Martin Luther King as head of the Southern Christian Leadership Conference, protested the event to call attention to the plight of the poor of the United States. He and 500 marchers of the Poor People's Campaign arrived at the Kennedy Space Center to contest the meaning of the moon launch.[28] The protesters held an all-night vigil as the countdown proceeded and then made a march with two mule-drawn wagons as a reminder that while the nation was spending significant money on Project Apollo, poverty ravaged the lives of many Americans. As Hosea Williams said at the time, "We do not oppose the Moon shot. Our purpose is to protest America's inability to choose human priorities."[29]

This protest pointed more effectively than anything else to the collision of NASA's high-tech moon program with the ever-present problems of America's racial divide. Abernathy asked to meet with the NASA leadership. Thomas

O. Paine, James Webb's successor, agreed to a meeting on July 15, 1969, the day before the launch. Paine later recounted the incident:

> We were coatless, standing under a cloudy sky, with distant thunder rumbling, and a very light mist of rain occasionally falling. After a good deal of chanting, oratory and lining up, the group marched slowly toward us, singing "We Shall Overcome." In the lead were several mules being led by the Rev. Abernathy, Hosea Williams and other leading members of the Southern Christian Leadership Conference. The leaders came up to us and halted, facing Julian [Scheer] and myself, while the remainder of the group walked around and surrounded us. . . . One fifth of the population lacks adequate food, clothing, shelter and medical care, [Rev. Abernathy] said. The money for the space program, he stated, should be spent to feed the hungry, clothe the naked, tend the sick, and house the shelterless.

Abernathy said that he had three requests for NASA: that some of his group view the launch; that NASA "support the movement to combat the nation's poverty, hunger and other social problems"; and that NASA technical people work "to tackle the problem of hunger."

Paine responded with an answer that suggested that the space agency was moving toward a greater social conscience. He invited Abernathy and a busload of his supporters to view the Apollo 11 launch from the VIP site with other dignitaries. He also said how difficult it was to apply NASA's scientific and technological knowledge to the problems of society. "I stated that if we could solve the problems of poverty in the United States by not pushing the button to launch men to the Moon tomorrow," Paine said, "then we would not push that button." He continued:

> I said that the great technological advances of NASA were child's play compared to the tremendously difficult human problems with which he and his people were concerned. I said that he should regard the space program, however, as an encouraging demonstration of what the American people could accomplish when they had vision, leadership and adequate resources of competent people and money to overcome obstacles. I said I hoped that he would hitch his wagons to our rocket, using the space program as a spur to the nation to tackle problems boldly in other areas, and using NASA's space successes as a yardstick by which progress in other areas should be measured. I said that although I could not promise early results, I would certainly do everything in my

own personal power to help him in his fight for better conditions for all Americans, and that his request that science and engineering assist in this task was a sound one which, in the long run, would indeed help.

When Abernathy held a prayer meeting later that day with his protestors, Paine asked that they "pray for the safety of our astronauts." As Paine recalled, Abernathy "responded with emotion that they would certainly pray for the safety and success of the astronauts, and that as Americans they were as proud of our space achievements as anybody in the country."[30]

Paine concluded that unfortunately the social problems of the United States could not be solved entirely by shifting resources from NASA to other initiatives. He agreed that the problems of society were much more complex and defied resolution using the tools, knowledge, and resources that NASA had used to accomplish Project Apollo. While it might be tempting to generalize from the experience of NASA during the 1960s that its success might be duplicated in other domains of US society, that was not the case. As one observer commented: "NASA's effective implementation of the Apollo mission shows that anything we set our minds to can be done, provided all the conditions are met. Unfortunately, there will be few areas in American life where such will be the case. Nevertheless, Apollo will serve as an everlasting precedent to which optimists will be able to point."[31]

There were no easy fixes for NASA's difficulties with establishing racial equality in its workforce. Complying with federal law, the agency established its Equal Employment Opportunity office at NASA Headquarters in September 1971. The NASA equal opportunity head, Ruth Bates Harris, pushed hard for equitable hiring practices and efforts to create a meaningful path to a successful diversity program. She faced an agency that had little interest in making changes. When Harris arrived at NASA, its 5 percent minority workforce representation was the lowest of all federal agencies. That diversity rate was even lower in science and engineering positions, where only 4.5 percent of administration were people of color by 1975. A 1973 report on NASA's hiring practices condemned NASA's Equal Opportunity Program as a "near-total failure." It took years of effort thereafter to begin to change this situation, and efforts to achieve equitable representation in NASA's workforce are still far from complete.[32]

The Huntsville School of History and the Interpretation of Project Apollo

The problem of interpreting the Apollo program also had an important southern connection. In 1991, Rip Bulkeley christened a particular celebration of space history as the Huntsville School, so named because of where it originated.[33] Wernher von Braun and the German rocket team had moved to Huntsville, Alabama, in 1950 and made it their home for more than two decades. White southerners in Huntsville welcomed them; they became celebrities in the community to the extent that to this day their image is jealously protected and anyone who casts aspersions on it is maliciously attacked by the keepers of the German rocketeers' legacy. For years, von Braun and his confidants and followers dominated the study of space history, and their perspectives remain prominent in space historiography. Von Braun's most significant contribution to this writing was the massive illustrated *History of Rocketry and Space Travel*, which went through three editions from 1967 to 1986 and has been influential in shaping popular conceptions about space flight.[34] Von Braun associates Ernst Stuhlinger, Walter Dornberger, Frederick C. Durant III, Frederick I. Ordway III, and Mitchell R. Sharpe have extended the approach von Braun pioneered in their historical writings about space exploration.[35] A younger generation of historical writers have also promulgated the message of the Huntsville School.[36]

There are key elements in the work produced by the Huntsville School of space history. First, and most important, von Braun and the German rocketeers who built the V-2 rocket and then came to America at the end of World War II are presented as far-sighted visionaries with an integrated space exploration plan that fostered a future of great discovery on the "final frontier." The Huntsville School also minimized the wartime cooperation of von Braun and his rocket team with the Nazi regime in Germany and maximized the team's role in the development of American rocketry and space exploration. Both are conscious distortions of the historical record that were promulgated for specific personal and political reasons. Even today, few Americans realize that von Braun was a member of the Nazi Party and an officer in the SS or that the V-2 was constructed using slave labor from concentration camps. People also do not understand that the United States had developed a very capable rocket technology at the Jet Propulsion Laboratory, in the National Advisory Committee for Aeronautics, in the Air Force, and in private corporations.[37] The result has been both a whitewashing of the less savory aspects of the careers of the German rocketeers and an overemphasis of their influence in American rocketry.

A core question is why the von Braun rocket team has been so celebrated while other experimenters are nearly forgotten. Clearly the German team was enormously successful in developing the V-2 ballistic missile during World War II, but why were the less savory aspects of its working for the Nazis swept under the rug?[38] As M. G. Lord wrote in a study of the Jet Propulsion Laboratory rocketeers led by Frank Malina from the 1930s through the 1950s, "With accomplishments like these you'd expect to find [Malina] enshrined in history, alongside Goddard, the quirky collector of rocketry patents who did not work well with others, or von Braun, the oily ex-Nazi who very much did."[39] The reason, it turns out, is because Malina had ties to leftist organizations in the 1930s, although he always denied being a Communist. His pedigree was not one of unabashed anti-communism, and in the Red Scare era of the 1940s and 1950s, it would not do for the keepers of this new and remarkably powerful technology of rocketry to have any Marxist leanings. Always conscious of the horror of rockets in war, Malina believed that his work in World War II was appropriate only to rid the world of a great evil. After World War II, he left the jet propulsion laboratory and accepted employment at UNESCO headquarters in Paris, a post he held from 1947 to 1953. In 1952, the US government indicted Malina and revoked his passport for failing to disclose his association with communism. Although the indictment had little power as Malina was already living abroad, revoking his passport forestalled any chance that he would return to the United States.[40] Fraser MacDonald amplifies this story with declassified CIA records showing the intentional sidelining of Malina's team.[41] Lord said it best: "To a country that viewed international politics as a clash between teams—us and them, right and wrong, good and evil—Communists were them, wrong and evil. What is more, people who had expressed curiosity about communism in the 1930s were not allowed to reconsider. Regardless of the way their sympathies may have evolved, they were inexorably tainted."[42] Malina and the accomplishments of his "suicide squad" have been largely omitted from the history of spaceflight. It is the Huntsville School's systematic pursuit of a selective history of spaceflight that is most interesting here.

Less nefarious perhaps but equally biased, the Huntsville School also downplayed the role of another equally worthy team in spaceflight, the Pilotless Aircraft Research Division/Space Task Group led by Robert R. Gilruth at the National Advisory Committee for Aeronautics (NACA) at Langley Research Center in Hampton, Virginia, which relocated to Houston in the early 1960s. During the latter part of World War II, leaders at NACA, the predecessor to NASA, became interested in the possibilities of rockets and the future of spaceflight. To pursue that technology, NACA created the Pilotless Aircraft Research Division (PARD), under the leadership of the young and promising Gilruth.[43]

Gilruth, as much as anyone and far more than Wernher von Braun, is the godfather of human spaceflight in the United States. After his significant role at PARD, he went on to lead the Space Task Group for NASA that accomplished Project Mercury, then he served as director of the Manned Space-craft Center—renamed the Johnson Space Center in 1973—which supervised Project Gemini and Project Apollo. His organization recruited, trained, and oversaw the astronauts and the human spaceflight program throughout the heroic age of spaceflight. Yet his name is much less well known than others associated with these projects. He was a contemporary on a par with Wernher von Braun, the former Nazi rocketeer who led the effort to build the first launcher that sent spacecraft into orbit for the United States.

Gilruth established a rocket test site on Wallops Island in Virginia. The first launch from there took place on June 27, 1945. While that launch was inauspicious, it was only the beginning, and from 1945 to 1950, Gilruth's team flew 386 rockets. Thereafter, PARD went on to evaluate multistage, hyper-sonic, solid-fuel rockets: a four-stage rocket on October 14, 1954, and the first five-stage rocket in August 1956. The latter rocket reached a speed of Mach 15. These strides in rocket technology positioned Gilruth's organization as critical to the American quest for space in the 1950s. Gilruth championed the idea of human space exploration: "When you think about putting a man up there . . . there are a lot of things you can do."[44] As NASA was being estab-lished in 1958, PARD was beginning testing on larger rockets that launched early capsules capable of sustaining astronauts in space. A series of tests of the Little Joe rocket carrying these capsules proved that they were spaceworthy. Little Joe underwent eight launches in 1959–1960 from Wallops Island to as-sess a launch escape system as well as the Mercury capsule. Rocket testing at Wallops Island has evolved since that time, and it remains a central site of space operations to the present.

It seems a tragedy that Gilruth is not mentioned in the same breath as Wernher von Braun. He was just as central to the Apollo effort, and it could be argued that he was more important because of his role in overseeing the program and managing the astronauts. Could it be that he did not possess the same charisma and charm that everyone acknowledged von Braun exuded? Probably. Could it also be that he did not toot his own horn, write his own history, and encourage others of like mind in that regard? Absolutely. Our understanding of the history of Project Apollo would have been quite differ-ent had there not been a Huntsville School of space history, and the place of a southern aspect of Apollo would have been far different without it. That is a fundamental agenda item for future historians seeking to reinterpret the his-tory of this singular event of the twentieth century.

So What?

This is the fundamental question in any historical investigation. Why is the subject and the historian's handling of it important and what might we learn from it? In the case of the Apollo story in the American South, three major themes are at play and are deserving of serious consideration.

First, the fact that NASA's major human spaceflight centers—Kennedy Space Center, Marshall Space Flight Center, and Manned Spacecraft Center/Johnson Space Center—were located in the South ensured that a large body of NASA employees and contractors were endemic to the region. This gave Apollo a unique hue; it might be partly about the accents of mission controllers and astronauts on the radio but more important it lent privilege to the engineering styles and approaches to problem solving learned at the great universities of the South such as Virginia Tech, Georgia Tech, and Texas A&M. Many of the engineers who made Apollo successful arose out of the red clay of the American Southeast. Often they were the first members in their families to attend university, and they brought with them a drive and zeal to succeed that came from years of struggle. Without making too much of this, let me suggest that there is room for a prosopography of the engineers who made Apollo successful that includes their origins and education. Did those factors affect the engineering solutions they adopted? Using the approach of the social construction of technology, we may well find that their decision making was not exclusively dictated by hard data but may have included non-engineering considerations such as regional perspectives.[45]

The second area of significance deserving sustained investigation is the place of Apollo in the larger efforts of the administration of Lyndon B. Johnson. Like the New Deal objectives on which Johnson cut his political teeth, the Great Society initiatives of the 1960s were aimed at alleviating economic conditions in areas of the country where they are most evident. This was especially true in the American South. I have laid out this case above, but I invite others to explore it further. Most especially, what was the thinking of southern Democrats in supporting NASA and Apollo, and to what extent did mitigation of economic hardship come into the consideration of budgets and spending priorities? Useful studies might explore this in the context of the individual NASA human spaceflight centers, in the demographics of jobs and contractors, and in the considerations of infrastructure needed in the South to support Apollo. Another topic is the powerful social history of how race and class related to Project Apollo. We know that NASA was committed to exporting its approach to problem solving from Apollo for use on other urgent national problems. They had only modest success but might have realized much more with a more diligent effort. Are there ways to expand on this area in the context of the history of the American South?[46]

Finally, I remain fascinated with the peculiarity of the Huntsville School of space history and its assignment of heroic status to Wernher von Braun and the German rocket team versus a more rounded interpretation of Project Apollo. Like the possibility of a uniquely southern version of engineering that I raised above, was there a peculiarly German version of engineering that dominated the development of the Saturn V? I believe there was. Many have complained of von Braun's conservative engineering approach: each rocket had to have the strength of a bridge that would last for many years even though it only needed to fly once. But the celebration of the rocket team's success is most assuredly out of proportion to its demonstrated success. How and why did this occur? Only through much additional investigation will we get to answers to this question. I would call for historical excavation of how the public came to accept this story of German heroism during Project Apollo and how it originated and was transmitted over time from the early part of NASA's history to the present. This is more than debunking; it is really about the dialectic of historical understanding with each study extending, modifying, or overturning parts of what went before. That type of engagement with historians of differing perspectives offers the prospect of a mature encounter with the NASA past, something that has only now begun to take place.[47]

Notes

1 On American exceptionalism, see Charles Lockhart, *The Roots of American Exceptionalism: Institutions, Culture and Policies* (Basingstoke, UK: Palgrave Macmillan, 2003); David W. Noble, *Death of a Nation: American Culture and the End of Exceptionalism* (Minneapolis: University of Minnesota Press, 2002); Godfrey Hodgson, *The Myth of American Exceptionalism* (New Haven, CT: Yale University Press, 2010); and Andrew J. Bacevich, *The Limits of Power: The End of American Exceptionalism* (New York: Henry Holt and Co., 2008).

2 I have discussed these responses preliminarily in Roger D. Launius, "American Spaceflight History's Master Narrative and the Meaning of Memory," in *Remembering the Space Age*, ed. Steven J. Dick (Washington, DC: NASA, 2008), 353–385.

3 James E. Webb, NASA administrator, to E. F. Buryan, July 18, 1961, Webb, James E. Papers, Harry S. Truman Presidential Library & Museum, Independence, MO (hereafter Webb Papers). See also James E. Webb to George R. Herbert, president, Research Triangle Institute, July 10, 1961; James E. Webb to John I. Snider, chairman, United States Industries Inc., February 7, 1962; James E. Webb to John J. Burke, January 3, 1963; James E. Webb memorandum to Mr. Lingle, October 17, 1962; James E. Webb to Senator Jennings Randolph, October 7, 1962; and James E. Webb to Thomas H. Eliot, chancellor, Washington University, December 26, 1962, all in Webb Papers.

4 James E. Webb memorandum for program officers, headquarters, and directors, NASA centers and installations, July 5, 1961, Webb Papers. See also James E. Webb to Chase, June 8, 1961; NASA Office of Advanced Research and Technology, "Extract of Remarks by R.L. Bisplinghoff," December 9, 1964; and James E. Webb to President John F. Kennedy, May 31, 1961, both in Webb Papers.

5 James E. Webb to Marvin Kelly, February 17, 1955; James E. Webb to Lloyd V. Berkner, May 27, 1957; James E. Webb to Homer E. Newell, November 17, 1961, all in Webb Papers; Clayton Anderson, "Science: Oklahoma's New Frontier," Oklahoma Today, January 1956, 8, 17.

6 JFK to LBJ, July 29, 1963; LBJ to JFK, July 31, 1963, White House Famous Names, LBJ Presidential Library, Austin, TX (hereafter LBJ Library); Walter A. McDougall, The Heavens and the Earth: A Political History of the Space Age (New York: Basic Books, 1985), 322–323, 376, 389–396.

7 Lyndon B. Johnson, "Remarks at the University of Michigan," May 22, 1964, The American Presidency Project, https://www.presidency.ucsb.edu/documents/remarks-the-university-michigan.

8 Daniel Walker Howe, The Political Culture of the American Whigs (Chicago: University of Chicago Press, 1979), 20.

9 See Julian E. Zelizer, The Fierce Urgency of Now: Lyndon Johnson, Congress, and the Battle for the Great Society (New York: Penguin, 2015); Randall B. Woods, Prisoners of Hope: Lyndon B. Johnson, the Great Society, and the Limits of Liberalism (New York: Basic Books, 2016); and John A. Andrew III, Lyndon Johnson and the Great Society (Chicago: Ivan R. Dee, 1998).

10 James Webb to LBJ, December 20, 1963, WHCF:GEN:FG260:NASA; and Webb to LBJ, October 29, 1965, WHCF:EX:FG260, LBJ Library.

11 See Edward C. Welsh's weekly NASA reports to LBJ for 1964 in WHCF:EX\OS, LBJ Library.

12 Webb-LBJ Briefing, October 14, 1968, from which the quotes are drawn, James Webb Papers, NASA Historical Reference Collection, NASA History Office, Washington, DC.

13 Robert A. Divine, The Johnson Years, Volume Two: Vietnam, the Environment, and Science (Lawrence: University Press of Kansas, 1987), 235; LBJ, interview with Walter Cronkite, July 5, 1969, LBJ Files, NASA Historical Reference Collection; Lyndon Baines Johnson, The Vantage Point: Perspectives of the Presidency, 1963–1969 (New York: Holt, Rinehart, and Winston, 1972), 285–286.

14 "George A. Smathers, United States Senator, 1951–1969," Oral History Interviews, August 1 to October 17, 1989, 85, U.S. Senate Historical Office, Washington, DC; McDougall, The Heavens and the Earth, 361–363, 373–376.

15 NASA, Conference on Space, Science, and Urban Life (Washington, DC: NASA, 1963).

16 NASA, Conference on Space, Science, and Urban Life, 218.

17 Eugene S. Uyeki, "Review of Conference on Space, Science, and Urban Life," Technology and Culture 5 (Autumn 1964): 626–628.

18 Henry R. Hertzfeld, "Space as an Investment in Economic Growth," in Exploring the Unknown: Selected Documents in the History of the U.S. Civil Space Program, vol. 3, Using Space, ed. John M. Logsdon (Washington, DC: NASA, 2000), 385–401.

19 Roger W. Hough, "Some Major Impacts of the National Space Program," June 1968, 1–2, 19–22, 36, NASA Historical Reference Collection.

20 "Economic Impact of Stimulated Technological Activity," Final Report, Midwest Research Institute, Contract NASW-2030, October 15, 1971, 1–11, NASA Historical Reference Collection.

21 Lee D. Saegesser, "High Ground Advantage," n.d., NASA Historical Reference Collection.

22 Arthur M. Schlesinger Jr., *Robert Kennedy and His Times* (Boston: Houghton Mifflin Company, 1978), 335–336.

23 James A. Webb to Wernher von Braun, June 24, 1963, Minority Groups folder, NASA Historical Reference Collection; Marshall Space Flight Center Manpower Office, "A Chronology of the Equal Employment Opportunity Program at MSFC," February 1971, 5.

24 Andrew J. Dunar and Stephan P. Waring, *Power to Explore: A History of Marshall Space Flight Center, 1960–1990* (Washington, DC: NASA, 1999), 116–128, quote on 118–119.

25 Margot Lee Shetterly, *Hidden Figures: The American Dream and the Untold Story of the Black Women Mathematicians Who Helped Win the Space Race* (New York: William Morrow, 2016); Richard Paul and Steven Moss, *We Could Not Fail: The First African Americans in the Space Program* (Austin: University of Texas Press, 2015).

26 Charles L. Sanders, "The Troubles of 'Astronaut' Edward Dwight," *Ebony*, June 1965, 29–36; Frank White III, "The Sculptor Who Would Have Gone to Space," *Ebony*, February 1984, 54–58; Betty Kaplan Gubert, Miriam Sawyer, and Caroline Fannin, *Distinguished African Americans in Aviation and Space Science* (Westport, CT: Oryx Press, 2001), 113–117.

27 Ben A. Franklin, "Wallace Is Given a NASA Warning; He Is Told on 'Truth' Tour to Liberalize His Policies," *New York Times*, June 9, 1965; Paul and Moss, *We Could Not Fail*, 203–205, 214–215; Robert Stone and Alan Andres, *Chasing the Moon: The People, the Politics, and the Promise that Launched America into the Space Age* (New York: Ballantine Books, 2019), 154–159.

28 Duke Newcome, "Not Everyone Came to Praise the Launch," *Florida Today* (Orlando), July 17, 1969.

29 *Titusville Star-Advocate*, July 15, 1969; Bernard Weinraub, "Hundreds of Thousands Flock to Be 'There,'" *New York Times*, July 16, 1969.

30 Thomas O. Paine, NASA Administrator, Memorandum for Record, July 17, 1969, NASA Historical Reference Collection.

31 Thomas P. Murphy, "The Moon and the Garbage of New York," *Review of Politics* 34 (April 1972): 271–273.

32 Eric Fenrich, "Détente and Dissent: Apollo-Soyuz, Ruth Bates Harris, and NASA's Rhetoric of Cooperation," *Quest: The History of Spaceflight Quarterly* 22, no. 1 (2015): 4–15.

33 Rip Bulkeley, *The Sputniks Crisis and Early United States Space Policy: A Critique of the Historiography of Space* (Bloomington: Indiana University Press, 1991), 204–205, 208.

34 Wernher von Braun, Frederick I. Ordway III, and Dave Dooling, *History of Rocketry & Space Travel* (New York: Thomas Y. Crowell Co., 1966).

35 Representative examples of these works include Frederick I. Ordway III and Mitchell R. Sharpe, *The Rocket Team* (New York: Thomas Y. Crowell, 1979); Ernst Stuhlinger and Frederick I. Ordway III, *Wernher von Braun: Crusader for Space*, 2 vols. (Malabar, FL: Robert E. Krieger Company, 1994); Frederick I. Ordway III and Randy Liebermann, eds., *Blueprint for Space: From Science Fiction to Science Fact* (Washington, DC: Smithsonian Institution Press, 1992); Frederick C. Durant III, ed., *Between Sputnik and the Shuttle: New Perspectives on American Astronautics* (San Diego, CA: American Astronautical Society, 1981); Walter Dornberger, *V-2: The Nazi Rocket Weapon* (New York: Viking, 1954).

36 For example, see William B. Breuer, *Race to the Moon: America's Duel with the Soviets* (Westport, CT: Praeger, 1993); and Marsha Freeman, *How We Got to the Moon: The Story of the German Space Pioneers* (Washington, DC: 21st Century Science Associates, 1993).

37 Clayton R. Koppes, *JPL and the American Space Program: A History of the Jet Propulsion Laboratory* (New Haven, CT: Yale University Press, 1982); James L. Johnson, "Rockets and the Red Scare: Frank Malina and the American Missile Development 1936–1954," *Quest: The History of Spaceflight Quarterly* 19, no. 1 (2012): 30–36.

38 Many have studied this story, but the standard work on the V-2 program is Michael J. Neufeld, *The Rocket and the Reich: Peenemünde and the Coming of the Ballistic Missile Era* (New York: Free Press, 1995).

39 M. G. Lord, *Astro Turf: The Private Life of Rocket Science* (New York: Walker and Co. 2005), 66.

40 See James L. Johnson, "Frank Malina: America's Forgotten Rocketeer," IEE Spectrum, July 31, 2014, https://spectrum.ieee.org/frank-malina-americas-forgotten-rocketeer; Dr. Frank J. Malina (1912–1981), Jet Propulsion Laboratory, California Institute of Technology, https://www.jpl.nasa.gov/who-we-are/faces-of-leadership-the-directors-of-jpl/dr-frank-j-malina-1912-1981.

41 Fraser MacDonald, *Escape from Earth: A Secret History of the Space Rocket* (New York: Hachette Book Group, 2019).

42 Lord, *Astro Turf*, 95.

43 Much of this section is based on Roger D. Launius, "Godfather to the Astronauts: Robert Gilruth and the Birth of Human Spaceflight," in *Realizing the Dream of Flight: Biographical Essays in Honor of the Centennial of Flight, 1903–2003*, ed. Virginia P. Dawson and Mark D. Bowles (Washington, DC: NASA, 2005), 213–256.

44 Robert R. Gilruth, interview with Linda Ezell, Howard Wolko, and Martin Collins, National Air and Space Museum, Washington, DC, June 30, 1986.

45 I am struck by the concept of heterogeneous engineering here, which recognizes that technological issues are simultaneously organizational, economic, political, and cultural. See John Law, "Technology and Heterogeneous Engineering: The Case of Portuguese Expansion"; and Donald MacKenzie, "Missile Accuracy: A Case Study in the Social Processes of Technological Change," both in *The Social Construction of Technological Systems: New Directions in the Sociology and History of Technology*, ed. Wiebe E. Bijker, Thomas P. Hughes, and Trevor J. Pinch (Cambridge, MA: The MIT Press, 1987), 111–134 and 195–222, respectively.

46 See Roger D. Launius, "Managing the Unmanageable: Apollo, Space Age Management, and American Social Problems," *Space Policy* 24 (August 2008): 158–165.

47 I sought to wrestle with these issues, at least somewhat, in Roger D. Launius, "Historical Dimensions of the Space Age," in *Space Politics and Policy: An Evolutionary Perspective*, ed. Eligar Sadeh (Dordrecht: Kluwer Academic Publishers, 2002), 3–25.

2

Lyndon Johnson, NASA, and the South

Andrew J. Dunar

From the launch of Sputnik in October 1957 until he left the presidency in January 1969, Lyndon B. Johnson held key positions that he used to influence the American space program in its formative years. Historian Robert Dallek has argued that LBJ "deserves to be remembered as the elected official who did as much, if not more, for space exploration than any other American political leader in [the twentieth] century."[1] As Senate majority leader in the late 1950s, Johnson chaired the two most important Senate committees involved in space-related activities: the Senate Armed Services Committee's Subcommittee on Preparedness, which investigated the nation's missile program, and the committee that wrote the legislation that established the National Aeronautics and Space Administration. As vice-president during the Kennedy administration, Johnson chaired the National Aeronautics and Space Council (usually known as the Space Council), which had the potential to influence the nation's space policy. He served as president during crucial years of the American space program, when NASA and his administration faced dramatic challenges: fulfilling President Kennedy's pledge to place a man on the moon and return him safely to Earth by the end of the decade; overcoming the most serious setback of the Apollo years, the 1967 Apollo fire that killed three astronauts; funding the program in the context of the expensive domestic agenda of the Great Society and the debilitating war in Vietnam; mitigating the ongoing contention over the division between civilian and military space programs; addressing interservice rivalry over responsibilities for space within the military sector; supporting an international agreement to prevent the weaponization of space without curtailing other military involvement; and planning the next step in space after the Apollo program.

From October 1957 to November 1963, Johnson's ties to the South were linked to his role in the American space program. This was less the case after he became president, and indeed his decision to support a strong civil rights bill after the Kennedy assassination signaled a break with the South that would not be repaired. But following the launch of Sputnik in 1957, Johnson depended on southern support to maintain his leadership in the Senate. This period coincided with the rise of the civil rights movement and with a competitive contest for the 1960 Democratic presidential nomination. Maintaining southern support while demonstrating his capacity for national leadership was a challenge even for someone with Johnson's considerable political instincts.

Johnson's involvement with NASA had more impact on the South than on any other region. The location in the South of the major NASA centers involved in Apollo made this inevitable. But other factors were involved. There were echoes of Franklin Roosevelt's Tennessee Valley Authority in the economic impact of Apollo on the South. And NASA furthered other national goals, including modernization, advancing civil rights, and ameliorating poverty. Executive office attention to some of these issues predated Johnson's presidency. For example, the Kennedy administration pressured Johnson, as Space Council chair, to pressure the NASA centers in the South to increase the hiring of African Americans with the threat that projects would be transferred elsewhere if they did not do so.

Lyndon Johnson's relationship with the South was complicated. Journalists and historians have pondered the relationship, especially in evaluations of the Civil Rights Act of 1964 and the Voting Rights Act of 1965. When Johnson became president after President Kennedy's assassination, the question of Johnson's sectional loyalty was a concern for many Americans, especially those concerned about civil rights. Civil rights activist John Lewis recalled reacting with trepidation because Johnson was "a Southerner, a Texan."[2] But did Johnson consider himself to be a southerner? Journalist James Reston wrote that "if you said he was from the South, he would say he was from the West, and the other way around."[3] James Rowe, a close political advisor, said of Johnson: "He lived in both worlds, and could go back and forth."[4] Even Johnson's wife, Lady Bird, questioned the southern component of her husband's identity, calling him a "southwesterner" and claiming that she was more southern than he was.[5] She had been raised in East Texas but spent summers with relatives in Alabama as a young woman.

Johnson identified with the South when it suited his purposes to do so. Historian William Leuchtenburg labeled his section identity "fungible."[6] While southern support was vital to Johnson as majority leader and in his campaign for the presidency in 1960, his support for the South varied with circumstances. James Parker, an African American who was a driver for LBJ, recalled that Johnson often called him "nigger" in front of guests and that Johnson "especially like to put on a show for Senator Bilbo," the Democrat from Mississippi who was probably the most racist member of the US Senate. "I used to dread being around Johnson when Bilbo was present," Parker recalled, "because I knew it meant that Johnson would play racist."[7] Hubert Humphrey, who served in the Senate while Johnson was majority leader before becoming his vice-president, said that LBJ "was on good working relationships with every southerner, but he wasn't quite southern. . . . Johnson was never a captive of the southern bloc. He was trying to be a captain of them, not a captive."[8]

During his rise to power, Johnson often nurtured ties to powerful older men, including President Franklin D. Roosevelt and Speaker of the House Sam Rayburn. Rayburn, a fellow Texan, was a bachelor who led a lonely private life. Johnson took him into his family, having him over for dinner so frequently that the Johnson daughters called him "Uncle Sam." Johnson developed a similar relationship with Senator Richard Russell, the Georgia Democrat who was the leader of the Southern Caucus. Russell initiated the relationship in 1949, when he invited Johnson to attend a meeting of the caucus at a time when he feared that President Truman planned to initiate civil rights legislation. Johnson was a freshman senator, and Russell assured him that if he did not attend, no one would know the invitation had been extended. In the years that followed, Russell became a mentor to Johnson. Like Rayburn, Russell was a bachelor, and the Johnson family extended the same welcome to Russell that they did to Rayburn; Russell became "Uncle Dick." Russell and Johnson often met for breakfast in the Senate dining room in the Capitol, and they often met Rayburn for dinner at a Washington seafood restaurant.[9]

Johnson's relationship with Russell, which connected him to southern Democrats, paid dividends in the years that followed. The southern Democrats wielded power beyond their numbers in the 1950s and 1960s. Seniority overrode all other criteria in the selection of chairs for Senate committees. Since Reconstruction, southern Democrats had had such dominance in elections that the region became known as the Solid South. Whoever won the Democratic primary in southern states had a virtual lock on victory in the state's senate election and on reelection when their term expired. Thus, they accumulated seniority that earned them chair positions in most of the key senate committees. As such, they shaped debate on important issues and had the power to prevent legislation on issues such as civil rights from reaching the senate floor for a vote. By the time he became majority leader in January 1955, Johnson had an excellent working relationship with the southern

Democrats. He had worked with them as minority leader in the preceding Congress, and his ties to Richard Russell continued to serve both men well.

In 1957, two issues came before Congress that tested the nation and provided a measure of Johnson's leadership ability. Civil rights, the first of these matters, had become the major domestic issue, one that demanded federal attention. And the Russian launch of Sputnik in October shattered the nation's confidence in its technological dominance and its role of international leader. Lyndon Johnson's response to these two crises relied on his ties to the Southern Caucus, enhanced his power in the Senate and placed him as a leading contender in what promised to be a strong field for the Democratic nomination for president in 1960.

Civil rights had been looming over American politics for decades, but it became central after the Supreme Court ruling in the 1954 case *Brown v. Board of Education of Topeka, Kansas* that segregation in public schools was unconstitutional. However, southerners had decades of experience in opposing changes in civil rights, usually by insisting that civil rights matters were to be decided at the state level, not by the federal government. Although a Supreme Court decision was not so easily brushed aside, southerners still sought to oppose it. In 1956, 101 southern congressmen and senators signed the Southern Manifesto, which labeled the Court's decision a "judicial abuse of power" and promised to oppose the decision by any legal means available. Lyndon Johnson chose not to sign, as did most of the Texas delegation. Southerners pressured him to sign, but Russell gave him cover, reasoning that as leader of all Democrats in the Senate, it would not be proper for him to sign.[10]

In 1957, the issue became more difficult for Johnson when President Eisenhower proposed a civil rights bill centered on voting rights. This was familiar ground for southern Democrats, who were accustomed to swatting down such proposals by burying them in committees, staging filibusters, and applying the states' rights argument. This was more difficult for Johnson, particularly because as majority leader his actions would be under national scrutiny. If he opposed the bill, he would be praised in the South but vilified in the rest of the nation. It would likely end any possibility of gaining a presidential nomination. If he supported the bill, he stood to lose the support of the South that he had so carefully cultivated. James Rowe offered sage advice, suggesting that although his friends and enemies both believed that "this is Lyndon Johnson's Waterloo," he would be better served by fashioning a reasonable approach that would satisfy southerners and yet offer a modest achievement that would mollify Eisenhower and the North.[11] Southern Democrats knew that the only candidate from the South who could become president was Lyndon Johnson and that the only way he could do so was by supporting some form of civil rights legislation. Southern senators who supported him expected that he would give them enough sweeteners in the bill to make it palatable. Many considered his support for any civil rights legislation to be a betrayal, but he had some room for maneuver.

Johnson's argument to the South was that there was broad support for civil rights across the nation and that although the issue would remain controversial, something had to go forward because it was a proposal from the president. However, he knew the South had to be appeased. A fight over civil rights could have torn the Democratic Party apart and surely would have ended Johnson's presidential aspirations. It was a difficult task, but not a lost cause. Biographer Robert Caro argues that Johnson had the ability to simultaneously convince people on both sides of an issue that he was on their side. Since he "was the only bridge between the two sides," he had to keep himself as the negotiator between them.[12] And he had the invaluable support of Richard Russell, who convinced his angry colleagues not to stage a filibuster that might have ended a chance for a compromise. Ultimately, South Carolina senator Strom Thurmond executed a solo filibuster that lasted just over twenty-four hours, setting a record. But the real victor in the filibuster battle was Russell, who kept the rest of the Southern Caucus together. Johnson allowed revisions that weakened the bill enough to temper opposition from southern Democrats. The Senate passed the bill, which was watered down to the point that it did little to protect voting rights. However, it did establish the Civil Rights Commission, which gave the bill a small measure of credibility. Johnson, as the only senator who could have engineered such a compromise, emerged from the ordeal with his reputation as majority leader enhanced and with his credentials as a national leader established.

The next test of LBJ's national leadership ability was not long in coming. On October 4, 1957, the Soviet Union announced that it had placed a satellite in orbit around the earth. Americans had come to assume that the United States was the world's technological leader, but the launch of Sputnik instantly changed that image. Johnson was at his ranch in Texas when the news broke. In his memoir, he recalled taking a walk with his guests along the road that led to the Pedernales River, looking to the sky, hoping to catch a glimpse of the satellite.[13] Gerald Siegel, who was one of those guests, remembered that walk and recalled that when they returned to the ranch house, Johnson decided to call Senator Russell.[14]

Russell was at home in Georgia, fielding phone calls from anxious colleagues who were urging him as chair of the Senate Armed Services Committee to hold hearings to consider how to respond. Most adamant among the callers was Stuart Symington (D-Missouri), who had served as the first secretary of the air force during the Truman administration. Symington had already expressed interest in running for president and Russell worried that he would use the committee to attack the president (as he had already been doing) and that he would push defense-oriented responses. Russell was concerned that he no longer had the energy to lead hearings that were bound to be full of controversy and partisan disputes. He decided instead that the Preparedness Investigating Subcommittee, a subcommittee of the Senate Armed Services Committee, ought to stage hearings. That committee had been inactive since the end of the Korean War. But now Russell believed that the committee chair, Lyndon Johnson, would be the ideal senator to lead such an investigation, so he decided to reactivate it. He reached Johnson just as Johnson was preparing to call him.[15]

When the Preparedness Subcommittee was formed in 1950, during the Korean War, its mission was to monitor government spending, much as the Truman Committee had done during World War II. Chairing this committee was a plum assignment for Johnson, who was a freshman senator in 1950. At the time, President Truman took Johnson aside and told him that his success in the committee would depend on his relationship with the ranking Republican and that he was fortunate to have Styles Bridges (R-New Hampshire) in that position. Truman said that Bridges would watch him and that if he "dealt off the top" of the deck, Bridges would do the same.[16] In 1957, Bridges was still the ranking Republican on the committee. So when Johnson's aide George Reedy advised to run the committee in a bipartisan fashion, the advice fell on fertile ground.

The senators on the committee were already in place, but Johnson had to hire a staff. The Democrats on the committee included two senators who were likely rivals for the party's presidential nomination, Stuart Symington and Estes Kefauver (D-Tennessee), and one of the members of the Southern Caucus, John Stennis of Mississippi. Johnson hired Ed Weisl as chief counsel to the committee. Weisl brought along two lawyers from his law firm, his son Ed Jr. and Cyrus Vance. At Russell's recommendation, Johnson also hired Eilene Galloway, a national defense analyst who had served on the staff of Russell's Senate Armed Services Committee. A few months earlier, she had written a report entitled "Guided Missiles in Foreign Countries."[17]

Johnson's closest advisors in the months after the Sputnik launch had diametrically opposing recollections of the senator's actions during the two weeks after October 4. Reedy, Johnson's key advisor on a wide range of policy issues, and LBJ were both in Texas—Johnson at his ranch and Reedy in Austin, fifty miles east. Reedy recalled that "for about two weeks I merely let the thing vegetate. Senator Johnson had so many problems on his mind that I doubt whether he devoted too much attention to it. . . . I sort of put the space thing on the back burner."[18] Johnson contacted Eilene Galloway on October 7. She remembered that "LBJ reacted swiftly and became the most energized leader I have ever beheld in galvanizing the Congress, the Pentagon, industry and the scientific community, to take decisive action to achieve U.S. preeminence in outer space."[19] Galloway and Reedy participated in a NASA-sponsored conference in 1992 and both commented on Johnson's actions in the two weeks after October 4. The difference between the two accounts takes on added significance because of a memo that Reedy wrote to Johnson on October 17 that outlined suggestions for Johnson's response to Sputnik, which is discussed below. That memo was the first contact between Reedy and Johnson since the Sputnik launch, and Reedy assumed that the memo energized LBJ, whom Reedy believed had not taken an interest in the American response to that point. In an interview in 1982, Reedy explained, "I had been a very ardent advocate of the whole space program. In fact, I'm pretty sure I'm the one that sold Johnson on it."[20] Galloway offered the most plausible explanation for the difference between her account and that of Reedy. She surmised that Johnson had not contacted Reedy earlier, so Reedy assumed that Johnson had not been interested in the issue. She suggested that "Reedy was describing his own reactions, and ascribing them to the senator."[21] Another member of the staff of the Preparedness Investigating Subcommittee, Glen P. Wilson, agreed with Galloway. "There have been some people who said Johnson was slow to get started," Wilson said, "and the record shows that that was anything but the truth. The truth is that he was the first man to do this."[22]

While he was still in Texas, Johnson began to plan for committee sessions and put his staff to work gathering information. On October 7, Solis Horowitz, an attorney on Johnson's senate staff, asked for a report on the missile and rocket work of each of the services; all aspects of the Vanguard program, including budgets and contracting; and details about the Soviet missile and rocket programs.[23] LBJ followed up when it seemed that the Defense Department was taking too much time. He called Secretary of Defense Neil McElroy, explaining that it was the committee's responsibility to maintain "surveillance" over the Defense Department. He asked for all relevant information on the satellite, including "why we did not have it first." McElroy said he had already turned over some material to Horowitz and would send a report by the end of the week.[24]

Reedy's memo of October 17 may not have been the trigger that first stimulated Johnson's interest in space policy, but it was the best advice Johnson received concerning how to proceed. Reedy had worked with Johnson since 1951. He had advised Johnson on several policy matters, including civil rights. The impetus for the space memo had southern roots. Its genesis was a meeting Reedy had with Charles Brewton, formerly an aide to Senator Lister Hill (D-Alabama). Shortly after the Sputnik launch, Brewton flew to Texas to confer with Reedy. Brewton argued that Sputnik offered a chance to change the nation. It would help the Democrats "clobber" the Republicans, lead to technological advances, and elect Lyndon Johnson as president. Reedy said Johnson was not interested in running for president. (This is one thing about Reedy's recollections that Eilene Galloway agreed with.[25]) Brewton said he'd settle for clobbering the Republicans. This was the impetus for Reedy's October 17 memo to Johnson, in which he stressed the need for bipartisanship but also made more specific suggestions. He urged the senator to avoid bringing in air power advocates such as Curtis LeMay, commander of the Strategic Air Command, who would be supported by Senator Symington and who would argue that the solution to Sputnik should be a military response that would center on ballistic missiles and favor the air force. Instead, Reedy argued, bring in scientists. Treat Sputnik as a challenge similar to Pearl Harbor. The nation needed to beat the Russians in space, and Sputnik could inspire American technology. Reedy counseled Johnson on the importance of this opportunity, which offered an unusual chance to combine politics and statesmanship.

Meanwhile, Senator Symington contacted the White House, calling for a special session and for an investigation of how American scientific and technological capabilities compared with those of the Soviet Union, what the pace of the ballistic missile program was, and what limits there were on defense spending. President Eisenhower had downplayed the importance of the Sputnik launch, insisting that the American missile program was in good shape and that Sputnik posed no threat to national security. He brushed aside Symington's concerns, replying that he didn't see a need for a special session and that he had sent copies of Symington's request and his rejection letter to Senator Johnson. Eisenhower's aide, Bryce Harlow, passed the copies to Johnson's staff.[26]

On November 3, the Soviet Union launched a second Sputnik into Earth orbit, carrying a dog named Laika. This proved that the Soviets were not a one-shot wonder. This time the capsule carried a larger payload than the first Sputnik, which had been an aluminum sphere 22 inches in diameter that weighed 184 pounds. Sputnik II weighed 1,100 pounds and had instrumentation to cool the capsule and monitor Laika's vital statistics. Laika died on the fourth orbit when the temperature inside the capsule rose to 109 degrees, but the Soviet accomplishment, especially the weight of the payload, only added to the American panic.[27]

There was no time to waste. Johnson met with President Eisenhower on November 6 and assured him that the Preparedness Subcommittee hearings would be bipartisan. Eisenhower and Johnson had worked together often in the preceding years, particularly on foreign policy issues. In that domain, Eisenhower was closer to Democratic positions than he was to those of the conservative isolationist wing of the Republican Party. One of the benefits of bipartisanship in this case was the common goal of limiting Symington's role: Johnson saw Symington as a rival for the 1960 Democratic nomination and Eisenhower wanted to avoid having the hearings focus on a demand for more money for the military. For Johnson, there was the added advantage that limiting Symington's role satisfied Russell (and the Southern Caucus). Eisenhower knew Johnson well enough to be wary, but after meeting with Johnson he told his secretary that Johnson had said the right things. He concluded, "I think today he was being honest."[28] But he also told Bryce Harlow to begin an investigation into missile development during the Truman years in case the Democrats began to blame his administration for a failure to develop a missile program.

Johnson also gave his pledge of bipartisanship to Styles Bridges. While the hearings were in session, he kept his word. Ed Weisl helped maintain bipartisanship through the unusual arrangement of serving as counsel to both the Democrats and the Republicans on the committee. He later recalled that Johnson "was always a consensus man; he wanted to get unanimous agreement from the committee on every issue, from both Republicans and Democrats; and we succeeded in getting that."[29] Johnson appointed a southerner, Senator John Stennis (D-Mississippi) as his vice-chair. Stennis told Johnson, "This is so vital a matter that nothing short of your guidance will give it the necessary prestige and force."[30]

Johnson called a planning session for the afternoon of November 22. At that meeting were Senators Stennis (the vice-chair), Symington, Bridges (the ranking Republican), and Ralph Flanders (R-Vermont), who had made his mark three years earlier when he was one of the first Republicans to denounce Senator Joseph McCarthy on the floor of the Senate. Johnson established his control of the hearings. He would introduce each witness and each senator would have ten minutes to question each witness in the first round. He announced nine witnesses who had already been scheduled. When Symington named several individuals he would like to have appear before the committee, Johnson asked that Symington and any other senators who wanted to suggest witnesses do so in a letter to the committee counsel. Symington asked if he could bring his assistant, Edward Welsh, to the hearings, since Welsh had responsibilities related to missile development. Johnson turned him down, saying that if he allowed Symington to bring a member of his staff, everyone else would want to do so.[31]

The hearings, which were held on the third floor of the Senate Office Building, began on November 25. Johnson took control from the start. He emphasized that the hearings would be bipartisan, as he had stressed from the beginning. Then he compared the Sputnik crisis to Pearl Harbor, suggesting (without saying so) that the Eisenhower administration had left the nation unprepared. Johnson seized the moment, as he intended to do. Glen Wilson remembered that it "put him up-front in the public eye as being 'Mr. Space,' an image he held for many years as the one politician who was truly interested in the space program and its implications for the future of the United States and its place in the future development of mankind."[32]

Johnson then introduced the first witnesses, doing so in a way that enabled him to further establish control over the proceedings. The first witnesses were scientists, which kept Senator Symington from turning the hearings into a forum for demanding larger budgets for the military. While this satisfied the president's desire to limit Symington's role, Johnson had another motive. No one could complain that the scientists were partisan, but inevitably they would say something about how the Russians had seized the moment, catching the Americans unaware, perhaps not blaming the president but certainly not defending him either.

The first witness, Edward Teller, was a nuclear physicist who had gained fame as the man who led the development of the hydrogen bomb. He described the frightening prospect of Russian ballistic missiles armed with thermonuclear bombs. Next came Vannevar Bush, who had trained as an electrical engineer but who had had a varied career in academia and public service. He had served on the National Advisory Committee for Aeronautics (NACA), the predecessor to NASA, and had pushed for the establishment of the Manhattan Project that developed the atomic bomb. His contribution to the hearings was principally to promote science education, an argument that broadened the discussion beyond consideration of developing military missiles. Jimmy Doolittle, who also had NACA credentials, and had become its chair in 1956, shifted the focus to the military and warned that the United States risked falling even further behind the Russians if it did not fully fund missile development.

The second day produced the most alarming revelations of the hearings. In a top-secret session, CIA director Allen Dulles and Herbert Scoville, the CIA's assistant director for scientific research, set out details of Soviet nuclear research and missile development that gave context to the Sputnik launches. The CIA had conducted a series of flights over Soviet territory with U-2 aircraft, which could fly at an altitude of 70,000 feet, which at the time was too high for the Soviet Union to shoot down. The CIA also had an advanced radar system in eastern Turkey that yielded information on the frequency and range of Soviet missile tests. The combined data indicated that perhaps as early as 1958 the Soviets could have a missile that could carry a one-ton nuclear weapon 5,000 miles. Furthermore, the CIA claimed, they could feasibly have as many as 500 intercontinental ballistic missiles by mid-1960.[33]

After the first three days of testimony, Reedy warned Johnson that the pattern that was emerging was "the extreme difficulty in pinning down lines of authority in the missile and satellite programs" and said that the committee would have to "inquire very carefully who is running what."[34] That issue became clearer as the committee moved into the next phase, when representatives of the armed services appeared. Confusion became apparent as each of the services had their say, opening issues that had worried Eisenhower. Each of the services had its own intermediate range ballistic missile program: the army had Jupiter, the navy had Polaris, and the air force had Thor. The Vanguard program, created under the auspices of the navy, suffered a humiliating launch failure on December 6, while the hearings were under way, giving credence to the military's complaints of inadequate funding. When William Holaday, the special assistant to the secretary of defense for guided missiles, admitted that he was not an expert on missiles, it added to the appearance of mismanagement. For the military men, the hearings were a chance to air their grievances, mainly about unnecessary duplication of effort and lack of adequate funding. Historian Robert A. Divine referred to this portion of the hearings as a "military sour grapes session."[35]

Nevertheless, the press considered the hearings a triumph for Johnson, who received praise for his statesmanship and bipartisanship. While he did not criticize the Eisenhower administration, he didn't have to, since witnesses before the committee did so throughout, even during the testimony of the science experts. His performance established him as the legislator with the strongest record on space in the Congress. But Johnson's proclivity to exaggerate also characterized his approach. His frequent comparisons between Sputnik and Pearl Harbor were just one example of Johnsonian hyperbole. In a speech on January 7, 1958, which was something of a "pre-bate" to Eisenhower's state of the union address two days later, Johnson argued that "control of space means control of the world." Critics referred to his committee as the "Johnson Earth Control Clinic."[36] Robert Caro argues that his concern was "less preparedness than publicity."[37]

Despite these criticisms, when Congress moved forward to construct the architecture of a new space agency, Johnson was at the peak of his powers as

Senate majority leader and well established as the Senate's expert on space. The House and the Senate each organized a special committee to craft bills to meet President Eisenhower's goal of creating a civilian agency. The Senate voted 78–1 on February 6 to create a Special Committee on Space Aeronautics.[38] To no one's surprise, LBJ was elected to chair the committee. He selected senators he knew would support him as committee members. The task before Congress was to recommend legislation to form a civilian space agency that would be responsible for space science research but would also accommodate military space research. The House produced its bill on May 20 and sent it to the Senate on June 2. The Senate thus had the House bill, the results of the Preparedness Investigation Subcommittee hearings, and additional information from its own hearings conducted in May. The Senate bill used much of the language from the House bill but broadened the statement on the scope of the agency, added a policy board, and inserted a statement on international cooperation.

The White House objected to the policy board, since President Eisenhower believed it might be an overly bureaucratic organization that could undercut the role of the president in setting space policy. Johnson and Senator Bridges met with the president on July 7, mainly to discuss the policy board. Eisenhower stated his objections and Johnson reminded him that he had initially supported the idea. Johnson suggested that the president should be made president of the board, which would remove Eisenhower's objections. Eisenhower agreed, and the board became part of the enabling legislation for the new agency. As it turned out, this agreement was fateful for Johnson's future with the Space Council. Eisenhower never warmed to the idea of serving as president of the council and never met with it. When John Kennedy became president, he also wanted to avoid such hands-on involvement in space policy and passed the chair on to his vice-president—Lyndon Johnson.

After Johnson's meeting with Eisenhower, the remaining issues were quickly resolved. The House established a standing Committee on Science and Astronautics on July 21 and the Senate created a standing Committee on Aeronautical and Space Sciences two days later. President Eisenhower signed the bill creating NASA on July 29, fixing October 1 as the date when it would officially begin business. On that date, the National Advisory Committee for Aeronautics would end operations and NASA would absorb its functions. The new organization was to plan, direct, and conduct aeronautical and space activities. The act needed adjustment as time went on. For example, the Space Council underwent modification and the Civilian-Military Liaison Committee (to resolve differences between NASA and the military) had a short life span. But the framework proved to be fundamentally sound.

Lyndon Johnson had played a prominent role, and he would continue to do so for his remaining months as majority leader. He and the first NASA administrator, T. Keith Glennan, conferred often on issues such as finances and organization. LBJ's control of his committee remained strong, bolstered by his connection to the senior southern senators who chaired committees. Glennan once visited Senator Stennis to urge him to talk to Johnson about a matter of NASA legislation that Glennan wanted to move along. Afterward, he noted in his journal that "it is interesting to note the way in which these senior senators still defer to the majority leader."[39]

Years later, when Johnson was president, Marianne Means, a reporter for the Hearst press, recalled an occasion when she rode in a limousine to Camp David with Senator Russell and President Johnson. Russell and Johnson reminisced about the aftermath of Sputnik and the establishment of NASA. "Johnson went on and on, and as the story went on his role got bigger, and pretty soon he had done everything and Eisenhower wouldn't have done anything without him. And Russell, who knew all the facts of the case, of course, listened to him for a long time and finally said, 'That's the joy of being president; you get to rewrite history.'"[40]

The formation of NASA required the establishment of new centers devoted primarily to human space flight. One of the responses to Sputnik was the organization of the Space Task Group (STG), the most direct successor of NASA's predecessor, the National Advisory Committee for Aeronautics. Two questions arose early in the process of determining how the STG would operate. Would it be located with an existing center or be an independent center? And where would it be located? Early discussions considered a merger with Lewis Research Center, either at the Lewis location near Cleveland or at Langley, or locating STG at the Goddard Space Flight Center in Greenbelt, Maryland. NASA administrator T. Keith Glennan, his assistant Hugh Dryden, and Abe Silverstein, who had served as director at Goddard in its early months, became convinced that locating STG at an existing center was unworkable. STG had to be an independent operation that was responsible to NASA headquarters, not to an existing center.

Texas also had a space-related political juggernaut. During Glennan's term as NASA administrator, LBJ was senate majority leader and the chair of the two major senate committees related to space. When Johnson left the Senate to run for vice-president, his friend Robert Kerr (D-Oklahoma) took over as chair of the Senate Committee on Aeronautical and Space Sciences. As vice-president under Kennedy, Johnson chaired the Space Council. In the House of Representatives, Albert Thomas (a Texas Democrat whose congressional district included Houston), chaired the subcommittee of the Appropriations

Committee that dealt with NASA appropriations and Olin Teague chaired the Subcommittee on Manned Space Flight. Teague and Bob Casey both sat on the Committee on Science and Astronautics. Both Teague and Casey were also Texas Democrats. Another Texan, Sam Rayburn, was speaker of the House.

Soon after Glennan became NASA administrator in the fall of 1958, Albert Thomas began calling him regarding his desire to place a NASA lab in Houston. Glennan informed him that a site in Beltsville, Maryland, had already been approved and that construction had already begun. Thomas replied that he had been absent at the time of that decision and he had not approved that site. After back-and-forth in other phone calls, Thomas played his trump card: "Now look here, Dr., let's cut out the bull. Your budget calls for $14 million for Beltsville and I am telling you that you won't get a God-damned cent of it unless that laboratory is moved to Houston."[41] Glennan didn't yield, but the decision carried over to the Kennedy administration, when Glennan was no longer NASA administrator. The governor of Ohio, who knew Glennan from Glennan's years as president of Case Institute of Technology, contacted Glennan and asked for help in proposing Ohio as site for the new NASA center. Glennan laughed, and told him, "You know, I suppose there are 25 states doing just this at the present time, and I'll lay you a year's salary that the Center is going to Houston."[42]

When Johnson joined the Kennedy administration as vice-president, many expected that he would take lead in space-related matters. As a result of LBJ's negotiations with President Eisenhower, the Space Act of 1958 designated the president as chair of the Space Council. Kennedy had no more desire to take on this responsibility than his predecessor had, and LBJ's involvement in space matters made him a logical candidate. Johnson became chair of the Space Council and the administration sponsored legislation that made this post permanently a responsibility of the vice-president. On April 20, 1961, Kennedy did ask LBJ to use the Council to make "an overall survey of where we stand it space," and that report gave substantial support to a lunar landing program—although some suggested that it merely endorsed a decision Kennedy had already made. Nonetheless, it was an excellent report that included the opinions of a wide range of experts about both the technical and financial aspects of the proposed program. This was the last major enterprise of the Space Council under Johnson, however. It never became the operational organization some expected, and Kennedy gave responsibility for NASA operations to the NASA administrator.

Johnson's role in the space program as vice-president changed radically from what it had been as Senate majority leader. His relationship with the South with regard to space also changed. While he still had contact with members of the Senate's Southern Caucus and with Senator Russell, he no longer could draw on their support or expect it. He still had a role in matters of space-related activities that impacted the South, however. Early in the administration he played a decisive role in the appointment of the new NASA administrator, a "southwesterner" like he was. Likewise, although he was not the key decision-maker, he influenced the designation of Houston as the site of the new NASA center. And almost inevitably, the civil rights movement crossed paths with the space program, drawing Johnson into a controversy.

In between his election and Inauguration Day, President Kennedy gave Johnson a central role in the selection of the next NASA administrator. Johnson asked Kenneth Belieu, the staff director of the Senate Space Committee, to lead the search. By late January, Belieu and his search committee had contacted twenty-five individuals without success. It was the highest administration position that remained vacant and Kennedy was impatient to fill it. He told LBJ and his science advisor Jerome Wiesner that if they didn't find someone, he would do so himself. Among the reasons others had rejected the job were the apparent lack of direction of the space program under Kennedy and a reluctance to work for LBJ, who had a reputation as an unrelenting taskmaster and who people expected to head the space program under Kennedy. Johnson talked to Senator Robert Kerr (D-Oklahoma), whom Johnson had recommended to head the Senate Space Committee after his departure. Kerr suggested James E. Webb, a businessman who had served as head of the Bureau of the Budget (the predecessor of the Office of Management and Budget) and as undersecretary of state during the Truman administration. He had also worked for Kerr for five years in business. Wiesner had thought of Webb too, so he and Johnson concurred.

Webb had strong administrative and policy experience but little background in science and technology. He met with Johnson and told him that he didn't think he was suited to the position. Johnson wouldn't hear of it and applied some of his legendary arm-twisting. "There wasn't any doubt in my mind that he wanted me to say yes that day," Webb recalled years later. Webb agreed to meet with the president that afternoon. Kennedy told him that he considered policy experience to be more important to the NASA administrator than technical expertise and said that he wanted Webb for his policy chops. Webb later insisted that "I've never said no to any President who has asked me to do things."[43]

Webb believed he could work with Kennedy, but he was concerned that Johnson would use his position as chair of the Space Council to interfere in NASA business. He and Kerr addressed the issue of "how do you handle Lyndon Johnson?" Webb recalled that "Kerr performed a great service for me.

He told Johnson this story about my independence, how I wouldn't be kicked around, which meant, Johnson never tried."[44]

When Webb took command as NASA administrator on February 14, 1961, the site for the location of the STG remained undetermined. Webb set up a site selection team in August 1961. The team developed specific criteria for the site of the proposed manned spaceflight laboratory that included a moderate climate, access to water for barge transportation for large rockets, nearby industrial capability and labor supply, an institution of high learning, and at least 1,000 acres of land. Proposals came from Florida, Louisiana, Texas, California, Massachusetts, Rhode Island, Virginia, and Missouri. Houston offered the San Jacinto Ordnance Depot and later added sites offered by Rice University and the University of Houston. A NASA team toured twenty-three sites and narrowed the field to three finalists. MacDill Air Force Base in Tampa (where the air force planned to shut down its Strategic Air Command activities) ranked first, followed by the Houston site and the Benicia Ordnance Depot in the San Francisco Bay area.

Although MacDill, which was close to Cape Canaveral, was the preferred site, Houston had a lot to offer. It met virtually all of the search criteria. Its warm climate offered advantages for a program that demanded year-round development. Rice University had reserved 3,800 acres of land for a research facility, and the nearby Houston Ship Channel offered the required facility for transporting rockets to Cape Canaveral. The Texas congressional delegation still held key space-related committee assignments, and now Johnson was vice-president and chair of the Space Council. NASA was committed to placing the new center in Tampa, but the ground suddenly shifted when the air force decided to hold on to MacDill. So the door opened and Houston moved into first place.

Webb presented the NASA decision to President Kennedy. Even though Kennedy's home state had been in the running and Massachusetts governor John Volpe had pushed for Boston, Kennedy accepted Webb's recommendation. Kennedy phoned Albert Thomas and told him that he needed assistance in passing three bills. Thomas hesitated, "Well, Mr. President, I don't know about this." Kennedy replied, "You know Jim Webb is thinking about putting this center down in Houston." Thomas said, "In that case, Mr. President—," and the deal was struck.

Years later, when an interviewer asked Webb about Lyndon Johnson's involvement in the selection of Houston as the site for the Manned Spacecraft Center, he said, "Mr. Johnson was always interested in Texas. A good deal of the time he was Vice President, he still sort of thought of himself as a Senator from Texas. He was interested in everything that went on in Texas. . . . The basic fact of the matter is that as we studied these things, I informed President Kennedy and not Mr. Johnson that we were moving in our thinking towards Houston."

Others had a different opinion. Senator George Smathers (D-Florida) said that he and Johnson argued about it and LBJ "tried to act like he didn't know . . . It never has made any sense to have a big operation at Cape Canaveral and another big operation in Texas. But that's what we got, and we got that because Kennedy allowed Johnson to become the theoretical head of the space program."[45]

The selection of Houston completed assignment of the major facilities for NASA's human space flight centers. The centers, all located in the South, formed a crescent that swept from Cape Canaveral on Florida's Atlantic coast to the Manned Spacecraft Center in Houston. Cape Canaveral (now the Kennedy Space Center), on Merritt Island off the Florida coast at Titusville, was the launch site for the Mercury, Gemini, and Apollo projects and for the Space Shuttle. In Huntsville, Alabama, Wernher von Braun had served as the technical director of the Army Ballistic Missile Agency at Redstone Arsenal. In July 1960, that facility became NASA's Marshall Space Flight Center with von Braun as director. It was responsible for rocket development, especially the Saturn V, the launch vehicle for Project Apollo. The Mississippi Test Facility at Pearl River, which Marshall Space Flight Center operated in the early 1960s, later became the Stennis Space Center. Finally, the Manned Spacecraft Center in Houston (which became the Johnson Space Center in 1973) was responsible for developing the capsules for Mercury, Gemini, and Apollo and for training the astronauts.

President Kennedy told Webb that he had received complaints from other states, including "Michigan, Pennsylvania, and the eastern states" about the fact that NASA was concentrating its spending in California and the South.[46] NASA had other centers, including some that had been established under NACA. NASA also realized that it would have to involve other parts of the country in order to garner the political support it needed. The administration sent Richard Callaghan to serve as Webb's legislative affairs director and asked him to help Webb "to arrange for a more equitable distribution of contracts . . . and to find out whether Kerr and Johnson were pulling strings for their friends as NASA." Callaghan reported back that he didn't find any evidence of such interference in the awarding of contracts. Some historians have argued that the reason so much of the human spaceflight organization was centered in the South was to aid the economic development of the region.[47] Historian Walter MacDougall has argued that it was a Second Reconstruction that created a New South. Indeed, it was an important factor for the Texans who lobbied for the Houston center.

By June 1963, civil rights had become the major domestic issue confronting the nation. In May, brutal police suppression of a civil rights march in Birmingham, Alabama, prompted President Kennedy to take action. On June 11, speaking from the Oval Office, Kennedy gave his most important address on civil rights. For the first time he referred to civil rights as a moral issue and proposed legislation that became the Civil Rights Act of 1964. Kennedy's new aggressive stance on civil rights put pressure on Congress and on Vice-President Johnson.

In addition to his role as chair of the Space Council, Johnson chaired the President's Committee on Equal Employment Opportunity (PCEEO). Civil rights impacted both of Johnson's committees, never more dramatically than at a meeting of the PCEEO on June 18, one week after Kennedy's speech. The PCEEO became a target of criticism for the meager progress it had made in Black employment. A survey of 35,000 companies with government contracts revealed that 25,000 of them had no Black employees. In some cases when the PCEEO claimed that some companies had increased Black employees by 100 percent, it turned out that they had done so by hiring one additional person.

Attorney General Robert Kennedy and President Kennedy had discussed the shortcomings of the PCEEO. Robert Kennedy and Lyndon Johnson had long been bitter rivals, and the attorney general was quick to put the blame on LBJ. With civil rights attorney Burke Marshall in tow, he burst into the June 18 meeting of the PCEEO. Kennedy launched into questions of Black employment in Birmingham, defense industry employment, and PCEEO action. Johnson tried to respond and asked James Webb to report on the fine work on Black employment in NASA. Webb, caught off guard, gave a general answer that didn't satisfy. Kennedy shot back, asking Webb why NASA had only two people working on equal employment to service $3.5 billion in contracts. Kennedy persisted: "Mr. Webb, I just raised a question of whether you can do this job and run a Center and administer its $3.9 billion worth of contracts and make sure that Negroes and nonwhites have jobs . . . I am trying to ask some questions. I don't think I am able to get the answers, to tell you the truth." Kennedy also attacked Hobart Taylor, a Texan who was close to Johnson and was serving as special counsel to the PCEEO, for the lack of Black employment in Birmingham. One of those present said that Kennedy left only after "completely humiliating Webb and making the vice president look like a fraud." Kennedy's outburst embarrassed Johnson and exacerbated an enmity that already had deep roots and that became perhaps the most bitter intraparty feud in American political history.[48]

Webb shared Johnson's sense of shame after the encounter, and both men recognized the need to respond. Once again, Johnson found himself at the juncture of civil rights and space policy. As civil rights emerged as the most pressing domestic issue of the 1960s, the location of NASA's flagship centers in the South embroiled the agency in contention over federal hiring practices. Kennedy's assault may not have been diplomatic, but it hit a vulnerable target. NASA's record on hiring minorities, particularly African Americans, was abysmal. The administration had few Black employees, and what few it had were clustered predominantly in unskilled positions, most commonly in menial jobs such as janitorial or food service duties.

Alabama had become the chief battleground in the campaign for civil rights reform. Rosa Parks and Martin Luther King emerged as national figures in the Montgomery Bus Boycott in 1955. In 1961, freedom riders were beaten in Anniston and Birmingham. Eugene "Bull" Connor, Birmingham's commissioner of public safety, led the attack on protestors in May 1963. And on the same day that Robert Kennedy gave his verbal tongue-lashing to Johnson and Webb, Alabama Governor George Wallace staged his "stand in the schoolhouse door," promising "segregation today, segregation tomorrow, segregation forever."

It was Webb who responded to Robert Kennedy's challenge, invoking Johnson's authority when he did so. Huntsville, in northern Alabama, was the site of NASA's Marshall Space Flight Center. Webb told the center's director, Werner von Braun, that "the Vice President has expressed considerable concern over the lack of equal employment opportunity for Negroes in Huntsville, Alabama." Webb suggested that some of Marshall's work would be shifted elsewhere if the situation did not improve. The community responded. A new business group, the Association of Huntsville Area Contractors, worked with Marshall and the Chamber of Commerce to address Huntsville's segregation in education and public facilities. They urged Webb to consider that the situation in northern Alabama was markedly better than it was in Birmingham and Montgomery. Von Braun acknowledged problems in hiring because of Alabama's image: it was hard to attract Black engineers to Dixie when there were alternative jobs in the North and West. And the problem was not only in Alabama; other centers in the South had similar issues, and indeed all of NASA fared poorly in relation to other federal agencies.[49]

Lyndon Johnson's relationship with the South thus impacted his actions on space throughout his tenure as Senate majority leader and in less direct fashion during his vice-presidency. His close ties to Senator Richard Russell gave him entree to the Southern Caucus, and Russell's support placed him in a position to use the nearly defunct Preparedness Committee as a vehicle for controlling the post-Sputnik debate on space policy. The issue of civil rights was never far removed from this relationship. In the late 1950s, his adept handling of the passage of the Civil Rights Act of 1957 enabled him to use his position as majority leader to emerge as the leading space advocate in the Senate and to be at the center of the crafting of the Space Act of 1958.

When Johnson became president, he continued to be involved in the space program. He was president during some of the great achievements of Project Apollo, although he left office months before the moon landing in July 1969. He served during what was at the time the greatest disaster in NASA's history, the Apollo fire that killed three astronauts undergoing training for an Apollo mission. One of his greatest achievements, the Civil Rights Act of 1964, severed his relationship with the South. He said at the time that he was afraid that the act would cost the Democratic Party support of the region, where his support for civil rights seemed to be a great betrayal. He understood the South, and he was right.

Notes

1 Robert Dallek, "Johnson, Project Apollo, and the Politics of Space Program Planning," in *Spaceflight and the Myth of Presidential Leadership*, ed. Roger D. Launius and Howard E. McCurdy (Urbana: University of Illinois Press, 1997), 69.

2 John Lewis with Michael D'Orso, *Walking with the Wind: A Memoir of the Movement* (San Diego: Harcourt Brace & Company, 1998), 246.

3 Johnson quoted in William E. Leuchtenburg, *The White House Looks South: Franklin D. Roosevelt, Harry S. Truman, and Lyndon B. Johnson* (Baton Rouge: Louisiana State University Press, 2005), 382.

4 Reston quoted in Leuchtenburg, *The White House Looks South*, 382.

5 Rowe quoted in Leuchtenburg, *The White House Looks South*, 382.

6 Leuchtenburg, *The White House Looks South*, 382.

7 Robert Parker with Richard Rashke, *Capitol Hill in Black and White* (New York: Dodd, Mead & Company, 1986), 23.

8 Leuchtenburg, *The White House Looks South*, 255.

9 John A. Goldsmith, *Colleagues: Richard B. Russell and His Apprentice, Lyndon B. Johnson* (Washington, DC: Seven Locks Press, 1993), 13–15.

10 Leuchtenburg, *The White House Looks South*, 251–252; Goldstone, *Colleagues*, 51–52.

11 Leuchtenburg, *The White House Looks South*, 259–261.

12 Robert A. Caro, *Master of the Senate*, vol. 3 of *The Years of Lyndon Johnson* (New York: Alfred A. Knopf, 2002), 954.

13 Lyndon B. Johnson, *The Vantage Point: Perspectives of the Presidency, 1963–1969* (New York: Holt, Rinehart and Winston, 1971), 272.

14 *Legislative Origins of the National Aeronautics and Space Act: Proceedings of an Oral History Workshop, April 3, 1992,* 18, Monographs in Aerospace History number 8 (Washington, DC, NASA History Office, 1998), https://history.nasa.gov/40thann/legislat.pdf.

15 Caro, *Master of the Senate,* 1021–1022.

16 Transcript of George E. Reedy, oral history interview with Michael L. Gillette, August 28, 1988, 4, Discover LBJ, https://www.discoverlbj.org/item/oh-reedyg-19880828-23-99-56.

17 *Legislative Origins of the National Aeronautics and Space Act of 1958,* 2, 57; Caro, *Master of the Senate,* 1022.

18 *Legislative Origins of the National Aeronautics and Space Act of 1958,* 8.

19 "Additional Comments by Eilene Galloway," Appendix B of *Legislative Origins of the National Aeronautics and Space Act of 1958,* 57.

20 Transcript of George E. Reedy, oral history interview with Michael L. Gillette, September 25, 1986, Discover LBJ, https://www.discoverlbj.org/item/oh-reedyg-19860925-20-99-53.

21 "Additional Comments by Eilene Galloway," 57–58.

22 *Legislative Origins of the National Aeronautics and Space Act of 1958,* 22.

23 Solis Horowitz to Secretary of Defense, October 11, 1957, box 355, folder Hearings on Preparedness Subcommittee on Satellite and Missile Programs, Papers of Lyndon Baines Johnson, United States Senate, 1949–1961, Committee on Armed Services, LBJ Presidential Library (hereafter Preparedness Subcommittee folder).

24 Transcript of telephone conversation, LBJ with Neil McElroy, October 21, 1957, box 355, Preparedness Subcommittee folder.

25 "Additional Comments by Eilene Galloway," 58.

26 Robert A. Divine, *The Sputnik Challenge: Eisenhower's Response to the Soviet Satellite* (New York: Oxford University Press, 1993), 62-64.

27 Divine, *The Sputnik Challenge,* xiii; Yanek Mieczkowski, *Eisenhower's Sputnik Moment: The Race for Space and World Prestige* (Ithaca, NY: Cornell University Press, 2013), 96.

28 Divine, *The Sputnik Challenge,* 64.

29 Transcript of Edwin L. Weisl Sr., interview with Joe B. Frantz, May 13, 1969, 12, Discover LBJ, https://discoverlbj.org/item/oh-weislsr-19690513-1-74-267.

30 Stennis quoted in Mieczkowski, *Eisenhower's Sputnik Moment,* 139.

31 Minutes of Meeting of Senate Armed Service Preparedness Subcommittee, November 22, 1957, Preparedness Subcommittee folder.

32 Glen P. Wilson, "How the U.S. Space Act Came to Be," Appendix A of *Legislative Origins of the National Aeronautics and Space Act of 1958,* 50.

33 William I. Hitchcock, *The Age of Eisenhower: America and the World in the 1950s* (New York: Simon & Schuster, 2018), 384–385.

34 George Reedy, memorandum to LBJ, November 26, 1957, LBJ Senate Papers, LBJ Presidential Library.

35 Divine, *The Sputnik Challenge,* 67.

36 Mieczkowski, *Eisenhower's Sputnik Moment,* 140–141, 143.

37 Caro, *Master of the Senate,* 1028.

38 Mieczkowski, *Eisenhower's Sputnik Moment,* 141.

39 T. Keith Glennan, *The Birth of NASA: The Diary of T. Keith Glennan*, ed. J. D. Hunley (Washington, DC: NASA, 1993), 201.

40 Cited in Mark K. Updegrove, *Indomitable Will: LBJ in the Presidency* (New York: Crown Publishers, 2012), 226.

41 Glennan, *The Birth of NASA*, 14.

42 Glennan quoted in Walter A. McDougall, *The Heavens and the Earth: A Political History of the Space Age* (New York: Basic Books, 1985), 374.

43 Transcript of James E. Webb, oral history interview with T. H. Baker, April 29, 1969, 5, Discover LBJ, https://discoverlbj.org/item/oh-webbj-19690429-1-74-266.

44 Webb quoted in W. Henry Lambright, *Powering Apollo: James E. Webb of NASA* (Baltimore, MD: The Johns Hopkins University Press, 1995), 103.

45 Robert Dallek, *Flawed Giant: Lyndon Johnson and His Times, 1961–1973* (New York: University Press, 1998), 22.

46 California had more money from NASA prime contract awards that any other state during the 1960s. In FY 1961, it had 39.1 percent of NASA procurement money. See Jane Van Nimmen and Leonard C. Bruno with Robert L Rosholt, *NASA Resources 1958–1968*, vol. 1 of *NASA Historical Data Book* (Washington, DC: NASA, 1974), 200 Table 5-17, "Distribution of NASA Prime Contract Awards by State, FY 1961–FY 1968."

47 John M. Logsdon, *John F. Kennedy and the Race to the Moon* (New York: Palgrave Macmillan, 2010), 135–136. The second quotation is from a memo from Frederick G. Dutton to Webb, August 15, 1961, quoted in Logsdon, 125.

48 Jeff Shesol, *Mutual Contempt: Lyndon Johnson, Robert Kennedy, and the Feud that Defined a Decade* (New York: W. W. Norton & Company, 1997), 80–87; Arthur M. Schlesinger Jr., *Robert Kennedy and His Times* (Boston: Houghton Mifflin Company, 1978), 335–336.

49 Andrew J. Dunar and Stephen P. Waring, *Power to Explore: A History of Marshall Space Flight Center, 1960–1990* (Washington, DC: NASA History Office, 1999), 115–125.

3

Black Education in the Shadow of Apollo

Dr. Richard D. Morrison and Alabama A&M College

BRIAN C. ODOM

The continuation of barriers to Black education through the 1950s closed off training for most technical careers and kept new opportunities in the aerospace industry beyond the grasp of most Black students and graduates. Generations of resource deprivation in Black education created a training gap. Administrators at Historically Black College and Universities (HBCUs) found it difficult to develop curriculum in the emerging technology fields. They were hesitant to forgo traditional professional training programs within the Black community such as those in education and agriculture and expand into new areas when there were questions about whether those careers were indeed open to their students. At the same time, HBCU administrators performed a delicate balancing act between the forces of social revolution sweeping the country and close surveillance on the part of white funding agents at the local and state levels. Change would come only through an integrated national commitment from prominent Black educators backed by extensive funding for the space program during Project Apollo. This chapter examines that transformation by examining the relationships that developed in Huntsville, Alabama, home of NASA's Marshall Space Flight Center (MSFC). In Huntsville, Alabama A&M College, executives at MSFC and for aerospace companies, the Association of Huntsville Area Contractors (AHAC), and civil rights activists formed dynamic new relationships.[1] It does so by exploring the evolving thinking of Black educators in the early Cold War years, the actions of Alabama A&M College president Dr. Richard Morrison, and the brokerage provided by AHAC.

The white perspective of this era dominates Huntsville's historical narrative. Dr. Ernst Stuhlinger, the chief of science at MSFC, helped construct this narrative by arguing that the administration at Alabama A&M is what stood in the way of that institution's technical engagement with NASA. In Stuhlinger's account, President Morrison continuously rebuffed Wernher von Braun's offers of NASA technical contracts. Stuhlinger claims that Morrison snubbed him and told him that A&M College was an "agricultural college" and that what he was attempting to achieve at the university was to "make better farmers." In Stuhlinger's version, Morrison believe that "this new-fangled technology at the Marshall Center is not for us."[2] Morrison's actions at A&M College and his consistent, lucidly articulated educational philosophy contradict Stuhlinger's claims. The framing of Morrison as uncooperative served to support the decision to push for state funding to establish a new research institution that became the University of Alabama in Huntsville instead of investing funds at Alabama A&M. This chapter challenges Stuhlinger's narrative by highlighting the contours of the civil rights movement in the city and exploring Morrison's actions at A&M College.

Black Technical Education at the Beginning of the Cold War

Black educators had long been aware of the changing nature of education in the United States in response to emerging markets and the new technological careers associated with increased automation and expanded research and development. As early as January 1935, Lewis K. Downing, the dean of engineering and architecture at Howard University, described the new opportunities for Black students in the field of engineering.[3] Downing argued that the "importance of technological service to the life, safety, and economic welfare of society" was evident from the billions in federal funding appropriated across the country each year for public works projects. Downing, who understood that many students continued to question the role of racial prejudice in their competition with white graduates for jobs, asked, "Just where does the Negro come in?" To rectify the lack of Black representation in fields such as engineering, Downing called for the organization of Black firms of engineers, architects, and contractors in large population centers in the country. In the meantime, Downing thought there was enough work with educational, church, and business institutions to "absorb all of the technical graduates our schools can supply for many years to come."[4]

When Black soldiers returned from World War II, they found increased opportunities for college training through the GI Bill and other programs. But as Gavin Wright points out, southern members of Congress inserted language in the legislation that decentralized its application and stripped it of any federal oversight and enforcement.[5] Despite these challenges, Black educators increasingly turned to the federal government to support the development of Black state and private institutions. The arrival of the new Cold War economy marked an important shift from agricultural jobs to jobs in technological fields. In 1946, Dr. Thomas Turner, a retired botany professor at Hampton Institute, argued that the fact that HBCUs continued to suffer from limited facilities and resources had led to the "compartmentalization" of Black students into a few areas, such as teacher training and agricultural sciences, that led to limited opportunities. This narrow focus resulted in the establishment of a limited science curriculum that left Black graduates ill prepared for advanced studies in the physical sciences. Turner explained that this neglect of the physical sciences "ignored the trends pointing to the largest field of job activity in the future."[6] Turner's arguments were part of a larger discourse that called on HBCU administrators to embrace the changing job market and shift their curricula away from the limited occupational training of the past and toward the widening opportunities in the technical and scientific professions.

In 1949, Ambrose Caliver, a longtime leader in Black education, argued that while Blacks preferred not to be "singled out for any special consideration by their government," they had come to realize "through long and bitter experience" that they could not leave their progress to "good will among men" or the "natural process."[7] The situation required the "strong persuasive arm" of the national government. Caliver argued that enacting strong educational reforms was made more difficult when state leaders were calling the granting of equal education facilities to Black children a "fatal blow to the Southern way of life."[8] Black educational institutions in Deep South states with a long history of Jim Crow discrimination faced further obstacles to their long-term success. Reviewing the state of Black education in Alabama after World War II, historian W. E. Anderson noted that despite the state's pledge to ensure that its school system would not be discriminatory, there was "objective evidence of educational inequalities in the administration [of] and support" for Black schools—inequalities that fell outside the state law.[9] The statistics were staggering. During the 1929–1930 school year, the average expenditure per pupil in primary education was $10.09 for Black students and $36.43 for whites. During the 1944–1945 school year, the numbers were $22.85 per Black student and $56.31 per white student. In higher education, $10,275,000 was available for all state higher education institutions in Alabama. Of that amount, only $487,500 went to Black schools, and only $285,238 of that came from state contributions. On a positive note, Anderson relayed that between 1930 and 1945, school surveys, curriculum revision programs and state equalization laws had brought about positive action on many of the major challenges Black educators faced.[10] However, many problems persisted, such as lower pay for

Black faculty and fewer funds for facilities on HBCU campuses.[11] The problem continued to be the refusal of state agencies to provide public HBCUs with the funding equivalent to what their white counterparts received.

The *Brown v. Board of Education* decision of 1954 was a key turning point in the long battle for equal opportunity. The following year, the Court ordered that the desegregation of public education in the country should proceed with "all deliberate speed." The decision made Black education a key topic in the Cold War battle against communism. According to historian Mary Dudziak, the *Brown* decision provided the federal government with the "counter to Soviet propaganda it had been looking for" because the State Department used it to present a "rosy picture of racial equality" in the country.[12] In 1955, Walter Crosby Eells, a former advisor on higher education to the Supreme Commander for the Allied Power in postwar Japan, noted that he received many queries from international students on the topic of segregation and discrimination against Blacks in American higher education. Eells argued that while conditions in higher education were gradually getting better, "good news travels much more slowly than bad."[13] The situation made it difficult for the United States to counter communist propagandists who skillfully compared the situation in US higher education unfavorably to the racially open policies in Asian and African universities.

If HBCU administrators were to take advantage of increasing opportunities brought about by increased federal investment in research and development and by geopolitical considerations, they would need to either shift the resources they had away from traditional curricula or locate additional funding for expansion. In early 1956, education professor Edward K. Weaver of Atlanta University addressed hesitation about diversifying the curriculum, arguing that there was no difference between a proper curriculum for Black science education and white science education and that the problems Black colleges were experiencing in establishing a rigorous science program were the same as those small universities and poor white universities in the South faced. A general lack of financial investment, a "faltering administrative outlook," a lack of skilled instructors, a poor occupational outlook, and "clouded uncertain standards and values" combined to limit the ability of Black college leaders to institute modern science education programs.[14] Jim Crow segregation and discrimination had adversely affected growth in the Black education sector just as it had in "every other aspect of Negro American life."[15] Weaver noted that in the quest to meet the educational needs of all students, struggling Black colleges had developed an "elaborate compartmentalization of programs," by which he meant that most HBCUs used their limited resources to implement programs geared toward professions within the Black community.[16] The few training programs in science that existed in Black institutions lived with a fundamental contradiction. They worked diligently with limited resources to prepare students to participate in a democratic society that discriminated against them socially and politically while striving to educate them in scientific fields where jobs were closed to them.

Following the Soviet launch of Sputnik 1 in October 1957, many leaders in Black education noticed a potential for progress. In the spring of 1958, W. Montague Cobb, a professor at Howard University's College of Medicine, wrote that there was no longer a need to discuss the poor representation of Blacks in the science professions or demonstrate their many significant accomplishments in those fields. Instead, Black students only required the proper inspiration and motivation to take advantage of the growing opportunities for Black graduates in the era of automation. He also argued that HBCU administrators and faculty must "put our shoulders to the wheel, search out our gifted young," and "motivate them with high goals."[17] These increased opportunities created a unique environment at Black colleges, where the interests of traditionally conservative institutions for Black elites started to intersect more and more with the social revolution in the Black community that affected more than W. E. B. Du Bois's "talented tenth."

The 1960s opened upon a segregated world in which Black students saw no hope of pursuing professional careers in technology fields. They could be confident that however many new jobs were created in the aerospace industry, those jobs would inevitably go to whites. As historian David Wharton points out, of all the degrees granted to Black students in 1962 and 1963, over half were in education. Fields such as engineering remained a "white, male enclave." Black educators realized that change would come only through massive government intervention.[18] Administrators at publicly funded Black universities in the South were challenged with balancing engagement with the social revolution sweeping the country without alienating state and local funding agencies that were committed to maintaining Jim Crow segregation. At HBCUs, desegregation offered the potential for a new funding dynamic and created more space in the public sphere for students to critique the larger society. For example, Fisk University's Diane Nash argued that if Black institutions could only provide students with equal access to a quality education, "maybe someday a Negro will invent one of our missiles."[19] HBCU presidents were forced to chart a course on a perilous new terrain that included the forces of social revolution, adversarial funding agents at the state level, and rapidly expanding opportunities in science and technology careers for students. Coordinating those efforts with the massive Cold War–era federal investment in research and development and technology programs would be critical.

Richard D. Morrison, Alabama A&M College, and the Civil Rights Context

In Huntsville, Alabama, a transformation in Black education took place at Alabama A&M College when the presidency transitioned from Dr. Joseph F. Drake to Dr. Richard D. Morrison in March 1962. From his appointment as president of Alabama A&M College in 1962 to the end of the decade, Morrison engaged in a sustained, comprehensive campaign to diversify an overwhelmingly agricultural and teacher training curriculum into one that incorporated technical training programs capable of producing graduates for new space-age careers. Despite limited funding, a staff and student body that was hesitant to embrace change, and increasing white surveillance of the college's activities related to the civil rights movement, Morrison and the university administration added new programs, improved facilities, added specialized faculty, and established cooperative agreements with government facilities and other universities (fig. 3.1). At the end of the decade, Morrison stood at the helm of a university that had weathered desegregation, expanded its technical and scientific curriculum, and secured avenues for employment and advanced studies for Alabama A&M students.

Figure 3.1. In November 1968, Alabama A&M University president Dr. Richard D. Morrison (*right*) and NASA Marshall Space Flight Center director Wernher von Braun (*left*) signed a cooperative agreement between the center and university. Courtesy of NASA.

Dr. Richard David Morrison was born January 18, 1908, in Utica, Mississippi. He began his education at Tuskegee Institute in 1927 at the request of George Washington Carver and earned a Bachelor of Science in agriculture in 1931. Following five years as a member of the faculty at Talladega College, Morrison came to Alabama A&M College in 1937 to direct its agricultural department. Morrison continued his education, receiving a master's degree from Cornell University in 1941 and PhD from Michigan State University in 1954. On March 1, 1962, the Board of Trustees named Morrison to replace the outgoing A&M president, Dr. Joseph F. Drake. During his first decade in office, Morrison walked the fine line between Black elites and civil rights activists who demanded intellectual leadership from the area's primary Black institution of higher education and those who controlled funding in federal, state, and local government, white elites, and aerospace executives. During that decade, Morrison used the small funds at his disposal to improve the facilities at Alabama A&M College and expand the college's technical training in higher-level science programs such as physics and hands-on programs such as drafting and computer science. Morrison accomplished his delicate task by employing a public rhetoric characterized by a politically moderate discourse that was palatable to those who advocated for a top-down progressive movement that best represented by Whitney Young Jr. and the National Urban League.

Civil rights protest formed the contextual backdrop of Morrison's initial efforts at reform. January to July 1962 was the critical period of the civil rights movement in Huntsville. On January 13, the headline in the *Huntsville Mirror*, the local Black newspaper, was "13 Sit-Inners, Stand Inners are Arrested— A&M College Students Lead Demonstrations." Working with Congress of Racial Equality (CORE) representatives Hank Thomas and Richard Haley, students from Alabama A&M College and William Hooper Councill High School were arrested throughout the week on trespassing charges at segregated lunch counters in the city. The story reported that the students conducted themselves in a "very submissive and orderly" way. They had been schooled in this procedure by CORE representatives who were "said to be active in directing these demonstrations."[20] In the early stages of the civil rights movement in the city, Huntsville activists followed a pattern similar to that of the larger national movement such as the sit-ins at Nashville, Tennessee, and Greensboro, North Carolina, both of which began in February 1960. This scene of Black students and local youths arrested for sitting in continued throughout the first half of 1962, and many students were arrested multiple times. Over the course of 1962, student participation in the Huntsville sit-in movement grew from "eight or ten students" to "something like 250 to 300."[21]

Alabama A&M College faculty proved critical to the Huntsville movement and its expansion beyond the city. This activism garnered the attention of Huntsville aerospace investors. For example, it was A&M professor Randolph Blackwell's idea to target the New York Stock Exchange and the Midwest Stock Exchange in Chicago as a way of bringing national attention to the Huntsville movement. According to Sonnie Hereford III, Blackwell provided the Community Service Committee (CSC) with "thousands of ideas" acquired from his work with the civil rights movement in North Carolina. On May 18, 1962, protesters allied with the Huntsville movement converged on the New York Stock Exchange, where they asked potential investors to avoid investing in aerospace companies with Huntsville operations, including IBM, Brown Engineering, and Chrysler. During the protest at the stock exchange, supporters from the Cashin family of New Jersey wore signs reading, "Don't invest in Huntsville, Alabama, it's bad business" and "I ordered a hamburger and was served a warrant in Huntsville, Alabama."[22] Blackwell's strategies not only proved critical for coordinating and strengthening the Huntsville movement but also showed how that movement could borrow from the larger national civil rights movement while maintaining its distance. When Huntsville leaders failed to take action toward desegregation following initial protests, Blackwell and Rev. Ezekiel Bell organized another protest at the Midwest Stock Exchange in June 1962. At Chicago, the group of Huntsville demonstrators passed out 2,000 handbills that said, "Don't Invest in Huntsville, Alabama Corporations. It's bad for business."[23] The handbills included a brief overview of the sit-in movement in Huntsville, called on recipients to write to both Mayor Robert B. Searcy and President John F. Kennedy to ask for redress, and asked the question, "Can democratic defenses be built in an undemocratic city?"[24] The strategy brought national attention to the movement in Huntsville and gained the attention of city leaders and aerospace executives who wanted to avoid any negative publicity. Funding agencies tasked Morrison with limiting faculty participation in civil rights activism.

Many within the leadership of both A&M College and the civil rights movement questioned Morrison's elevation to the position of president. Some members of the faculty thought Morrison was much too conservative while others supported more senior candidates at the college. During an introductory informal faculty meeting, Morrison noted that he wanted to push past initial resistance and pledged to lead the development of A&M to an institution capable of producing graduates that both the state and the nation needed—well-rounded, technically trained individuals.[25] Morrison maintained a comfortable distance from what white elites perceived as the more radical social elements of the civil rights movement while also pursuing a

revolutionary agenda of educational reform that aggressively sought new avenues of funding and new opportunities for graduates in space-age careers.

From the perspective of civil rights leaders including Dr. John Cashin, the
removal of Alabama A&M College's previous president, Joseph F. Drake, was
an attempt by the state of Alabama to "cut into the CSC's troop strength." The
state viewed Morrison as someone who was "committed to disciplining any
students who participated in the sit-ins."[26] This move had the intended effect
and soon CSC frontline support from university students had been reduced
to only the "hardcore faithful."[27] Dr. Sonnie Hereford III noted that he was not
surprised by Morrison's appointment because Alabama governor John Patterson had chastised Drake, telling him that he was "gonna have to keep the kids
straightened out up there." Drake's inability to control the students' participation in the downtown sit-ins had upset Governor Patterson.[28] His removal
echoed the removal of President H. Council Trenholm at Alabama State University three months earlier for allowing student demonstrations on campus.
Patterson claimed that a "consistent operation of the college was lacking" as
the reason for his dismissal of Drake.[29] In addition, by depriving institutions
such as Alabama A&M of equitable funding, Governor Patterson and other
state leaders placed Black administrators in a vulnerable position: they could
be held personally responsible for situations far beyond their control.

Administrators at Black universities had long been vocal about the impact of the civil rights movement on education. In November 1963, President
Samuel Proctor of North Carolina A&T of Greensboro argued that college
presidents could no longer "appease the Negro Press one minute and Southern Governors the next."[30] Black educators had long battled to overcome the
educational resource gap at small, underfunded Black high schools. Proctor
pointed out that the leaders of the civil rights movement had been educated
in southern institutions. He contended that it was not the job of the colleges
to integrate quickly; their job was to continue to improve the education they
offered students in a world where funding for HBCUs remained heavily reli-
ant on local industries and white donors.[31]

In the decade after *Brown*, Black educators, including Proctor, grappled
with the implications of that decision. At issue was how to navigate the end of
segregated schools and how to best position their own institutions to take advantage of the new opportunities Cold War–era industries offered. The Space
Age offered the possibility that the current generation of Black graduates
could rise to a level of equality with their white counterparts. This rhetoric
of uplift was more palatable that more radical civil rights rhetoric to whites,
who heard its conservative tone and could see it as a sign of assimilation.
What emerged at the end of the period was a consistent statement from Black

administrators throughout the country that had been coordinated through academic literature.

Many civil rights leaders and Black university administrators argued that the movement was conservative in nature while others felt the protest movement was a liability for Black institutions. Florida A&M University president George W. Gore relayed how his campus had been in an "uproar" since 500 students had participated in a CORE-led demonstration in Tallahassee, violating an injunction against more than eighteen people picketing at one time. After Gore expelled the two key student leaders, civil rights leaders waged a "systematic campaign of vilification" against him in May 1963. Gore insisted that the purpose of the university was education, not social revolution.[32] Southern University president Felton Clark had also expelled students from the university for their participation in the March 1960 sit-in movement in Baton Rouge, Louisiana. Morehouse president Benjamin Mayes explained that "to be president of a college and white is no bed of roses," but to be president and Black "is almost a bed of thorns." Tennessee Agricultural & Industrial president Walter Davis concluded that "there is no question about it, mass demonstrations put the college president in a tough spot."[33] Alabama A&M College wrestled with this dilemma when CORE visited its campus in January 1962 and started the sit-in movement in downtown Huntsville. After Richard Morrison became president, the university appeared to be leaning toward a more moderate stance toward civil rights activities in Huntsville.

Scholar Eddie Cole argues that college presidents were actually a "driving force behind many of the social changes, initiatives, and struggles" during the civil rights movement because they "actively, though often quietly, shaped racial policy both inside and outside of the educational sphere."[34] Cole points out that leaders of Black institutions of higher education, including Martin Jenkins at Morgan State College, worked diligently to "maintain control of their institutions, secure money from white legislators, fight back against racists, and balance student demands with other demands."[35] For Cole, exploring the "quiet," offstage efforts of Black presidents is critical in understanding how they shaped "racial practices beyond the gaze of white state legislators, governors, and trustees" and in the process "sacrificed their reputations and accepted public criticism without being able to explain the full breadth of how they supported desegregation."[36] Over the course of the 1960s, that is the situation Richard Morrison found himself in.

One of the most significant obstacles to progress for many HBCUs was gaining and maintaining a positive status with accreditation agencies. Given the inequitable distribution of funding from state appropriations and limited supplemental funding, loss of accreditation was a constant fear. The drive for accreditation at A&M certainly hung over all discussions of updating the math and science curricula. The previous effort to achieve accreditation had ended in December 1961, when the Commission on Colleges and Universities had denied the university membership in the Southern Association of Colleges and Schools (SACS). The rationale included an insufficient number of faculty with doctoral degrees, "lack of sufficient library space and titles," and low standards for admissions. When Morrison took over as president in 1962, the university immediately reapplied for SACS accreditation. Following a new round of inspections in April 1963, the SACS committee reported that it was "favorably impressed" with the steps taken at A&M that included upgrading the faculty and course offerings. However, a new library, facility upgrades, and additional PhD faculty were still necessary if the university had any hope of maintaining membership. The SACS Special Study Committee also informed the university that an increase in association standards that would take effect in 1965 threatened its tentative hold on membership.[37] This evaluation period created an opportunity to reassess many university priorities. What new programs could the A&M administration successfully integrate into its curriculum? How did the promise of additional funding for the space program play into that integration? Going forward, the opportunities offered by the space program and the quest for accreditation provided the framework for the university's changing curriculum as well as its interaction with the civil rights movement in Huntsville.

President Morrison viewed the booming aerospace industry in Huntsville as an important foundation for his program of educational reform at Alabama A&M College and the source of additional resources to supplement education funds from Montgomery. At the 1963 meeting of the National Association of State Universities and Land-Grant Colleges in Chicago, Morrison contended that for the "first time in American history, Negro youth possess what appears to be at least the minimum preparation to participate and share in the development of a new age in the history of mankind." This had not been the case in the past, when "the dawn of each new era found the Negro ill-prepared to even adjust, much less participate in the dawn of a new age."[38] Now the challenge for Black youth in space research and development was to "keep up instead of catching up," something that was possible now due to the new emphasis on the role of the sciences in American society.[39] Morrison's language reflected a consistent strategy that recognized the changing opportunities for A&M students and a willingness on the part of A&M administrators to shift funds toward the creation of new technical training programs. In 1962, Morrison had developed two new technical education programs at the college, both of which were open to high school graduates. Part of the A&M Division of Mechanical Arts, these new programs included one in mechanical drafting and technology, which could lead to jobs as draftsmen, drawing checkers, or engineering assistants, and another in electronics technology, which could lead to careers as installation technicians, electronics repairmen, or engineering assistants.[40] These electronics and draft technician courses provided gave engineering graduates the foundation to "write [their] own ticket[s]" and optimistically promised "good pay—good working conditions with missiles contractors."[41] But to ensure that these changes were lasting, Morrison would need to secure assistance from the aerospace community.

Top-Down Brokerage in Educational Improvement

President Kennedy's March 6, 1961, executive order on equal employment opportunity met with little initial response from NASA Marshall and Huntsville's aerospace industry. It did, however, point to a mounting framework of federal expectations that government agencies and contractors would be forced to respond to. The explosion of violence in Birmingham in the spring of 1963 accelerated the president's program for progress. At a meeting of the President's Committee on Equal Employment Opportunity (PCEEO) on July 18, 1963, Attorney General Robert Kennedy blasted NASA administrator James Webb for making little tangible progress on the issue. Newly motivated, Webb redoubled efforts within NASA, focusing on Marshall because of its location and its previous poor performance in increasing diversity. The renewed demand for results in equal employment opportunity sparked further direct engagement with the local Black educational institutions that were capable of producing qualified job candidates.

A fair amount of coordination was already happening behind the scenes. Hobart Taylor Jr., executive vice-chair of the PCEEO, continued to work with federal government programs and departments to facilitate the implementation of the executive order and the associated Plans for Progress. Even before Robert Kennedy's verbal assault at the July 18 PCEEO meeting, Taylor understood the importance of engaging with HBCUs for meeting the terms of equal employment. In a July 24, 1963, letter to Taylor, Alfred S. Hodgson, NASA's principal compliance officer, reported that the agency had been "instructed to provide expert assistance to the President of Alabama A&M College and Tuskegee Institute" in the development of grant proposals that would "assist with the improvement of curriculum and faculty to provide better qualified graduates for employment with NASA at Huntsville."[42] Hodgson reported that a meeting with Richard Morrison had taken place on June 28, 1963, at which assistance in the form of a NASA grant was discussed. Establishing a cooperative education program and an intern program with Alabama A&M College were also discussed at that meeting. Hodgson also relayed plans for additional meetings with Morrison and other HBCU representatives from across the region, including North Carolina A&T, Fisk University, Miles College, and ten more HBCUs.[43] Taylor's instructions and Hodgson's report demonstrate that both the PCEEO and NASA understood that progress in equal employment opportunity would come only through direct and continual engagement with HBCUs.

The organization created to address the challenge of meeting President Kennedy's call for greater employment diversity was the Association of Huntsville Area Contractors (AHAC). This group was a collaborative institution that acted as an organizing force, a clearinghouse, and a coordinating agency to ensure compliance. Initially made up of representatives from fifteen aerospace firms, AHAC served as the mediation body between the white and Black communities in Huntsville. From this position, AHAC was an important point of contact for government compliance bodies, private funders, and Black educational policy makers and a vital information clearinghouse for Black students and job candidates. The association's stated employment objectives included plans to implement training programs, recruit on HBCU campuses, and place Black graduates in company personnel departments. AHAC outlined a plan to work with local Black educational institutions to develop increased offerings in technical training, a major part of which was the establishment of cooperative scholarships and summer student employment backed by funding from member corporations. The booming aerospace industry was an obvious opportunity to break into more professional and skilled craftsmen categories. However, as members of AHAC and Alabama A&M College could attest, accumulating skills and education was a "time-consuming process."[44] Past and continued discrimination proved difficult to overcome. As Herbert Northrup argued in his 1970 evaluation of Black employment in the aerospace industry, it was "expecting too much for a man who knew that he would always be a laborer ever to believe that he may be promoted if he takes training."[45] Administrators at HBCUs such as Alabama A&M's Morrison understood that dilemma all too well.

AHAC's educational reform efforts aligned with A&M College's ongoing upgrade of its curriculum and facilities. Participating aerospace firms conducted internal assessments to determine what in-house resources were available in terms of professional lecturers, equipment for loans, and facilities for tours and visitations. AHAC also worked by connecting universities with professional organizations including the National Urban League, the Urban Teacher Preparation Program, the General Electric Summer Guidance Fellowship Program, and the Olin Mathieson Chemical Corporation's Aid-to-Education Foundation. Even limited efforts could produce positive returns. An internal memorandum from Charles Grainger of local aerospace firm Brown Engineering to AHAC president Milton Cummings revealed that efforts at minority colleges throughout the Southeast were creating public relations victories, such as the announcement of beginning of the cooperative program at Southern University in fall 1963. Grainger also noted that visitors to Huntsville had "run into the problem of housing" and that inferior facilities at the "Negro schools" presented a major obstacle to equipping students with the proper training for space-age jobs.[46]

The view from NASA's Marshall Space Flight Center underscored the primary issues with integration. NASA leadership immediately grasped that the location of much of the space program's facilities in the Jim Crow South was deeply problematic in terms of enforcing equal employment policy. Historian Gavin Wright argues that federal legislation on equal employment was particularly effective in the South, where discrimination took the form of "vertical segregation" that excluded Black participation in entire professions in the aerospace industry, particularly above the most menial job level.[47] This problem persisted throughout the Apollo era. NASA failed to allocate portions of the substantial Apollo funds to Black education institutions that were developing employment opportunities for Black graduates through technical training. Throughout the 1960s, NASA's southern centers continued to fill lower-grade positions, including janitorial and clerk positions, with Black employees while the federal government failed to invest in training programs that already existed at nearby colleges, including Alabama A&M and Oakwood Colleges, that equipped Black students for new technical careers. The brief period from 1963 to 1965 was the high point of federal funding for Project Apollo and therefore a missed opportunity at reducing structural inequality through job training.

The person tasked with leading the day-to-day operation of the AHAC was Executive Director L. C. McMillan, an African American and a former administrator at Prairie View A&M College in Texas. Philosophically, McMillan was aligned with a moderate, top-down approach to civil rights reform. Over the course of the decade, he developed a comprehensive approach to the social and economic elements of the Huntsville civil rights movement. In the process, he became a liaison between the aerospace industry and Huntsville's Black communities. For McMillan, the situation in Huntsville was "paradoxical." The low cost of labor that originally drew industry to the South was now the "greatest burden in remaining in the region."[48] McMillan argued that the lack of training in the areas of physics, mathematics, chemistry, medicine, astronomy, and electronics in the region was not so much a race issue but the result of educational disparity between the South and other parts of the country. Black and white southerners found themselves with little hope of employment in the booming high-tech industries. While recruiting efforts could meet the immediate demand for skilled labor, long-term success required new directions in local workforce training.[49]

Precedent existed for McMillan's overall plan for ensuring compliance with federal guidelines on equal employment. In his position at Prairie View, McMillan had developed a cooperative project between Texas Instruments Incorporated and that HBCU that had created a committee of Texas Instruments personnel and college representatives dedicated to finding collaborative ways of connecting industry needs with an educational program at the college. McMillan maintained that only this type of partnership would promote good will and understanding and identify mutually beneficial areas of activity.[50] The program featured many elements that would appear in the Huntsville project, including a summer development program, scholarship programs, a program that offered students opportunities at Texas Instruments in Prairie View cooperative programs, and a clearinghouse for information on workforce development opportunities.[51]

Several assumptions lay at the heart of McMillan's strategy. First, his plan assumed that aerospace industries would continue to grow and that federal funding for the space program would keep pace. Second, McMillan assumed that government and industry would continue to increase their cooperative efforts in aerospace projects and would continue to compete with each other for the scarce supply of technically trained professionals. Third, missing an opportunity to provide "quality vocational training for Negro youth" would keep the South's Black population from equal economic participation, making them an "increasing burden rather than an asset."[52] Therefore, AHAC's project created mechanisms for informing teachers, employment services, and career counselors about opportunities to steer students toward technical training programs. Fourth, Black institutions of higher learning needed to develop their staff, their facilities, and a new curriculum that was capable of providing Blacks with technical education. Failure to do so would mean that Black students would continue to train for low-level careers. Fifth, Huntsville offered the best chance of success in a southern community. McMillan argued that Huntsville had demonstrated its willingness to explore new ways to work with industry and government partners to "alleviate the problems of minority employment."[53] Finally, McMillan recognized his plan's fundamental dependence upon larger support from government, private enterprise, labor, community leaders, and ordinary citizens for its success: "Negroes alone cannot crack through the vicious circle."[54] McMillan's blueprint was a well-conceived proposal for the reform of Black education and employment in the aerospace industry. It also was a major break from the token reforms that white community leaders had previously supported. McMillan, the only Black member of the AHAC's leadership, found himself standing between his desire for lasting reform and an organization dedicated to doing little more than remaining in compliance with federal equal employment guidelines.

New Directions and New Programs

In the years after the passage of the Civil Rights Act of 1964, AHAC remained committed to its educational development programs at Alabama A&M College. However, enforcement pressure from the federal level began to disappear and collaboration from NASA and Marshall began to recede. In this environment, funding for new programs would have to come from private or community organizations. In February 1967, new AHAC president Clinton Grace wrote to Dr. Marvin Feldman at the Ford Foundation with a proposal for a cooperative education program. Grace told Feldman about work that was under way at Alabama A&M under the direction of Dr. W. H. Hollins. That demonstration project, which had begun in the spring of 1966, included faculty training, consultation with industry, development of relevant curriculum, and outreach to secondary schools. Grace described Huntsville as "one of the most unusual labor markets in the world," one in which NASA collaborated with numerous aerospace firms in "exotic and far-reaching research programs that tend to stagger the imagination."[55] Any hope of producing a steady stream of qualified applicants from local educational institutions would require substantial supplementary funds. AHAC and Alabama A&M proposed a coordinated cooperative education program that would need $230,000 from the Ford Foundation for a three-year program. The successful implementation of this program at A&M enabled many Black students to work at Marshall while completing their degrees.[56] However, all of the effort expended to create and maintain these cooperative programs ultimately led to few permanent hires with federal government institutions and local firms.

The A&M cooperative college work-study program, which began in the spring of 1965, was an important bridge between A&M students and the aerospace industry. Initial support for the program came from the legal framework for President Lyndon B. Johnson's War on Poverty through the Economic Opportunity Act of 1964. Part of this legislation aimed to provide financial assistance to qualified students, "stimulate and promote worthwhile work experience," and enable more low-income students to attend college "without the necessity of incurring an unduly heavy burden of indebtedness" that would "seriously handicap them in their future careers."[57] Dr. W.H. Hollins explained how the program would join technical education with space-related industrial employment.[58] During the first year of the program, participating students would alternate periods of classroom and laboratory instruction with periods of job training at aerospace firms and federal facilities. Initially, twenty students chosen from A&M's electronics technology, mechanical drafting, and design technology programs filled positions with ten local industry partners.[59]

A constructive working relationship with Marshall Space Flight Center was an important element of the cooperative program. Marshall administrators understood the value of cooperative relationships with HBCUs. They were a way for the flight center to separate itself from the poor race relations that prevailed in Alabama. Marshall representatives, who had previous experience with HBCU cooperative work-study programs from its 1963 work with Southern University, met with Morrison on July 25, 1966, to discuss a new relationship. President Morrison said that A&M hoped to provide opportunities for students in mathematics and chemistry coursework and hoped that something could be accomplished later in physics. Marshall's offer included personnel and equipment, funding for faculty members, and hands-on experience for students in the form of summer employment at the center.[60] For Morrison, it was critical that this cooperative agreement form the foundation for future graduate-level math and science programs. Developing this close working relationship was vital for Morrison not only for the resources it put at his disposal but also for the opportunities it provided students who were willing to make the jump to the new technical programs.

One graduate of Alabama A&M College who participated in the cooperative work-study program at Marshall Space Flight Center was James Jennings. While completing his degree in mathematics (with a physics minor) at Alabama A&M, Jennings worked in Marshall's Computation Laboratory as a cooperative program student. Later, following service in the United States Army, Jennings returned to work at Marshall before moving on to Kennedy Space Center and eventually to NASA Headquarters in Washington, DC, where he served as an associate administrator. Jennings remembered his participation in the cooperative work-study program at Marshall as a smooth and rewarding process. Jennings recalled being "a little disenchanted with school" during his junior year as well as "tired of not having any money." Seeking a remedy for both, Jennings visited the Alabama A&M office and signed up.[61]

Another participant in the program at Alabama A&M was Charles Scales. As with many Black students of his generation, Scales understood how the lack of funding for Black education had created an imbalance in the types of careers open to Black graduates. Scales remembered how the segregated schools he had attended as a child in the 1960s had inferior equipment and outdated books and that in general, those schools had no science laboratories.[62] However, teachers who were HBCU graduates "worked very hard to try to teach as best they could with the materials they had and the background of their own experience," as they came from "schools that probably didn't have the resources they needed as well."[63] For Scales, the improvements Morrison had made at Alabama A&M by the time he arrived on campus represented a "quantum leap in facilities" over those of his prior experience. According to Scales, the saying "open while under repair" was indicative of the landscape at Alabama A&M as upgrades continued, noting that "while they were trying to get these things done, they still had a school to run. . . . You still had students to educate with the funding and facilities they had, while at the same time pushing for upgrades, pushing for the school of engineering—I think they did a tremendous job of balancing that."[64]

In 1966, Morrison developed another key relationship with the United States Army Missile Command at Redstone Arsenal. This cooperative agreement enabled Alabama A&M students and faculty to use the army's Physical Sciences Laboratory to compensate for the lack of similar facilities at the university. The agreement also allowed A&M faculty to supplement their small salaries and provided students with a work-study income.[65] The program had the reciprocal effect of creating an opportunity in which both students and faculty could stay informed on the most recent advances in solid-state physics. Morrison understood that having these programs in place enabled him to recruit the faculty he needed to add depth to the math and science programs. The ability to sustain these relationships, however, depended on government installations at Redstone Arsenal eliminating discrimination in employment, and Redstone's performance of that task varied over the decades. Nonetheless, cooperative programs with government partners and industry served not only as a bridge between graduates of Black institutions and employers but also as mechanisms for reinforcing the value of technical degrees inside those institutions.

The development of the Alabama A&M Computer Science Center in 1968 through the combination of state funding and a cooperative agreement with Marshall was another major milestone in curriculum development and demonstrated A&M's commitment to engaging with federal installations in new ways. The program included a four-year bachelor's degree in computer science, the first of its kind at any Alabama institution, white or Black. The director of the center, Clyde Foster, developed the program in the A&M Division of Technology while on a two-year leave from Marshall Space Flight Center. While on leave, Foster had familiarized himself with much of the new computing technology from companies such as IBM. The goal of the program was to expand scientific and educational research opportunities for A&M students; automate many of the administrative actions performed in the offices of the registrar and dean and the business administration office; and provide students with the opportunity to earn a bachelor of science degree in computer science.[66]

Foster pointed out that the new machines, including a third-generation IBM System/360 Model 30 mainframe, were "not intended to eliminate jobs but to speed up routine operations" and free personnel to concentrate on the many other problems of administration. The A&M student newspaper reported that the job outlook for the computer science field appeared "rosy" for A&M graduates as there was an "appalling lack of trained programmers, data processors, and operators of computer machines" in the current workforce. Formal training involved basic instruction in programming languages and required only a "limited mathematical background."[67] The space and missile industry was driving an "unlimited demand for competent, qualified computer science graduates."[68] The development of the new computer science program at A&M coincided with computer upgrades in Marshall's Computation Laboratory. The new equipment required new technicians trained in the various programming languages. Students from A&M now found themselves engaged in training programs with a high potential payoff. Both the cooperative program and the computer science program reflected the efforts of Morrison to use space program funding to develop a comprehensive technical curriculum.

One of the challenges to Morrison's plans for the university was a lack of faculty with advanced degrees and experience. Convincing qualified instructors to move to Alabama involved considerable effort. One of the most important new faculty members was Dr. Howard J. Foster, who arrived at A&M in 1965.[69] Morrison understood how the scarcity of Black physicists with PhDs was a formidable barrier to his plan to add a new technical curriculum at the university. Morrison noted how difficult it was to convince Foster to leave his position at the National Bureau of Standards to come to A&M and develop the physics department from the ground up. According to Morrison, although the university could not pay Foster enough to convince him to leave a "pretty good salary" to teach, Foster was "high over" the presence of NASA in the community.[70] The fact that the Oak Ridge National Laboratory in Oak Ridge, Tennessee, was nearby was a draw for Foster; it provided him with research opportunities with the Atomic Energy Commission.

One of the most successful programs of cooperative education was established between Alabama A&M College and the MIT Physics Department and Laboratory for Nuclear Science in the spring semester of 1970. Funded under Title III of the Higher Education Act of 1965, the cooperative agreement with MIT sought to allow juniors pursuing physics degrees at seven HBCUs to spend one semester at MIT taking classes and working in the laboratories.[71] The program was inspired by a similar program that began in September 1969 at Brookhaven National Laboratory known as the Brookhaven Semester. This program, which was funded by a National Science Foundation grant, also gave Black students and faculty from HBCUs an opportunity to spend one semester at the laboratory working in a hands-on environment with esteemed scientists.[72]

The initial proposal for the program reflected the larger philosophical foundations of what Morrison was attempting to accomplish at Alabama A&M. He argued that HBCUs were now engaged in a campaign to break the historical pattern of directing math and science graduates into the teaching profession and "other areas not optimally appropriate to students of high aptitude with desires for careers in the natural sciences and mathematics." One part of the program included a faculty exchange in which Dr. Howard Foster spent one semester at MIT and Dr. Robert Gilmore spent one at Alabama A&M. These educational exchanges were an incredibly rare opportunity for Black students to gain experience working directly on big science programs. The cooperative agreement also enabled Alabama A&M's fledgling science programs to benefit from the prestige of the alliance with MIT and from having its participants serve as examples to incoming students.[73]

During his tenure at Alabama A&M College, Howard Foster represented the aggressive flank of Morrison's more moderate approach and rhetoric on civil rights. Whereas Morrison was forced to maintain his restrained public persona, Foster could use more direct strategies. One example was Foster's falling out with Major General Edwin I. Donley at Redstone Arsenal in December 1970. The argument began with the army's July 1970 termination of its cooperative program between their Physical Sciences Laboratory and Alabama A&M, a program that dated back to 1966. Foster wrote to Donley from his temporary position as visiting professor at MIT in the hope of restoring the terminated cooperative agreement. He reminded Donley that Foster himself had participated in the program on a part-time basis, conducting work in solid-state physics. Foster began by noting the army's positive contributions to economic improvement and the educational advancement of Black students and then shifted to the army's poor local reputation for racial discrimination. Foster asked Donley directly whether reestablishment of the cooperative program might not "provide you with the positive mechanism of positive action in eliminating the racial image that presently exists within your command."[74]

On August 5, 1970, the A&M president wrote to Donley's superior in Washington, DC, the commanding general of the US Army Material Command, General Ferdinand Chesarek. Morrison detailed how critical the program had been to the success of university graduates, informed him that a total of seventy students had participated in the program, and said that the program had added an "incentive to strive for better academic preparation" on the part of the A&M administration and students. Morrison also added that the program had made it possible for the university to upgrade its facilities to support the technical curriculum.[75] Morrison also ensured that a copy of the letter was delivered to Major General Donley at Redstone Arsenal. This exchange is representative of Morrison's strategy for negotiating with white elites and speaks directly to scholar Eddie Cole's argument that although Black educational leaders received little in the way of public recognition, their actions "influenced educational policies and practices."[76] Morrison's letter presented a complimentary account of army and university cooperation. In truth, the letter's most important audience was Major General Donley, and its intent was to convince him that his ongoing negotiations with Black leadership would not occur in southern isolation.

Conclusion

Jim Crow segregation created an inequitable situation in terms of funding for Black education. This imbalance limited the technical training offered at HBCUs and therefore the career opportunities available to Black students when they graduated. Reflecting on the state of Black education in the era, Marshall's "Negro recruiter" Charles Smoot saw the situation as exceptionally bleak, as many Black students in the sciences continued to focus on careers in medicine, dentistry, or education. Smoot understood the delicate balancing act Richard Morrison undertook at Alabama A&M as he attempted to simultaneously project a moderate view of his own thoughts on civil rights reform while working diligently to expand Black technical education. Dr. Morrison walked a fine line between convincing white funding agencies of the need for improvements and expansion at A&M and more radical forces within Huntsville's Black communities that were calling for social justice and an immediate end to Jim Crow segregation. Smoot understood that if Morrison had aligned himself with or given the appearance of aligning with a more radical wing of civil rights, he risked losing the ability to institute the beneficial programs he planned for the university. Smoot recalled that Morrison was interested in keeping that school open, keeping funds coming in. [Morrison and the Alabama A&M administration] were limited in what they can do openly; they can do things behind the scenes to encourage change. But when they get ready to go down to Montgomery to get funds to operate that school, I'm sure in the back of their minds there are things I can't say, that I can't do and get the funds I need to do what I want to do over here. . . . And yet I feel I understand why he didn't, couldn't. That makes sense for me for him to not have Rap Brown come and tour A&M or to set up a chapter, a council there of the [Congress of Racial Equality] organization, of the [Student Nonviolent Coordinating Committee] organization, those organizations that were very active. To do that and survive would not have been wise. Now that I'm old, I didn't understand it when I was younger, but I do now.[77]

Throughout the 1960s, Morrison maintained a conservative stance in his relationship with Huntsville's civic and city leadership. Since his appointment by Governor John Patterson, Morrison had avoided speaking out on social revolution and had indeed prevented any major protests by A&M students. The advent of the Black Power movement and the continuation of racial unrest in urban centers across the nation forced Morrison to continually calibrate his statements in the public sphere and his actions at the college. By 1967, total enrollment at A&M College had expanded to 1,700 students and the school had a faculty of more than 150, all of whom held at least a master's degree. Twenty-five faculty had doctoral degrees from universities "mostly outside of the South."[78] Morrison argued that for the first ninety-one years of its history, A&M had been merely a "college *in* the community." The year 1967 was time for the university to become a "college *of* community—not just a college for Negroes, but a college where there is a community of scholars capable of learning, who are capable of teaching, who are capable of searching for truth and who are interested in humanity."[79] Morrison walked a moderate line on civil rights while engaging directly with NASA Marshall and local aerospace companies as he recrafted the technical curricula at A&M College.

Morrison's actions during the 1960s clearly challenge Ernest Stuhlinger's narrative of Alabama A&M's relationship with the aerospace industry. Stuhlinger's commentary reflects not only the power to shape the narrative but also the power that comes with controlling that narrative over time. Over the decade, Morrison and Alabama A&M College leadership transitioned the administration from that was one willing to engage openly with the forces of popular insurgency to one that was willing to play the long game so the college could expand its facilities and resources. This transition came about through Morrison's retreat from an open embrace of the protest movement in downtown Huntsville and his adoption of a strategy that sought to expand A&M College's facilities and develop a curriculum capable of preparing its graduates for space-age jobs that was concealed behind a conservative façade. Morrison leveraged cooperative agreements with NASA Marshall, federal directives on equal employment opportunity, and the fear of social revolution among white Huntsvillians to bring about lasting opportunities for graduates of the college.

At the end of the 1960s, the Black educational system was still at the crossroads of the South's Sunbelt modernity. The 1954 *Brown* decision had kicked off decades of ideological (and actual) battles in the public sphere about the role of education in reshaping the fortunes of educated Blacks and how those students would take advantage of the growing opportunities in the Cold War world. The desegregation of Huntsville's white institutions, including city schools and the University of Alabama in Huntsville, and the implementation of successful technical training programs at Alabama A&M under Morrison's direction were considerable steps forward. Blacks advanced through the twin forces of civil rights protest and a coordinated strategy of educational progress couched in a façade of conservatism that was palatable to white elites who hoped to project an image of the city as an oasis of progress.

Notes

1 On June 26, 1969, the Alabama State Board of Education officially changed the name of Alabama Agricultural & Mechanical College to Alabama Agricultural & Mechanical University. I will refer to that institution as Alabama A&M College throughout this chapter for consistency.

2 Ernst Stuhlinger and Fred I. Ordway III, *Wernher von Braun: Crusader for Space* (Malabar, FL: Krieger Publishing Company, 1996), 187.

3 Downing spent most of his professional career as dean of the School of Engineering and Architecture at Howard University in Washington, DC.

4 Lewis K. Downing, "The Negro in the Professions of Engineering and Architecture," *Journal of Negro Education* 4, no. 1 (1935): 60–70.

5 Gavin Wright, *Sharing the Prize: The Economics of the Civil Rights Revolution in the American South* (Cambridge, MA: Belknap Press of Harvard University Press, 2013), 68–69.

6 Thomas W. Turner, "Science Teaching in Negro Colleges," *Journal of Negro Education* 15, no. 1 (1946): 36–42.

7 Ambrose Caliver, "The Role of the Federal Government in the Higher Education of Negroes," *Phylon* 10, no. 4 (1949): 370–380.

8 Caliver, "The Role of the Federal Government in the Higher Education of Negroes."

9   W. E. Anderson, "The Education of Negroes in Alabama," *Journal of Negro Education* 16, no. 3 (1947): 311–316.

10 Anderson, "The Education of Negroes in Alabama."

11 Anderson, "The Education of Negroes in Alabama."

12 Mary A. Dudziak, *Cold War Civil Rights: Race and the Image of American Democracy* (Princeton, NJ: Princeton University Press, 2011), 107.

13 Walter Crosby Eells, "The Higher Education of Negroes in the United States," *Journal of Negro Education* 24, no. 4 (1955): 426, 433–434.

14 Edward K. Weaver, "Development of Science Curricula in Negro Schools," *Journal of Negro Education* 25, no. 2 (Spring 1956): 118–119.

15 Weaver, "Development of Science Curricula in Negro Schools," 118–119.

16 Weaver, "Development of Science Curricula in Negro Schools," 118–119.

17 W. Montague Cobb, "Not to the Swift: Progress and Prospects of the Negro in Science and the Professions," *Journal of Negro Education* 27, no. 2 (1958): 120–126.

18 David E. Wharton, *Struggle Worthy of Note: The Engineering and Technological Education of Black Americans* (Westport, CT: Greenwood Press, 1992), 97–98.

19 Quoted in Clayborne Carson, *In Struggle: SNCC and the Black Awakening of the 1960s* (Cambridge, MA: Harvard University Press, 1981), 13.

20 "13 Sit-Inners, Stand Inners are Arrested," *Huntsville Mirror*, January 13, 1962, 1. Among those arrested were Frances Sims, Nelson E. Wilkerson, Alonza Hinton, Dred C. McMillan II, Samuel F. Evans, Louis Hampton, Leon Felder, Holman Turner Jr., Melvin L. Haley, John Sims, William Pearson, Roland Dawson, and Robert Matthews. Founded in Chicago in 1942, CORE was one of the primary civil rights organizations dedicated to desegregation. CORE participated in a number of major events associated with the civil rights movement, including the Montgomery Bus Boycott, the sit-in movement, and the Freedom Rides in the summer of 1961.

21 Sonnie Hereford III, oral history interview, September 18, 1973, 7–8, Statewide Oral History Project, Alabama Center for Higher Education.

22 Sheryll Cashin, *The Agitator's Daughter: A Memoir of Four Generations of One Extraordinary African American Family* (New York: PublicAffairs, 2008), 147.

23 Cashin, *The Agitator's Daughter*, 10–11.

24 Theresa P. DeShields, "The Acquisition of Civil Rights in Huntsville, Alabama from 1962–1965" (MA Thesis, Alabama A&M University, 1965), Appendix A.

25 Richard B. Morrison, *History of Alabama A&M University, 1875–1992* (Huntsville: Liberal Arts Press, 1994), 150–179, 265–266.

26 Cashin, *The Agitator's Daughter*, 140.

27 Cashin, *The Agitator's Daughter*, 140.

28 Sonnie Wellington Hereford III and Jack D. Ellis, *Beside the Troubled Waters: A Black Doctor Remembers Life, Medicine, and Civil Rights in an Alabama Town* (Tuscaloosa: University of Alabama Press, 2011), 95.

29 "A&M Prexy Given Leave of Absence," *Huntsville Mirror*, February 17, 1962, 1, Digital Collections, Marshall Space Flight Center Archives, Huntsville, Alabama (hereafter MSFC Archives).

30 "Negro Colleges Face Challenge," *Huntsville Mirror*, November 16, 1963, 1, 8, Digital Collections, MSFC Archives.

31 "Negro Colleges Face Challenge."

32 "Negro Colleges Face Challenge," 8.

33 Davis quoted in Eddie Cole, *The Campus Color Line: College Presidents and the Struggle for Black Freedom* (Princeton: Princeton University Press, 2020), 63.

34 Cole, *The Campus Color Line*, 10.

35 Cole, *The Campus Color Line*, 11.

36 Cole, *The Campus Color Line*, 65.

37 "The Outlook for Accreditation: A Report by Dr. Richard D. Morrison, Ph.D., President of Alabama A&M College," *The Normal Index*, December 1963, 5.

38 *The Campus Intercom: A Weekly Bulletin*, November 15, 1963.

39 *The Campus Intercom: A Weekly Bulletin*, November 15, 1963.

40 "Alabama A&M College Organizes Two New Curricula in Technical Education," *Huntsville Mirror*, August 25, 1962, 2.

41 *Afro-American*, Fourth Annual Engineering Supplement, September 24, 1966.

42 Alfred Hodgson to Hobart Taylor, July 24, 1963, file Implementation of Executive Order 10925 (1963), MSFC Archives.

43 Hodgson to Taylor.

44 Herbert R. Northrup, *Negro Employment in Basic Industry*, vol. 1 of *The Negro in the Aerospace Industry*, (Philadelphia: University of Pennsylvania, 1970), 222–223.

45 Northrup, *Negro Employment in Basic Industry*, 222–223.

46 Northrup, *Negro Employment in Basic Industry*, 222–223.

47 Wright, *Sharing the Prize*, 121.

48 L. C. McMillan, "A Project Design to Create a Continual and Increasing Pool of Qualified Negro Applicants for New Employment Opportunities Available in Southern Industry and Business, Prepared for The Huntsville (Alabama) Contractors Equal Employment Opportunity Committee at the Request of Brown Engineering Company, Inc., Especially for Mr. Milton K. Cummings, President," August 1963, Association of Huntsville Area Contractors Collection, Alabama A&M University Archives, Huntsville, Alabama.

49 McMillan, "A Project Design to Create a Continual and Increasing Pool of Qualified Negro Applicants."

50 McMillan, "A Project Design to Create a Continual and Increasing Pool of Qualified Negro Applicants."

51 McMillan, "A Project Design to Create a Continual and Increasing Pool of Qualified Negro Applicants."

52 McMillan, "A Project Design to Create a Continual and Increasing Pool of Qualified Negro Applicants."

53 McMillan, "A Project Design to Create a Continual and Increasing Pool of Qualified Negro Applicants."

54 McMillan, "A Project Design to Create a Continual and Increasing Pool of Qualified Negro Applicants."

55 Clinton Grace to Dr. Marvin Feldman, February 9, 1967, Hobart Taylor, Jr. Papers 1961–1981, Bentley Historical Library, University of Michigan, Ann Arbor, Michigan.

56 Grace to Feldman.

57 *The Campus Intercom: A Weekly Bulletin*, March 19, 1965.

58 "A&M Fund to Help Train Negroes Here," *Huntsville Times*, April 22, 1965.

59 *The Campus Intercom: A Weekly Bulletin*, June 18, 1965, 2.

60 Meeting notes, July 25, 1966, Richard Morrison Collection, Alabama A&M University Archives, Huntsville, Alabama (hereafter Morrison Collection).

61 James Jennings, oral history interview, September 17, 2015, MSFC Archives, Huntsville, Alabama.

62 Charles H. Scales, oral history interview with Brian Odom, February 12, 2016. MSFC Archives, Huntsville, Alabama.

63 Scales oral history.

64 Scales oral history.

65 Howard J. Foster to Edwin I. Donley, December 17, 1970, Morrison Collection.

66 "Computer Science Center Offers New B.S. Degree," *The Normal Index*, n.d. [ca. 1967], 4.

67 "Computer Science Center Offers New B.S. Degree."

68 "Computer Science Center Offers New B.S. Degree."

69 A native of Gadsden, Alabama, Foster joined the military after completing only a seventh-grade education. After his military service, Foster completed his high school diploma, graduated magna cum laude with a master's degree in physics from Fisk University, and then earned a PhD at Catholic University of America in 1964. Following graduation, Foster found work as a solid-state physicist at the Institute for Material Research of the National Bureau of Standards. That is where he was working when Dr. Richard Morrison invited him to come to Alabama A&M and become the head of the Department of Mathematics and Physics. Unfortunately, Foster died in a car accident on January 23, 1973, while returning home from Gadsden, Alabama.

70 Richard Morrison Oral History with Bob Ward, July 23, 1997, Milton Cummings Collection, MSFC Archives, Huntsville, Alabama.

71 Bennett College, Fisk University, Hampton Institute, Norfolk State College, North Carolina A&T University, and Virginia State College for Science Education also participated in the program. Graduate student Ronald McNair, who would go on to be selected for NASA's astronaut corps, represented North Carolina A&T University in the program. McNair was killed in the *Challenger* disaster in January 1986.

72 Brookhaven National Laboratory Annual Report, July 1, 1969, Morrison Collection. Miles College, Tuskegee Institute, and Talladega College in Alabama; Grambling College in Louisiana; Tugaloo College in Mississippi; Langston University in Oklahoma; Prairie View A&M College, Texas College, and Jarvis Christian College in Texas; and Knoxville College in Tennessee also participated in the program.

73 "A Proposal to HEW from a Consortium Comprising Alabama Agricultural and Mechanical University, Bennett College, Fisk University, Hampton Institute, Massachusetts Institute of Technology, Norfolk State College, North Carolina A&T State University, and Virginia State College for Science Education at the Undergraduate Level," 1970, Morrison Collection. Michael Hartwell and Jimmie Jones represented Alabama A&M University for the first semester (spring 1970) were.

74 Howard J. Foster to Edwin I. Donley, December 17, 1970, Morrison Collection.

75 Foster to Donley.

76 Cole, *The Campus Color Line*, 69.

77 Charles Smoot, Oral History Interview, August 2, 2016, MSFC Archives.

78 "President R. D. Morrison Addresses the Metropolitan Kiwanis Club of Huntsville," *Normal Index*, March 1967, 16–17.

79 "President R. D. Morrison Addresses the Metropolitan Kiwanis Club of Huntsville."

4

Engineering with an Accent?

A Discussion of the *Challenger* Accident

STEPHEN P. WARING

After the space shuttle *Challenger* accident, blame fell on managers, southerners all, from NASA's Marshall Space Flight Center in Huntsville, Alabama. The Marshall managers oversaw the space shuttle's solid rocket motors. A joint in the segmented boosters leaked, destroying the space shuttle in an accident that killed the shuttle's astronauts. The Presidential Commission to Investigate the Space Shuttle Accident, commonly called the Rogers Commission, found that engineers and managers had discussed flaws in the joints for years and had underplayed the dangers of cold weather for the rubber O-ring seals in the joints.

The commission concluded that Marshall manifested a culture of "management isolation," used a "flawed" process of decision-making before the launch, and failed to convey "the seriousness of concern" about risk to top shuttle officials.[1] The *New York Times* described Marshall as a "secretive and authoritarian" organization that treated reporters with "pervasive suspicion, frequent anger, and widespread reluctance to talk."[2] Astronauts expressed anger that they had not known that risks such as the one associated with the accident had existed for years and that they had not known it because of Marshall's convoluted culture.[3] These statements about Marshall managers echoed negative stereotypes about southern culture and raised interesting questions. How (if at all) might the decisions before the accident have resulted from regional culture and how might the stories told after the accident show bias?

Historians have long used regionalism as a category of analysis. In the historical profession, social and cultural approaches have invigorated the study of regions other than nation-states and empires. Contemporary historians examine transnational areas for characteristics of language, knowledge, value, politics, gender, religion, social praxis, and economic modes.[4] Students of the US South have never stopped exploring southern society and culture.[5] Historians of NASA, including contributors to this volume, have explored the relationship of the agency to various changing social patterns in the South.[6]

Few historians have consciously applied southern regionalism as a category of analysis to technological events. In the study of NASA technology and engineering, historians and social scientists have used geography-free categories that transcend regional cultures and local knowledge. Typically, when assessing NASA's technical processes and events, they adopt the agency's internalist standards of management, communications, and engineering. The big question is such analyses becomes the historian's judgment of whether what the agency did worked. To scholars, the success of the hardware and of a mission means that the agency had made proper decisions and followed appropriate processes. However, in cases of failure, the focus of researchers shifts to tracing back to identify errors. The success/failure category of analysis is bright, clear, and internalist. This universal standard of technological merit mimics NASA's idealized vision of itself as a "perfect place" that exists above the limitations of politics and place.[7]

A regional perspective on NASA muddies the waters and inserts externalist factors from a wider context into discussions of technical culture. The categories of analysis become more slippery and the questions become more open. How might local concerns impact technological practices? Would local best practices lead agency experts to practice management and engineering with a southern accent? After technical failure, might provincial stereotypes blame regional cultural backwardness? How might stereotypical perceptions become written in interpretation and memory?

For this historian of technology, venturing into a study of regionalism means stepping outside scholarly comfort zones into a historiographical morass.[8] Historians of the South have a nuanced tradition of exploring the culture of and stereotypes about the region. In the middle to late twentieth century, they emphasized the exceptionalism of the region compared to the rest of the United States. They saw a "benighted South" with a backward and barbaric culture. They often confirmed northern stereotypes, such as the perception that the South was behind in education and business. In recent decades, after the civil rights era and the Sunbelt's spectacular growth, a younger generation of southern historians has reconsidered this analysis. Many see a "myth of southern exceptionalism." The recent history of the South, they argue, has not been substantially distinct from national experience and its culture has been inseparable from central patterns in the entire United States. They have also noted that a myth of southern exceptionalism, including the idea of the benighted South, endures in scholarship and popular memory. One reason such myths endure is that cartoonish exaggerations and oversimplifications can bring meaning to troublesome and persistent problems in American life.[9]

This chapter explores regional topics through events and perspectives that emerged following the space shuttle *Challenger* accident in January 1986. It has two main conclusions. Marshall did not engineer with a southern accent and no regional pattern caused the accident. However, in post-accident narratives about supposed managerial miscommunication and malfeasance, negative stereotypes associated with white southerners sometimes appeared. Such stereotypes were at most undertones in the report of the Rogers Commission. In some stories about the accident, however, images of a benighted South sometimes became explicit.

The essay has several sections. Biographies of key figures from NASA's Marshall Space Flight Center explore careers and work cultures to see if they engineered with a regional accent. A study of the events and dialogue that emerged during the hearings of the Rogers Commission checks whether they show evidence of regional rhetoric and regional bias. An examination of some scholarly and popular accounts seeks evidence consistent with regionalism. Finally, it comments on the utility of this approach to NASA's history.

Southern Engineering?

How "southern" were key Marshall engineers involved in the decision-making leading up to the accident? Did their engineering practices show a regional accent? Their careers and work culture at Marshall provide evidence that they were southerners but their engineering was not.

The Marshall engineers who took part in the decisions on the eve of *Challenger*'s fatal launch had southern roots. After receiving various criticisms in the Rogers Commission's report, all but one retired from NASA within months after its publication.[10] William Lucas, the Marshall center director, was born in Newbern, Tennessee, got a bachelor's degree in chemistry from what is now the University of Memphis, and earned a doctorate in chemistry from Vanderbilt University. George Hardy, the deputy director of science and engineering, was born in Russellville, Kentucky, and received a bachelor's degree in civil engineering from Georgia Tech. Stanley R. Reinartz, the manager of the Shuttle Projects Office, was a native of Ashland, Kentucky but grew up in Ohio and got his college degree in mechanical engineering from the University of Cincinnati. Judson Lovingood, the deputy manager of the Shuttle Projects Office, grew up in Birmingham, Alabama, and earned an engineering degree and later a doctoral degree in mathematics from the University of Alabama. Lucas, Hardy, and Reinartz started working in rocketry for the army at Redstone Arsenal in Huntsville in the 1950s and transferred to NASA with the establishment of the Marshall Space Flight Center in 1960; Lovingood started employment at the center in 1962. Lawrence B. Mulloy, the manager of the Solid Rocket Booster Project, grew up in Shreveport, Louisiana, and had a degree in mechanical engineering from Louisiana State University and a master's degree from the University of Oklahoma. On average, the five engineering managers had over twenty-eight years working in Huntsville's government rocketry organizations.[11]

When they moved to Huntsville in the 1950s, they helped transform the city from a struggling cotton-mill town to a booming "Rocket City." Like many cities in the region, Huntsville experienced a "nationalization" that removed or muted many of its "southern" cultural characteristics.[12] Early in their careers during the Redstone and Saturn programs, they took part in a federal approach to innovation called the "arsenal system." That system had several unusual features, but its primary characteristic was technological research, development, and even production by in-house civil servants. The arsenal system assisted in the success of the Saturn launch vehicles and Project Apollo. It stemmed from the army's arsenal approach to weapon development from the time of the early Republic. The arsenal system was reinforced by similar practices that V-2 missile engineers brought from the army in Nazi Germany.[13] The arsenal system owed nothing to the regional culture of the American (or even German) South.

The arsenal system, at first a defining feature of Marshall and a major difference from other NASA centers, had faded by the late 1960s because of changes in federal procurement policies. NASA imitated the air force and used contractors rather than civil servants to develop space hardware. To meet the goal of a lunar landing by the end of the decade, NASA turned to a nationwide network of aerospace contractors. Even before the conclusion of the lunar missions, NASA started slashing budgets. The cuts led to reductions in force at Marshall and essentially ended the arsenal system. The center came to rely on and manage contractors like other NASA centers and Marshall lost most of what had made it unique. In 1977, NASA even considered closing it and dispersing key personnel to other centers.[14]

The budget cuts undermined the prosperity of the communities in the NASA centers in the South. Faced with austerity, Marshall's leaders were not passive. As good citizens, they sought to preserve jobs and the local economy. Project Apollo had made Huntsville a boom town and center leaders felt a civil commitment to securing its prosperity during the Saturn bust. They responded with diversification, seeking contracts outside rocketry and establishing Program Development section to solicit extra work. The first manager of Program Development was William Lucas, who called it "the new business organization." The office was successful and led Marshall to work on such projects as scientific satellites, solar energy and coal mining, telescopes, and materials processing. Marshall's managers developed a mentality and praxis consistent with the tradition of industrial boosters of the New South.[15]

It is problematic, however, to conclude that Marshall's engineering decisions on the solid rocket boosters before the accident were warped with a regional bent. NASA had prolonged technical discussions about the best propulsion system for what became the space shuttle, and its teams chose to use solid rocket motors similar to those used in ballistic missiles. Center engineers had little expertise with solid-fueled rockets and regarded them with suspicion. In ground tests, solids were expensive and made the systems hard to verify; in flight, controllers could not throttle them back or shut them off. The center could hardly oppose solids for fear of losing responsibility for at least part of the propulsion work. Marshall supported NASA's choice of solid rockets because of the expectation of reusability and overall cost savings.[16]

Marshall managed the contractor for the solid rocket motor and recognized its expertise. NASA chose Thiokol, later Morton-Thiokol, as the contractor to develop the solid rockets and refurbish them after each flight. Marshall monitored the Utah-based company but mostly deferred to the expertise of Thiokol's engineers. The deference persisted even as Marshall engineers discovered problems with the solid rocket motor joints during tests. The motor cases had four segments in order to facilitate rail transport. The segments connected with a tongue-in-groove joint. These joints were complex and had two rubber O-rings designed to seal like a gasket in a garden hose. In hydroburst tests, during which the cases were filled with fluids and the joints were pressurized multiple times, the O-rings leaked drops of fluid. To a few low-level Marshall engineers, the leaks signified that the joint design was defective: the joints opened rather than closed under pressure and in the view of those engineers, the O-rings did not meet a safety requirement for redundancy. However, the Thiokol motor experts attributed the leaks to the repetitive pressurizations of the test. They pointed out that a launch would pressurize the motor only once and claimed that the joints and O-rings sealed as effectively as they had in static motor firings. Sharing Thiokol's consensus that the joint design was efficacious and redundant, Marshall's management exalted the expertise of Thiokol and disdained the localized knowledge of a few of its engineers. Even Marshall managers showed little trust in remnants of the arsenal system.[17]

Once the shuttle launches began and anomalies appeared in the joints, local knowledge developed, but not primarily at Marshall. Thiokol engineers discovered problems with the joints and seals during postflight inspections. Before the final *Challenger* launch, in about 4 percent of field case joints, they found that the hot gasses from the motor had burned away material of the primary rubber O-rings ("erosion") and in about 3 percent of field case joints they saw soot ("blowby") between the primary and secondary O-rings. Both Thiokol and NASA engineers understood the problems as a safety concern, but because flights succeeded despite the problems, they remained confident about the design. They studied the anomalies and worked to improve assembly and preflight inspection procedures. Despite their efforts, erosion and blowby grew more frequent. After an April 1985 flight that launched in 53F temperature had especially serious problems, NASA directed Thiokol to create an O-ring task force to study the problem and recommend a redesign. Some members of the Thiokol's task force circulated memos internal to the company that warned of catastrophe and waved a red flag about the resources devoted to the work. No one communicated the engineers' apprehension to Marshall or other NASA officials. In fact, the official line from Thiokol, one that Marshall and motor engineers at NASA Headquarters accepted at a major August 1985 briefing, was that the existing design was safe to fly and flight anomalies were "within [the] experience base."[18]

On the eve of *Challenger*'s ill-fated launch, Thiokol and Marshall discussed how cold weather might affect the O-rings and whether to delay the launch approved for the next morning. Previously, they had had superficial discussions about the effect of cold temperatures on the rubber seals, especially during the August 1985 briefing. When Thiokol engineers asserted that cold made the O-rings less likely to seal, Marshall managers Hardy and Mulloy challenged that view. In the teleconference, Thiokol engineers argued that the predicted freezing temperatures would make the rubber hard and stiff and recommended no launch until warmer weather. However, data in their faxed vu-graphs was contradictory and inconclusive. The key data compared joint anomalies from one cold-weather flight (the April 1985 launch) and one warm-weather flight. Marshall's managers pointed to that inconsistency and disputed Thiokol's conclusions about the risks of cold temperatures. They also made caustic remarks. Hardy stated that he was "appalled" at Thiokol's recommendation and Mulloy asked if they wanted to wait until "next April" to launch. Hardy also said he would not overrule a recommendation by a contractor to delay a launch. Because of the dispute, Thiokol managers in Utah asked to go offline for an internal caucus. In that private discussion, four Thiokol vice-presidents overruled the engineers and approved a cold-weather launch. The contractor did not share with Marshall its local knowledge—that Thiokol had an internal disagreement and that its managers had made the final decision. Thinking the launch recommendation had more support from contractor specialists than it did, Reinartz reported only to Lucas and not to shuttle officials outside Marshall. He reported Thiokol's final recommendation that *Challenger* was safe to launch in the cold but did not report Thiokol's initial recommendation or the conflict at the briefing.[19]

Investigating with an Accent?

Did the investigation show signs of regional stereotyping? After the accident, NASA and the Marshall officials lost control of the narrative of what happened. In the official investigation and in the media, stories emerged about the Marshall center and its managers that resembled negative stereotypes about southern culture. That narrative became dominant in the media and Marshall engineers reinforced it when they tried to refute it.

Although NASA had formal procedures for accident investigations conducted by agency experts, the Reagan White House appointed an independent presidential panel to direct the investigation. The Reagan administration did this for several reasons. NASA had a power vacuum because its administrator was on leave because of an indictment (later dismissed), and its deputy administrator was an inexperienced outsider. The White House wanted an investigation that handled the politics carefully in a way that would protect both NASA and the administration. Presidential policy, which Congress supported, had favored a rapid shuttle launch schedule to reduce the cost per flight and thus put pressure on NASA to meet a timetable. The shuttle was an important national symbol and the nation's prime carrier of scientific and military payloads. The *Challenger* mission had special importance because it carried Christa McAuliffe, the first member of the Teacher in Space Project, a Reagan administration initiative. After the accident, an uncorroborated urban legend developed at NASA and the Beltway that the White House had ordered the launch despite the cold weather, in effect putting politics over safety, so that President Reagan could talk with McAuliffe from orbit during his forthcoming State of the Union speech. William P. Rogers, a DC lawyer, a former secretary of state in Nixon's cabinet, and a Republican Party insider, headed the commission. In 1982, the president had selected Rogers to play the role of US president in a top-secret national security war game simulation.[20]

A combination of factors—NASA procedures, the Rogers Commission hearings, and media reporting—worked to transform depictions of the *Challenger* disaster from an accident into a scandal. In the aftermath of accidents,

NASA used a technique called fault-tree analysis, which examined all systems for failures, eliminating irrelevant factors while isolating relevant anomalies. NASA's practice was not to speculate on causes and to minimize public comment until the investigators isolated faults. But the media wanted stories immediately, and a flood of journalists headed to NASA centers. Kennedy Space Center went from issuing 150 press passes for a normal launch to over 1,100 after the accident. The media sought interviews of key managers and engineers, but those NASA personnel worked in the accident investigation and would not talk on the record. Journalists experienced NASA's behavior as stonewalling and sought sources outside the administration.

In initial presentations to the commission, Marshall engineers discussed a range of potential accident scenarios and did not focus on the anomalous flame coming from the side of the right solid rocket booster, a flame television news reports had noticed the evening of the accident. On February 6, in the commission's first public meeting, Judson Lovingood discussed the history of the joint problems, the concerns about cold weather, and Thiokol's approval of the launch and downplayed the anomalies as "having been thoroughly worked." Then, on February 9, the *New York Times* published a story from a former NASA budget analyst that exposed the long history of problems with the solid rocket booster joints and the fact that NASA Headquarter engineers had had serious concerns as early as the summer of 1985. The commissioners felt blindsided; they saw the story as an exposé of something dark at the agency and wondered if NASA and Marshall had sought to cover up how launches continued even with anomalous, dangerous hardware.[21]

Suspicion of Marshall intensified after a Thiokol engineer revealed the prelaunch teleconference on the cold temperatures. In a hearing on February 10, Lawrence Mulloy asserted that there was no statistical correlation between cold weather and O-ring problems and that Thiokol and Marshall had agreed that a launch in the cold was safe. Thiokol's Allan McDonald, the company's project manager for the solid rocket motor, asked to testify before the commission. For the first time, they learned from him that at the prelaunch teleconference, Thiokol had originally called for a postponement of the launch because of the cold. Also, he alleged, the company changed its recommendation under pressure from Marshall. Again, this helped convince the commission that NASA, especially Marshall, had management failings and was trying to hide embarrassing facts. The next day, Mulloy again testified that he saw no correlation between temperature and joint anomalies because problems had occurred in both warm and cold weather. Commissioner Richard Feynman, a Nobel Prize–winning physicist who was famous for his scientific demonstrations as a teacher, showed that when he removed a C-clamp from a section of O-ring submerged in ice water, deformations remained. This showed that a cold O-ring lost resiliency and likely would not seal the joint. As the commission continued its work, Thiokol engineers explained their version of the teleconference. Under the prompting of the commissioners, they said that Marshall managers' criticism of their no-launch rationale gave them the near-impossible task of showing it was unsafe to launch, the opposite of their usual job of showing it was safe to proceed. They experienced this as "pressure to launch" and said that when they could not prove it was unsafe, Thiokol's managers changed the launch recommendation. Not only did the commission express dismay at the dynamics of the teleconference, they also disapproved of how Marshall managers pressured Thiokol to change their recommendation and failed to report safety concerns to shuttle managers at NASA Headquarters or Johnson Space Center.[22]

To show the risk of cold temperature, a commission staff member made a bipolar chart that showed a statistical relationship between O-ring damage and temperature. The simple chart juxtaposed the problems seen in field joints and launch temperature.[23] It revealed that in the twenty-four flights before the *Challenger* launch, twenty missions had temperatures of 66F or above, and of these only three had thermal distress in field joint O-rings. In contrast, all four flights with temperatures below 63F had distress in field joint O-rings. The predicted temperature for the launch was far below the average launch temperature of 68.4F and hence was risky. A version of the chart appeared in the final report. For the commissioners at least, the chart confirmed that cold weather had caused the accident and that Marshall managers should have listened to the Thiokol engineers' warnings on the eve of the launch.[24]

Rogers informed President Reagan of its preliminary findings and on February 15 issued a public statement that said "in recent days, the commission has been investigating all aspects of the decision-making process up to the launch of the *Challenger* and has found that the process may have been flawed." The commission also changed its methodology. Its members sought to control the oversight of the investigation, bringing in more expert investigators and interviewers and removing several NASA managers who were taking part in the investigation, including Marshall Center director Lucas.[25]

In hearings after February 25, the commission narrowed its focus to the flawed launch process. Its hearings became more prosecutorial. Because "the presentation of the facts" would reveal pre-accident "human error," Rogers explained, the panel would provide "a right of reply" to anyone who believed they had been "unfairly criticized" or "inaccurately portrayed."[26] The first witnesses the commission called after February 25 were Thiokol engineers who had opposed a cold-weather launch or had provided warnings of joint problems. The commissioners asked them questions of fact about what had happened. Then the commission called the Marshall managers who it contended had failed to hear bad news, neglected to report bad news, or never understood reports of bad news but should have. The commissioners expressed their distrust by asking the managers counterfactual questions about supposed managerial errors and what they had not done. They also asked about why they had failed to heed clear warnings of danger, why they had not stopped "killing messengers," why they had not listened to those who reported risks, and so on.[27]

Presenting a chain of facts and logic, the commission sought to show that Marshall officials had underestimated risk and communicated ineffectively. The commission presented the occurrence of flight anomalies as warnings that the joint design was unsafe and said that booster managers wrongly interpreted them as "an acceptable risk." The Marshall booster managers, the commission contended, communicated in ways that obfuscated risks and deemphasized deficiencies. Such reports confused shuttle officials at Johnson Space Center and NASA Headquarters, who admitted under questioning that they had not understood the risk. In the commission's narrative, all of this culminated in the tragic preflight teleconference where Marshall managers pressured Thiokol to reverse its initial recommendation and failed to report it to Johnson Space Center and NASA Headquarters. In its final report, the commission argued that Marshall managers had not conveyed "the seriousness of concern" and that their organization suffered from "management isolation" and "a propensity of management . . . to contain potentially serious problems and attempt to resolve them internally rather than communicate them forward."[28]

In hearing testimony, the Marshall engineers sought to explain what had happened. They explained that before the ill-fated *Challenger* flight, Thiokol had defended the safety of their design, had approved it for flight, and had not shared company engineers' concerns that developed during the summer of 1985. Marshall engineers pointed out that they had initiated the prelaunch *Challenger* teleconference to discuss the risks of cold weather and pointed out that if they had not done so, the conversation would not have taken place. They argued that their challenges to the contractor's recommendation not to launch in the cold were normal efforts to criticize data and logic and not undue pressure to launch. Thiokol was the only source of solid rocket motors for the shuttle and was in no danger of losing its contract. The company's presentation of temperature data was contradictory and unconvincing. Thiokol did not tell Marshall that its vice-presidents had overruled their engineers during the offline caucus, so the Marshall Center engineers had assumed that its contractor had a consensus about launch safety. After the company reversed its no-launch recommendation, Marshall's Stanley Reinartz polled the participants of the teleconference by asking if there was more to discuss. No one from Thiokol's plant raised any objections.[29]

The commission's central effort to show the culpability of Marshall managers helped create a media maelstrom that was out of the managers' control. NASA Headquarters had ordered that no one comment on the ongoing investigation. The media descended on Huntsville and twenty-five and sometimes fifty reporters overwhelmed its press office. NASA required members of the press to have escorts and submit written questions. The reporters regarded the restrictions as part of an attempt to cover up a scandal.[30] They hung out in the cafeteria at the Marshall complex and camped outside the houses of Lucas, Lovingood, and Mulloy. Mulloy climbed out a rear window of his house to go to work. Reporters wrote stories without Marshall's input, and by February 19 had published reports that Marshall Center managers had withheld information from NASA about Thiokol's warnings. After Marshall's managers testified on February 25 to the commission about the teleconference, Lucas allowed press conferences to clear up misrepresentations.[31]

Mulloy, Hardy, Lovingood, Reinartz, and Lucas met separately with reporters from both wire services, all of the television networks, the *Washington Post*, the *Los Angeles Times*, the *Orlando Sentinel*, *Time*, *Newsweek*, and eight regional organizations. The managers tried to persuade the press that the launch process had not been flawed, but their words backfired. Mulloy said that the Thiokol engineers had not communicated their alarm about the safety of the field joints to NASA. "None of these [Marshall] engineers knew that level of concern existed." If serious concerns existed, Mulloy said, Thiokol engineers had several paths for transmitting information and did not require approval through Thiokol or Marshall's chain of command. Reinartz agreed, pointing out that Thiokol had not consistently communicated concern about cold weather and had not objected to launching in 40F weather on January 27, the day before the tragic launch. He also defended his decision not to report the teleconference to other NASA officials. Reinartz contended that "the pyramid approach says that everything cannot go to the top of an organization" and added that "decisions have to be made at every level." Lovingood said that the process of delegating technical decisions to the experts in NASA's field centers had existed since the beginning of the space program. All stated that given what they knew, they had acted appropriately. Mulloy said, "I think everything was done properly." Hardy stated, "I know of nothing now that I should have done differently."[32]

Lucas's meeting with the press was tumultuous. The media asked the Marshall director whether he agreed with William Rogers's statement that the launch decision process had been flawed. Lucas took pains to avoid controversy, saying that he didn't know what Rogers had meant. A reporter interrupted, yelling, "Goddamn it, answer the question!" Defending his people, Lucas said, "In my judgment, the process was not flawed." He explained that NASA must have a process that allows technical problems to be resolved at a level below the launch director and associate administrator. When reporters wanted to know why the Marshall managers had pressured Thiokol to launch or failed to report the teleconference, he refused to second-guess. A reporter then said, "You're ducking, you're dodging, you're slipping and sliding." Lucas responded, "You know, in this business you have to make decisions on the basis of information you have." Given "what my people say they knew and what they have testified they knew, I think it was a sound decision to launch" and not to report the initial decision to delay. He defended his center's engineering rigor, its long-standing commitment to safety, and its communication of the O-ring problem to all levels of the shuttle program. "No level of the Agency," he contended, "has not been aware of and has not considered the prospects of launch in view of the problems we've had on the O-ring seals." Lucas located the fault not in the process but in Thiokol's failure to communicate company engineers' growing concerns from the summer and fall of 1985 that may have led to mistakes in judgment in launching *Challenger*.[33]

The Marshall tactic of talking to reporters backfired. Media reports did not interpret the statements as important corrections. Rather, the stories interpreted the Marshall Center managers as arrogant, unrepentant, and unwilling to admit mistakes. Marshall, it seemed, sought to transfer blame to Thiokol and discredit the commission. The commission's response was even more critical. One commission member believed that the Marshall managers' justification of the flight readiness review process and their decisions were "totally insensitive." Commissioner Joseph F. Sutter believed that Marshall managers were "pretty defensive." According to the chief Marshall Center public affairs officer, commissioner Rogers "became infuriated with the remarks from Lucas" and demanded a tape of Lucas's comments. Marshall's director allowed daily press interviews for a few days. But then "someone chewed on Lucas" and he decided that being open with the media did more harm than good. Lucas never did another interview until his retirement in the summer.[34]

On March 6, the commission received an anonymous and vituperative letter written by a senior manager at the Marshall Space Flight Center. Signed "Apocalypse" and apparently addressed to commissioner Rogers, the letter had circulated at the center as early as February. Apocalypse blamed Lucas for the *Challenger* accident, alleging that the center director "more than any other individual in NASA is singly responsible for this terrible loss." The letter described Lucas as "a classic godfather" and the Marshall Center as "run by one man." Lucas ran the center as a "personal empire" with a "good old boy mafia" who were loyal to him. Since he made every major decision, everyone depended on his approval. In addition, he made all the decisions about technology and flight readiness, so all of NASA depended on his judgment. Apocalypse alleged that Lucas had "a psychopathic reaction to dissent" that inhibited the identification and solution of problems. Lucas had directed that "under no circumstances is the Marshall Center to be the cause for delaying a launch." In the flight readiness reviews, "for someone to get up and say that they are not ready is an indictment that they are not doing their job." Problems, the letter said, were "glossed over simply because we were able to come up with a theoretical explanation that no one could disprove." Marshall and its contractors would "basically 'snow' Headquarters with highly technical rationales that only the immediate experts involved can competently judge." The shuttle kept flying because of "the flawed management philosophy that if no one can prove the hardware will fail, then we launch." These practices corrupted Marshall and its contractors, leading to "ineptness and incompetence" and the flying of faulty hardware. Apocalypse claimed that after the accident, Lucas instituted a "cover-up" to obscure the center's responsibility. Officials presented information in as "highly technical" and "confusing" a way as possible. Then stories were "planted which would serve to shift blame away from MSFC to Thiokol and the contractors doing the processing at the Cape."[35]

The commission could not find Apocalypse. His letter contained internal inconsistencies. For instance, it mentioned a directive to never delay a launch and then described launches that Marshall had delayed. It ignored important facts, such as obscuring how NASA accident procedures, NASA Headquarters directives, and commission constraints had led Marshall to control information after the disaster. Despite this, the letter echoed many of the issues and concerns that had surfaced by mid-February, including poor communication, information delays, and pressure to launch. Apocalypse helped confirm many conclusions that the commission and the media had developed about the Marshall Center. The commission staff filed the letter under the filename "MSFC mismanagement."[36]

After four months of investigations and studies by NASA teams, the Rogers Commission published its report. It presented the narrative that the commission had established in its public hearings: the disaster stemmed from a flawed design of the solid rocket motor joints, a launch in cold weather, and a flawed decision-making process within NASA's shuttle program, especially the motor program directed by Marshall. Engineers had known of the danger and middle managers had not told top shuttle officials. NASA had failed to stop the launch, fix the problems, and prevent an accident. If Thiokol and Marshall's managers had communicated the bad news, the commission implied, wiser heads in Houston or at NASA Headquarters would have stopped launches and ordered a joint redesign. Calling the decision to launch *Challenger* "flawed," the commission argued that "those who made the decision" were "unaware" of key facts. "If the decisionmakers had known all the facts," the commission concluded, "it is highly unlikely that they would have decided to launch on January 28, 1986."[37]

Did the commission's process and narrative reflect a regional bias against southerners and the South? If regional bias meant explicitly provincial, consciously derogatory language, the answer is a resounding "no." There is no evidence of overt regionally inspired defamation of Marshall managers. And certainly the report found shared responsibility for the accident in decisions by non-southern managers at NASA Headquarters and Thiokol. It also found causes in a wider context. While exonerating the Reagan White House from the rumor that it had pressured NASA to launch *Challenger* on January 28 to coincide with the president's State of the Union speech, it also obliquely blamed the administration and Congress for policy and budgetary choices that constrained NASA's options.[38] Still, NASA's mistakes, mainly made by Marshall managers, were the real culprit rather than budget constraints.

Given that blame fell on Alabama, is it plausible that the commission had a disguised anti-southern bias that is manifest in regional stereotypes and tropes? Here the answer is "maybe." Regional prejudice was consistent with important patterns in the evidence of the hearings and final report. The "myth of the benighted South" has long conveyed negative images. Stereotypes often portrayed white southern men as backward, isolated, close-mouthed about their affairs, tribal in loyalty, intolerant of dissent, reluctant to admit fault, defensive of customs, inordinately proud, and defiant of central authority.[39] The commission portrayed Marshall and its managers in similar ways: as men who provided weak engineering, violated procedures, disallowed disagreement before the launch, stifled dissent after the launch, failed to allow problems to come to the surface, failed to admit guilt, and justified decisions even though they led to disaster. Its interpretation, which emphasized managerial malfeasance, had similarities to a narrative of regional backwardness. The commission's conclusions were consistent with notions that Marshall managed with a southern accent.

Congress offered an alternative narrative. In the three weeks after the commission published its report, the House Committee on Science and Technology and the Senate Subcommittee on Science, Technology, and Space had hearings on the accident. The members of these committees accepted the technical findings of the commission but complained about its handling of political and managerial accountability. Democratic members used the topic to embarrass the Reagan administration. They asked Watergate-style questions about who knew what when and what did they do about it, especially criticizing both NASA and Thiokol managers.[40]

The congressional reports, however, came from professional staff members rather than elected officials or allies of the president.[41] The staff included several with science, engineering, and social science backgrounds who downplayed "poor communications" and "inadequate procedures" and found "no evidence" that middle managers "suppressed information that they themselves deemed to be significant." Managers faithfully reported problems with the solid rocket motor joint, information was widely available, and that information showed that the problems were not serious. The congressional staffers found "no evidence to support a suggestion that the outcome" of *Challenger's* preflight meeting "would have been any different" had Marshall's Reinartz provided more information. NASA's top managers had a history of following the advice of technical managers who spoke for an entire team rather than to individual engineers.[42]

Rather than a narrative of malfeasance and miscommunication, the House report emphasized "poor technical decision-making" by NASA and Thiokol over many years. The motor's design specifications never matched the known cold weather of the Florida launch site. Thiokol's testing did not meet design specifications for redundancy or temperature. The "fundamental error" was accepting the joint problems. During development, some Marshall engineers identified the problem and recommended a redesign. NASA ignored their concerns because the joint worked even with anomalies. Rather than identifying O-ring erosion or blowby "as a joint that didn't seal, that is a joint that already failed, NASA, based on Thiokol's recommendations, elected to regard minor erosion and blow-by as 'acceptable.'" Based on faulty engineering analysis, NASA allowed Thiokol to continue to launch without a fix for four years. Once Thiokol engineers became concerned about cold temperature, they failed to call for a change in launch criteria and indeed they recommended the launch of the shuttle flight immediately before *Challenger* in cold weather.

Thiokol's teleconference presentation had contradictory and hence unpersuasive information. Regardless, Marshall should have accepted the engineering recommendation not to launch.

Why did the mistakes in engineering occur? Unlike the presidential commission, loyalty to a sitting president did not constrain the House committee staff. They argued that Congress and the president had set up a budget and a schedule that pressured technical managers and engineers to take risks. NASA sought to achieve the goal of twenty-four launches per year to reduce the costs per flight, changing itself from "an R&D agency into a quasi-competitive business operation." In effect, the space shuttle became a neoliberal launch vehicle operating with market incentives. Shuttle managers put schedules ahead of safety and they embedded this in the shuttle program. Shuttle contracts had penalties for catastrophic failure, not for hardware anomalies or weak engineering performance. The House committee report concluded that "operating pressures created an atmosphere which allowed" the accident to happen.[43]

Both the commission report and the House committee report influenced NASA. It reorganized the shuttle program after the accident. But in terms of memory, the commission's report proved far more influential than those of the two congressional committees. Because the commission's report was the first and was issued close in time to the accident, the media gave it lots of attention, but the congressional committee reports got lost in the shuffle. In terms of literary and stylistic qualities, the House committee report narrated a dull engineering plot with an engineering accent.[44] Its story attributed responsibility to flawed budgets, schedules, and engineering. In contrast, the Rogers Commission report had dramatic foreshadowing and a conflict between bumbling villains and bright heroes. Under the burden of responsibility from the accident as well as from the commission and the media, Lucas, Mulloy, Reinartz, and Hardy all retired from NASA by the end of 1986.

Narrating with an Accent?

How did some narratives spotlight "the South" and "southernness" after the accident? The *Challenger* accident and its investigations led to an explosion of publications and interpretations. Almost all the stories provided insights into the accident, shining a spotlight on a particular spot but also leaving other targets in the dark. Although explicit provincial diatribes did not develop, narratives that emphasized personal responsibility and blame sometimes used negative regional stereotypes.

After the accident, NASA and Marshall administrators took seriously the Rogers commission's narrative and implemented organizational reforms based on formal directives to improve communications. The reforms checked the power of the NASA centers by giving authority to NASA Headquarters and to astronauts. An administration-wide confidential communications channel allowed dissenters to report concerns. Marshall personnel received communications and management training designed to foster the flow of information and the dissemination of bad news.[45] By facilitating cross-checks and communications, NASA at least temporarily strengthened its technical culture and reined in negative regional peculiarities.

From the first weeks after the accident, a narrative of management failure has dominated, usually emphasizing poor communications and managerial misbehavior. The media interpreted Marshall officials' testimony as a cover-up and developed "scapegoating narratives" that cast blame on managers. Stories about culpability for the accident, while not the only stories, provided examples for the media to respond to the crisis with stock stories of bad leaders.[46] Scholars emphasized some variation of intentional suppression of bad news, unintentional misunderstandings, or tragic neglect of warnings.[47] At times, the stories had a pale regional tinge, with generalizations consistent with negative stereotypes of white southern men. Some, for example, discussed "group think" and toxic masculinity among middle-aged white male engineers of Marshall's solid rocket project.[48] Some, however, brought location to the forefront, as variations in two books show.

One book portrayed Alabama as a place desperately in search of status and funding from NASA. Malcolm McConnell, a professional writer and a journalist who at the time was covering space for *Reader's Digest,* was at the launch to cover the Teacher in Space Project. While his book had no citations or list of sources, it relied on extensive reading and interviews. McConnell's *Challenger, A Major Malfunction* had a subtitle that laid out its theme: *A True Story of Politics, Greed, and the Wrong Stuff.* He argued that "political intrigue and compromise" as well as "venality and hidden agendas" had led to the accident. In his book, he portrayed all key shuttle organizations and managers as corrupt and inherently political. Perhaps his most colorful chapter was "The Men from Marshall." He defined a "legacy of bitterness" at the center from years of success in technology without celebrity. Rivalry with the Johnson Space Center for control and contracts added to this feeling of slight and created a "Marshall-against-the-world mentality." In Director Lucas, this "coalesced into a garrison mentality." McConnell characterized the director as authoritarian, bureaucratic, rigid, and obsessed with maintaining a schedule. He wanted the center to always look good compared to other NASA centers. Relying on the "Apocalypse" letter, he portrayed Lucas's leadership as leading to secrecy, defensiveness, and isolation. Marshall managers tried to avoid delays and developed a "seemingly blasé attitude" to the booster joint problems and engaged in "blatant deceptions" to hide them. On the eve of the ill-fated launch, the atmosphere was one in which NASA managers easily overruled warnings from Thiokol and Rockwell about the cold in a desire to "press on" with the launch. In the end, McConnell concluded, "political compromise, bureaucratic deception, and corporate duplicity," led NASA to sacrifice safety and succumb to "production pressure" and "arbitrary operational goals."[49]

Another book portrayed Huntsville as an enclave of immoral habits from the military-industrial complex. Claus Jensen, a Danish professor of literature, took the bad management interpretation to its logical conclusions. For his 1996 book, he conducted no interviews and did no research in the United States. His title *No Downlink* referred to the loss of signal from the broken *Challenger* to the ground, a metaphor for what he described as a disconnect between engineers and managers. He described Marshall as an exemplar of the kind of organization where technology got out of control and managers failed to take responsibility, a phenomenon he called the "*Challenger* syndrome."[50] The pattern had developed in the military-industrial complex that Wernher von Braun and other German rocketry engineers had carried to Huntsville. They had developed the V-2 missile during World War II but took no moral responsibility for weapons of mass destruction or for the use of slave labor on their assembly line. Von Braun's team taught that same amoral discipline to American engineers and managers at Marshall Space Flight Center. NASA too inherited an apolitical, self-serving technocracy from the Cold War military-industrial complex. During the shuttle program, NASA sought to please the military, Congress, and the White House by staying on schedule.[51] Marshall managers, Jensen concluded, hid the problems of the solid rocket motors, making them NASA's "best-kept secret." He consistently put the worst possible spin on the facts and events about the motor joints and interpreted NASA actions as conspiratorial. The joints had "failed" on half the launches but "no one" had done anything "serious" to correct the problem. Marshall deliberately hid information about failures and suppressed conversations about problems. Since Thiokol engineers had been unanimous that the joints were dangerous, he portrayed Marshall managers as overruling contractor engineers' recommendation not to fly in the cold to stay on schedule. "A risk had been taken [with the motors], to keep the 'production line' moving." Marshall managers had not listened to bad news and had not taken individual responsibility. In short, the failures of communication and morality had created an organization that was short-sighted, self-serving, and out of control.[52]

Scholars who put the accident in a wider organizational context emphasized NASA's nationwide culture and political context rather than decisions made at Marshall Space Flight Center.[53] Social scientists in the 1980s devised "normal accident theory" and in the 1990s "high-reliability organization theory." These theories deemphasized factors like region in explaining disasters. They downplayed management error and played up the dysfunctions of social processes and breakdowns of organizations.[54] In 1996, sociologist Diane Vaughan blended these approaches in the award-winning book *The Challenger Launch Decision*. Although she only cited it a few times, her interpretation resembled the House report in its deemphasis of communications errors and rules violations. Instead, she emphasized how the accident happened because the engineers followed the rules and engineering practices. NASA and its shuttle organization had developed within daily work "bureaupathology" that accepted "mistakes" and the "normalization of deviance" from safe technology and engineering. Her interpretation had the effect of denying any responsibility for the disaster, whether personal or corporate. It also homogenized cultures among NASA contractors and centers in which overarching agency organizational and professional patterns mattered more than place. Indeed, in her outlook, professional aerospace engineering norms trumped local culture. While Vaughan's interpretation had other weaknesses, it eventually became very influential among academics and managers. After the second shuttle accident destroyed *Columbia* in 2003, her views influenced how its investigating committee interpreted the accident.[55] Ironically, Lawrence Mulloy, whom the Rogers Commission report had singled out for violating NASA rules, approved of Vaughan's account because it denied that Marshall had broken the rules and this absolved him of responsibility for the accident.[56]

Even so, in the lake of public memory, the managerial malfeasance interpretation had the biggest splash with ripples flowing toward backward southern managers. Protagonists fed the memory. Allan McDonald, who wrote a memoir that defended the Thiokol engineering team and its design of the solid rocket motor joint, attributed the accident to the cold weather and blamed NASA-Marshall (and company management) for the accident.[57] Obituaries of Thiokol engineers narrowed the timeline in ways that made them look good at the expense of Marshall engineers; they focused on dissent about the cold-weather launch and "whistle-blowing" to the commission after rather than the time before the launch that would reveal Thiokol's long-standing defense of the flawed joint design, failure to change launch temperature criteria, and poor communications.[58]

Movies painted a dark picture of Marshall. For example, *The Challenger Disaster* (2014) portrayed the commission's investigation through the character of Richard Feynman. The film depicted a fictitious trip of the physicist to Alabama. In dark lighting with ominous background music, he encountered cowardly and dim Marshall officials without an understanding of science or probability. In scenes about commission hearings, "Mulloy" was a hapless obfuscator; "Lovingood" was a central charlatan, a clueless Pollyanna. "Feynman" commented on the "crazy engineering" and "crap that goes down at Marshall."[59] Documentaries were often similar. The biggest was *Challenger: The Final Flight*, a four-part Netflix docuseries produced by J. J. Abrams. The series basically presented the plot of the Rogers Commission, showing Thiokol engineers as apologetic good guys and Mulloy and Lucas as the callous bad guys because they expressed no admission of guilt for wrongdoing.[60]

In conclusion, what can we learn from using the lens of place and regional culture as one tool to view a complex technological event? The best use of a regional lens for historians is to assess stories and memories. When regional tropes appear in storytelling, such narratives can serve political purposes and degenerate into regional prejudices. Alabama has a reputation (well deserved) as an undereducated and underperforming state when compared to urban and coastal areas. At times, this has led to a facile notion that the southern location of an important technology facility was nefarious. Marshall engineers and managers of course have worked on many successful projects such as the Saturn launch vehicles, Skylab, and the space shuttle main engines. But the center's southern location made a regional-type narrative likely when its programs competed for resources or problems developed. When Marshall has been involved in mistakes and accidents, negative southern stereotypes have helped make narratives easy to create and easy to believe. When telling a story about a tragedy involving multiple managers of the same regional background, an explanation based on malfeasance can look a lot like regional stereotyping.

Casting a penumbra of backwardness over Marshall's southern managers had political uses. The bad management interpretation appealed in part because it was consistent with preexisting negative regional stereotypes but required no overt provincial rhetoric. It helped Thiokol officials narrow the conversation to Marshall's mistakes in handling the cold-weather launch and it distracted from discussion of the contractor's engineering failures and naïveté as a prime contractor in human spaceflight. The Rogers Commission's interpretation also steered away from embarrassing implications of policy. Blaming civil servants was consistent with the era's neoliberalism that assumed that business knew best. While the interpretation that the proverbial buck stopped on NASA desks was correct, it obfuscated Thiokol's responsibility. The company was the prime contractor and had greater expertise with solid rocket motors than NASA did. It had developed a flawed design and continued to verify its safety despite those flaws. Finally, attributing responsibility to technical managers obscured the role of policymakers in Washington, DC, in increasing risk in the shuttle program. They had insisted that the shuttle program operate like a business, declared the launch vehicle "operational" like a commercial airliner, and pushed to increase the launch rate to reduce the cost per mission.

Understanding the decisions that led to the tragic *Challenger* accident requires viewing through many frames of reference. One helpful lens is a regional one.

Notes

1 *Report to the President by the Presidential Commission on the Space Shuttle Challenger Accident, June 6, 1986*, vol. 1 (Washington, DC: NASA, 1986), 200, 206, 86, https://sma .nasa.gov/SignificantIncidents/assets/rogers_commission_report.pdf.

2 Philip M. Boffey, "Zeal and Fear Mingle at Vortex of Shuttle Inquiry," *New York Times*, March 17, 1986.

3 Robert Reinhold, "Astronaut Assails NASA for Not Telling of Risk," *New York Times*, March 17, 1986; David Maraniss, "Astronauts 'Upset' at Learning of Booster Seal Problems," *Washington Post*, March 4, 1986.

4 C. A. Bayly, Sven Beckert, Matthew Connelly, Isabel Hofmeyr, Wendy Kozol, and Patricia Seed, "*AHR* Conversation: On Transnational History," *American Historical Review* 111, no. 5 (2006): 1441–1164; Ian Tyrrell, "Reflections on the Transnational Turn in United States History: Theory and Practice," *Journal of Global History* 4, no. 3 (November 2009): 453–474.

5 On regionalism in historiography, see "Bringing Regionalism Back to History," *American Historical Review* 104, no. 4 (1999): 1156; and Michael O'Brien, "On Observing the Quicksand," *American Historical Review* 104, no. 4 (1999): 1202–1207.

6 See, for example, Loyd S. Swenson, "The Fertile Crescent: The South's Role in the National Space Program," *Southwestern Historical Quarterly* 71, no. 3 (1968): 377–392; Andrew J. Dunar and Stephen P. Waring, *Power to Explore: A History of Marshall Space Flight Center, 1960–1990* (Washington, DC: GPO, 1999); Brian C. Odom and Stephen P. Waring, *NASA and the Long Civil Rights Movement* (Gainesville: University Press of Florida, 2019).

7 Garry D. Brewer, "Perfect Places: NASA as an Idealized Institution," in *Space Policy Reconsidered*, ed. Radford Byerly (Boulder, CO: Westview Press, 1989), 157–173.

8 In full disclosure, I am a native southerner—that is, a native of southern Nebraska—as well as a thoroughbred Yankee who has lived in Alabama for over three decades. In the 1990s, I co-authored a history of Marshall while I was under contract to the center. A contract controversy developed, mainly over the inclusion of stories about civil rights

and labor unions, so I experienced some provincialism from Marshall managers. See James R. Hansen, "Review Essay: A Battle over the Historian's 'Power to Explore,'" *Alabama Review* 55, no. 3 (2002): 192–199.

9   George Brown Tindall, *The Benighted South: Origins of a Modern Image* (Charlottesville: University of Virginia, 1964); C. Vann Woodward, *The Burden of Southern History*, rev. ed. (Baton Rouge: Louisiana State University Press, 1968); Matthew D. Lassiter and Joseph Crespino, eds., *The Myth of Southern Exceptionalism* (New York: Oxford University Press, 2009); Byron E. Shafer and Richard Johnston, eds., *The End of Southern Exceptionalism: Class, Race, and Partisan Change in the Postwar South* (Cambridge, MA: Harvard University Press, 2009); Laura F. Edwards, "Southern History as U.S. History," *Journal of Southern History* 75, no. 3 (2009): 533–564; Patrick Gerster, "Stereotypes," in *Myth, Manners, and Memory*, ed. Charles Reagan Wilson, vol. 4 of *The New Encyclopedia of Southern Culture* (Charlottesville: University of North Carolina Press, 2006), 170–175.

10 "NASA Rocket Official Retiring from Agency," *New York Times*, May 3, 1986; Dana Beyerle, "Former Shuttle Project Chief to Retire," UPI, November 27, 1986; Dana Beyerle, "Former NASA Rocket Boss Takes Early Retirement," UPI, July 17, 1986; "Embattled Director of NASA Rocket Center to Step Down: His Retirement Had Been Regarded as Inevitable Since Shuttle Disaster," *Los Angeles Times*, June 4, 1986; "NASA Figure in *Challenger* Explosion Retires at Space Center," AP News, February 22, 1988.

11 Biographical sketch of Stanley R. Reinartz, May 1986, Marshall Space Flight Center History Office, Huntsville, Alabama (hereafter MSFC History Office); biographical sketch of George Hardy, May 1986, MSFC History Office; biographical sketch of Lawrence R. Mulloy, May 1986, MSFC History Office; Wikipedia entry for William R. Lucas; EverybodyWiki entry for Judson Lovingood, https://en.everybodywiki.com/Judson_Lovingood.

12 Monique Laney, *German Rocketeers in the Heart of Dixie: Making Sense of the Nazi Past during the Civil Rights Era* (New Haven, CT: Yale University Press, 2015), 43–70; James T. Sparrow, "A Nation in Motion: Norfolk, the Pentagon, and the Nationalization of the Metropolitan South, 1941–1953," and Kari Frederickson, "The Cold War at the Grassroots: Militarization and Modernization in South Carolina," both in *The Myth of Southern Exceptionalism*, ed. Matthew D. Lassiter and Joseph Chrispino (New York: Oxford University Press, 2009), 167–189 and 190–209, respectively.

13 Merritt Roe Smith, "Military Arsenals and Industry before World War I," in *War, Business, and American Society: Historical Perspectives on the Military Industrial Complex*, ed. Benjamin Cooling (Port Washington, NY: Kennikat Press, 1977); David A. Hounshell, *From the American System to Mass Production 1800–1932: The Development of Manufacturing Technology in the United States* (Baltimore, MD: Johns Hopkins University Press, 1984); Thomas Charles Lassman, *Sources of Weapon Systems Innovation in the Department of Defense: The Role of In-House Research and Development, 1945–2000* (Washington, DC: Center of Military History, United States Army, 2009), 11–36; Roger E. Bilstein, *Stages to Saturn: A Technological History of the Apollo/Saturn Launch Vehicles*, NASA History Series (Washington, DC: NASA History Office, 1996).

14 Howard E. McCurdy, *Inside NASA: High Technology and Organizational Change in the U.S. Space Program* (Baltimore, MD: Johns Hopkins University Press, 1994); Dunar and Waring, *Power to Explore*, 39–144.

15 Dunar and Waring, *Power to Explore*, 135–270; Joan Lisa Bromberg, *NASA and the Space Industry* (Baltimore, MD: Johns Hopkins University Press, 2000); McCurdy, *Inside NASA*; Edward L. Ayers, *The Promise of the New South: Life after Reconstruction* (New York: Oxford University Press, 1992).

16 Dunar and Waring, *Power to Explore*, 225–270.

17 Dunar and Waring, *Power to Explore*, 339–350.

18 Dunar and Waring, *Power to Explore*, 350–370.

19 Dunar and Waring, *Power to Explore*, 370–379.

20 "Reagan's Nuclear War Briefing Declassified," National Security Archive, https://nsarchive.gwu.edu/briefing-book/nuclear-vault/2016-12-22/reagans-nuclear-war-briefing-declassified.

21 *Hearings February 6, 1986 to February 25, 1986*, vol. IV of *Report of the Presidential Commission on the Space Shuttle Challenger Accident* (Washington, DC: NASA, 1986), 91–92, 94–95, 99; "Pre-Launch Chill Was Discounted; Thiokol Consultation Brought a Go-Ahead, NASA Tells Probers," *Washington Post*, February 7, 1986; Philip M. Boffey, "NASA Had Warning of a Disaster Risk Posed by Booster," *New York Times*, February 9, 1986.

22 Richard S. Lewis, *Challenger: The Final Voyage* (New York: Columbia University Press, 1988), 97–121; Richard P. Feynman as told to Ralph Leighton, *What Do You Care What Other People Think? Further Adventures of a Curious Character* (New York, Norton, 1988), 141–153; T. F. Gieryn and A. E. Figert, "Ingredients for a Theory of Science in Society: O-Rings, Ice Water, C-Clamp, Richard Feynman and the Press," in *Theories of Science in Society*, ed. Susan E. Cozzens and Thomas F. Gieryn (Bloomington: Indiana University Press, 1990), 67–97.

23 Field joints were formed when motor segments were stacked in the Vehicle Assembly Building at Kennedy Space Center.

24 *Report to the President by the Presidential Commission on the Space Shuttle Challenger Accident*, 1: 145–146; Frederick F. Lighthall, "Launching the Space Shuttle *Challenger*: Disciplinary Deficiencies in the Analysis of Engineering Data," *IEEE Transactions on Engineering Management* 38 (February 1991): 63–74; John C. Macidull and Lester E. Blattner, *Challenger's Shadow: Did Government and Industry Management Kill Seven Astronauts?* (Coral Springs, FL: Llumina Press, 2007).

25 *Report to the President by the Presidential Commission on the Space Shuttle Challenger Accident*, 206–207.

26 *Hearings February 6, 1986 to February 25, 1986*, 709–710.

27 *Hearings February 6, 1986 to February 25, 1986*, 709–823; *Hearings of the Presidential Commission on the Space Shuttle Challenger Accident: February 26, 1986 to May 2, 1986*, vol. V of *Report to the President by the Presidential Commission on the Space Shuttle Challenger Accident* (Washington, DC: NASA, 1986), 825–955, 1505–1657.

28 *Report to the President by the Presidential Commission on the Space Shuttle Challenger Accident*, 86, 105.

29 Lewis, *Challenger*, 97–121. Marshall officials testified mainly on February 26 and 27 and May 2. See *Hearings of the Presidential Commission on the Space Shuttle Challenger Accident: February 26, 1986 to May 2, 1986.*

30 Charles Redmond, "Notes from MSFC Public Affairs Office on 51-L" and "Challenger Chronology Input from MSFC," n.d. (probably spring 1986), NASA Headquarters PAO 51-L Papers, NASA History Office, Washington, DC. See also William Lucas, "Remarks to Employees," January 30, 1986, MSFC History Office; Robert Dunnavant, "Marshall Tells Workers Not to Talk," *Birmingham News*, January 31, 1986; Robert Dunnavant, "Space Center Has Withdrawn into Its Shell," *Birmingham News*, February 2, 1986; Charles Fishman, "Questions Surround NASA Official," *Washington Post*, February 24, 1986.

31 Charles Redmond, Public Affairs Office, NASA HQ, "Challenger Chronology Input from MSFC," n.d. [probably fall 1986], NASA History Office.

32 All quotes in this paragraph are from Charles Redmond, Public Affairs Office, NASA HQ, "Challenger Chronology Input from MSFC."

33 Redmond, "Notes from MSFC Public Affairs Office on 51-L"; Charles Fishman, "5 NASA Officials Stand by Actions," *Washington Post*, March 1, 1986; J. Michael Kennedy, "NASA Officials Defend Methods, Score Engineers," *Los Angeles Times*, March 1, 1986; Martin Burkey, "Lucas: Launch Process Not Flawed," *Huntsville Times*, February 28, 1986; "Press Conference of William Lucas at MSFC," February 28, 1986, 27, Presidential Commission Records, National Archives, Washington, D.C.

34 Maura Dolan, "MSFC Key Officials Termed 'Insensitive,'" *Huntsville Times*, March 4, 1986; Redmond, "Challenger Chronology Input from MSFC."

35 "Apocalypse Letter," March 6, 1986, PC 167128, Records of the Presidential Commission on the Space Shuttle Challenger Accident, National Archives and Records Administration.

36 Records of the Presidential Commission on the Space Shuttle Challenger Accident, National Archives and Records Administration, Washington, DC; see also Martin Burkey, "Commission Wants to Talk to Letter Writer," *Huntsville Times*, March 12, 1986.

37 *Report to the President by the Presidential Commission on the Space Shuttle Challenger Accident*, 82.

38 *Report to the President by the Presidential Commission on the Space Shuttle Challenger Accident*, 164–177.

39 See, for example, Jim Goad, *The Redneck Manifesto: How Hillbillies, Hicks, and White Trash Became America's Scapegoats* (New York, Simon & Schuster, 1997); Patrick Huber, "A Short History of Redneck: The Fashioning of a Southern White Masculine Identity," *Southern Cultures* 1, no. 2 (1995): 145–166.

40 US Congress, House of Representatives, *Investigation of the Challenger Accident: Hearings before the Committee on Science and Technology, House of Representatives, Ninety-Ninth Congress, Second Session* (Washington DC: GPO, 1986); US Congress, Senate, *Space Shuttle Accident: Hearings before the Subcommittee on Science, Technology, and Space of the Committee on Commerce, Science, and Transportation, United States Senate, Ninety-Ninth Congress, Second Session, on Space Shuttle Accident and the Rogers Commission Report, February 18, June 10, and 17, 1986* (Washington, DC: GPO, 1986).

41 US Congress, House of Representatives, *Investigation of the Challenger Accident.*

42 US Congress, House of Representatives, *Investigation of the Challenger Accident,* 4–5, 148, 29, 172.

43 US Congress, House of Representatives, *Investigation of the Challenger Accident,* 62, 22–23, 123.

44 Thanks to my colleague Andrew J. Dunar for this phrasing.

45 J. M. Logsdon, "Return to Flight: Richard H. Truly and the Recovery from the Challenger Accident," in *From Engineering to Big Science: The NACA and NASA Collier Trophy Research Project Winners,* NASA SP-4219 (Washington, DC: NASA, 1998), 345–364; *Report to the President: Implementation of the Recommendations of the Presidential Commission* (Washington, DC: NASA, 1988); Bromberg, *NASA and the Space Industry.*

46 Christopher T. Caldiero, "Crisis Storytelling: Fisher's Narrative Paradigm and News Reporting," *American Communication Journal* 9 (Spring 2007), http://ac-journal.org/journal/2007/Spring/articles/storytelling.html.

47 Examples include Patrick Moore, "Intimidation and Communication," *Journal of Business and Technical Communication* 6, no. 4 (1992): 403–437; Patrick Moore, "When Politeness Is Fatal," *Journal of Business and Technical Communication* 6, no. 3 (1992): 269–292; Phillip K. Tompkins, *Organizational Communication Imperatives: Lessons of the Space Program* (Los Angeles: Roxbury Publishing, 1993); Maureen Hogan Casamayou, *Bureaucracy in Crisis: Three Mile Island, the Shuttle Challenger, and Risk Assessment* (Boulder, CO: Westview Press, 1993).

48 G. Moorhead, R. Ference, and C. P. Neck, "Group Decision Fiascoes Continue: Space Shuttle Challenger and a Revised Groupthink Framework," *Human Relations* 44, no. 6 (1991): 539–550; J. W. Messerschmidt, "Managing to Kill: Masculinities and the Space Shuttle Challenger Explosion," in *Masculinities in Organizations,* ed. Cliff Cheng (Thousand Oaks, CA: Sage Publications, 1996), 29–53.

49 Malcolm McConnell, *Challenger: A Major Malfunction: A True Story of Politics, Greed, and the Wrong Stuff* (Garden City, New York: Doubleday, 1987), x, 55, 79, 82, 112, 122, 167, 252, 256.

50 Claus Jensen, *No Downlink: A Dramatic Narrative about the Challenger Accident and Our Time* (New York: Farrar, Straus, Giroux, 1996), xii–xiii.

51 Jensen, *No Downlink,* 29–38, 50–55, 184–85, 194, 202.

52 Jensen, *No Downlink,* 206, 225–244, 286, 303, 319, 361–362.

53 See McCurdy, *Inside NASA.*

54 Charles Perrow, *Normal Accidents* (New York: Basic Books, 1984); Todd R. LaPorte and Paula M. Consolini, "Working in Practice But Not in Theory: Theoretical Challenges of 'High-Reliability Organizations,'" *Journal of Public Administration Research and Theory* 1 (January 1991): 19–48; Karlene H. Roberts, "Some Characteristics of High-Reliability Organizations," *Organization Science* 1 (1990): 160–177; Gene Rochlin, "Defining 'High Reliability' Organizations in Practice: A Taxonomic Prologue," in *New Challenges to Understanding Organizations,* ed. Karlene H. Roberts (New York: Macmillan, 1993), 11–32; Karl E. Weick, and Karlene H. Roberts, "Collective Mind in Organizations: Heedful Interrelating on Flight Decks," *Administrative Science Quarterly* 38 (September 1, 1993): 357–381.

55 Diane Vaughan, *The Challenger Launch Decision* (Chicago: University of Chicago Press, 1996); Stephen P. Waring, "Losing the Shuttle (or Nearly): Accidents and Anomalies," in *Space Shuttle Legacy: How We Did It and What We Learned*, ed. Roger D. Launius, John Krige, and James I. Craig (Reston, VA: American Institute of Aeronautics & Astronautics, 2013), 215–242.

56 Martin Burkey, "Manager: I Was Playing by the Rules," *Huntsville Times*, January 26, 1996, A1.

57 Allan J. McDonald, *Truth, Lies, and O-Rings: Inside the Space Shuttle Challenger Disaster* (Gainesville: University Press of Florida, 2009).

58 Douglas Martin, "Roger Boisjoly, 73, Dies; Warned of Shuttle Danger," *New York Times*, February 4, 2012; Clay Risen, "Allan McDonald Dies at 83; Tried to Stop the *Challenger* Launch," *New York Times*, March 9, 2021.

59 *The Challenger Disaster*, TV movie, directed by James Hawes, produced by BBC Films, Erste Weltweit Medien, Moonlighting Films, Pictureshow Productions, Science Channel, and Open University, 2013. Feeling defamed, Lovingood sued about the depiction; see "Opinion," *Lovingood v. Discovery Commc'ns, Inc.*, February 7, 2020, no. 18–12999, https://casetext.com/case/lovingood-v-discovery-commcns-inc-3.

60 *Challenger: The Final Flight*, documentary series, produced by Bad Robot, Zipper Bros Films, Sutter Road Picture Company, released September 2020.

Part 2

Colonizing the Crescent

5

Oranges and Rockets

The People and Technology of Cape Canaveral

RACHAEL KIRSCHENMANN

Apollo 16 astronauts John Young and Charlie Duke rested in the Lunar Module Orion after becoming only the ninth and tenth people to walk on the Moon. Young felt his stomach turn as the steady intake of orange juice during the trip to the Moon caught up to him.

"I'm gonna turn into a citrus product is what I'm gonna do," he complained.

Mission control in Houston piped back, "Oh, well, it's good for you, John."[1]

After the Apollo 15 mission, during which astronauts were afflicted with heart irregularities attributed to low levels of potassium, NASA physicians added electrolytes and potassium to the crew's food, most noticeably in the orange juice they were instructed to drink often. Earlier in the mission, while the crew prepared for the lunar landing, Duke had his own run-in with the drink when the valve on the drink bag in his helmet accidentally opened and he found himself with a helmet full of orange juice.[2] A fruit that was entangled with the space program's launch site had made its way to space.

When it was established in 1961 on North Merritt Island in Florida, the National Aeronautics and Space Administration's Launch Operations Center (soon to be renamed the Kennedy Space Center) stood in sharp contrast to its surroundings. The local communities and surrounding environments that had existed on the island and in the neighboring area presented an antithetical image to the one that the nation and the world would come to associate with the Space Coast.[3] Whereas the promise of human space exploration represented for many the next step in technological progress, local communities on Cape Canaveral sent their children to one-room schoolhouses, picked oranges in citrus groves, and made their living by fishing in lagoons. Viewed as a backwater area with more alligators than people, the land had remained largely underdeveloped. It was marshy and, as many of the histories of the Kennedy Space Center are apt to point out, mosquito ridden.[4] Beyond this initial image, however, is a much more complex story of an area with a rich local memory and a tradition of agricultural technology and development.

The unique location of Merritt Island showcases a dramatic juxtaposition between technology capable of landing humans on the moon and the everyday technology of agricultural laborers. This chapter will examine the contrast between the powerful institutional technology emerging from the Kennedy Space Center in the early 1960s and the community that existed both before and after the center was established. While the US space program brought great changes to the Cape Canaveral area, rocketry constitutes only one of many technologies found on Merritt Island throughout its history. While Project Apollo was an unquestionably impressive example of technological and organizational achievement, its operations on Merritt Island did not exist in a vacuum. For older residents, the role of the space center takes a backseat to memories of local activity and industry, in particular the technology and practice of cultivating citrus.

Accounts from citrus grove workers illustrate the ingenuity and adaptation they needed in order to ensure successful harvests each year. While Florida orange juice reached the homes of millions of Americans during this time, no one has written the biographies of the orange growers and pickers who made that possible. NASA's ability to seize the land and groves on Merritt Island was based on the technology that the space administration wielded and the government's decision to privilege that technology over others. The Kennedy Space Center had a profound impact on the economy and the population of Merritt Island and the surrounding area, yet the way the center is written about largely takes the preexisting community out of the conversation. When the story centers on the local culture that existed alongside the space program, a clearer picture emerges of how some technologies are remembered and some are taken for granted.

Citrus on North Merritt Island

The land that is now Brevard County can trace evidence of its first inhabitants to around 8000 BCE, when early hunter-gatherers began to establish semi-permanent villages along waterways like the Indian River.[5] In the early 1500s, the indigenous Ais groups living in the area encountered Spanish explorers searching for gold and setting up trading routes along the coast. Spanish writings from 1605 document the Ais settlements on Merritt Island, describing several villages near what is now the Haulover Canal and fishing camps on the Banana River.[6] By the nineteenth century, however, a series of wars between colonial powers and Native Americans in Florida had scattered and all but decimated the population of Ais.[7] With the passage of the Armed Occupation Act in 1842 and the Homestead Act in 1862, the United States made it possible for settlers to claim land, and by 1900, the island had grown to include around two dozen white communities. However, while towns across the Indian River like Titusville and Mims were gaining access to new railroad services, electricity, and other modern conveniences, Merritt Island remained inaccessible except by boat and was not connected to the electricity grid.[8] Families continued to rely heavily on their own hunting and fishing for survival, and tourists from outside Florida flocked to the area for the chance to experience the climate and to take advantage of opportunities for wild game hunting and excellent fishing.[9]

While white homesteaders made up the majority of the population of North Merritt Island, after the Civil War, formerly enslaved African Americans also sought land. During this time, "freedom colonies" became enclaves of hope and freedom for Black Americans, and on Merritt Island, the small towns of Clifton and Allenhurst, located in the northernmost part of the peninsula, have their roots in this movement. On the 1900 census, a total of 358 lived on the Merritt Island precincts of Canaveral, Merritt, and Haulover; 77 were African Americans and 281 were white.[10] Oral histories give some indication of mixing between these groups, and one former white resident reported that when she first arrived in the area, inhabitants attended church together in an integrated service because it was the only church in the vicinity.[11] Another account describes the homestead of Butler Campbell, a former officer in the United States Colored Troops during the Civil War. At a time when it was rare for a white-owned company to buy land from African Americans, the Indian River Fruit Company bought property from Campbell, showing evidence of interaction and cooperation, particularly in the context of the citrus industry.[12] This industry and the people who participated in it would shape the island in the years to come.

Perhaps no industry is as synonymous with Florida in the public imagination as the citrus industry, and Brevard County during the nineteenth and twentieth centuries was a model for successful citrus production. An installment of *Florida: A Guide* (1938), part of the 1938 Federal Writers' Project, described a "pastoral Florida of bucolic groves and colorful farmers."[13] The history, which traces the journey of the orange from early European expansion throughout the American south from Florida to California, boldly states that "one may safely say there is no fruit which is so closely woven into American expeditionary history as the orange."[14] In the subsequent decades, as American efforts extended upward to include space exploration, the orange had another front-row seat to the country's expeditionary efforts.

Along the Indian River, a modest number of orange groves had popped up by the 1830s, including that of Douglas Dummett, whose grove lay on a piece of land south of the Allenhurst Canal. Dummett used wild sour orange trees as the base for his grove and grafted sweet orange budwood onto the existing trees. While grafting was not an innovation at the time, Dummett is reported to have been the first person to use the technique on citrus in Florida, where most growers until the 1880s preferred starting trees from seed.[15] Grafting, however, offered a variety of advantages and is an agricultural technology that has been practiced for millennia; descriptions of grafting dating back to ancient Greece and Rome.[16] When a catastrophic freeze hit Florida in February 1835, killing nearly every orange tree in the territory, Dummett's trees were by most accounts the only sweet orange specimens left undamaged. A combination of the hardier sour orange species used as rootstock and the favorable location of the grove showed the benefits of budding. The survival of this Merritt Island grove ensured that orange growing in Florida could be revived through budding and reseeding, and it secured the island's reputation as a bastion for citrus.[17]

The popularity of the Indian River citrus label grew as the century progressed, and in northern markets, oranges from Dummett's grove fetched a dollar per box more than his competitors.[18] Other groves around the state branded themselves as "Indian River" in order to boost sales, and the area's reputation for citrus was so widespread as the decades progressed that references to the region can be found in newspapers across the country, from an 1873 article published in a Maine paper describing it as "the finest fruit that grows" to a newspaper in Davenport, Iowa, that extolled the virtues of the fruit in 1917.[19]

In 1966, when popular writer and journalist John McPhee visited the area to write a piece for the *New Yorker*, NASA had already acquired the land on North Merritt Island, including the old Dummett grove. One of Dummett's grandsons, Robert Hill, however, continued to manage a grove on an area of the island south of the space center. McPhee evocatively described how the "narrow roads wind through Merritt Island between high walls of orange trees, which are interspersed with numerous houses of growers."[20] Many of Hill's own orange trees had been planted by his father and grandfather, but his son, who worked for RCA, had little inclination to carry on the family

business. With the expansion of NASA and need for more residential housing, McPhee correctly predicted that within a few years the majority of North
Merritt Island groves like the one owned by Hill would be wiped out by development.

Work in citrus groves was a labor-intensive and precise practice. McPhee
interviewed an employee of the Silver Springs Land Company, who reported
that growing oranges "is no dead level of monotonous exertion, but one that
affords scope for the development of an ingenious mind."[21] Yet like much of
the history of agricultural labor, recognition and documentation of the lives
of workers involved has not always been so precise. For example, in McPhee's
account of Dummett's grove, the entrepreneur is described as putting in a
minimal amount of work to achieve his success, instead spending most of his
time "fishing or hunting wildcats on the mainland with his pack of dogs."[22]
There is no mention of how many people he employed or who those workers
were, but some clues paint a picture of the workers in his and other groves.
McPhee is one of the few writers to cover the citrus industry who remarks on
the demographics and lives of workers, who at the time were largely African
American.[23]

Pickers are rarely mentioned in the *Florida Star*, which was published in
Titusville in the early 1900s, and only in unnamed groups affiliated with a
named grove owner. For example, in the October 8, 1909, issue of the newspaper, a lengthy account is dedicated to the proceedings of a meeting of various
local citrus associations, and in a much smaller blurb further down the page,
wedged between a death notice and an account of a fever sufferer, readers
learn about an orange buyer and his "large force of orange pickers and pack-
ers."[24] Two months later, readers learned a bit more about the grower's workforce, which numbered about sixty and was able to pack as many as 1,000
boxes of citrus per day.[25] Countless promotions for orange buds dotted the
pages of the paper, including a 1909 announcement by a grower on Merritt
Island who offered buds for 25 to 35 cents.[26] Conspicuously absent from advertisements and from announcements and anecdotes found in papers of the
time is mention of the people who would have been using the ladders, applying the fertilizer, and budding the trees.

The answer to why that omission might have been made lies in the demographics of citrus pickers, a group that was majority nonwhite, poor, and
undereducated. McPhee, who interviewed pickers in the groves, paints an an-
imated picture of the lives of several workers. One man, an "amiable but un-
talkative, hardworking orange picker" nicknamed Bird Man, worked alongside his wife and their three-year-old child, who played nearby. Together, the
couple gathered an average of about eighty boxes of oranges a day, which

earned them about twenty dollars a day. On Merritt Island, while both white and Black citrus workers lived on the island, African Americans were much more likely to be employed in the citrus industry, and a far larger percentage depended on grove work for their livelihood. The 1940 census for the precincts covering the towns of Haulover and Wilson recorded a total of 208 residents, and of the eighty-three white residents with occupations listed, 29 percent were associated with citrus farming. Their occupations included "proprietor," "farmer," "nurseryman," "laborer," "packer," and "tractor driver."[27] Conversely, of the twenty African Americans with listed occupations, 45 percent had ties to citrus jobs. Unlike their white counterparts, these workers were listed exclusively as "laborer" or "farmer." The African American presence in the citrus industry on the island can be traced at least as far back as 1866, when Andrew Jackson, a formerly enslaved man from Georgia, began working for Douglas Dummett. In a 1926 interview about the history of the grove, Jackson did not comment on the nature of Dummett's workforce, which consisted of slave labor until just before the Civil War.[28] It is unclear how many enslaved workers and laborers worked for Dummett before the Civil War, but it was enough to ensure a relatively leisurely life for the grower. Instead of inquiring into these issues, the focus of the journal article was Dummett's skill as an innovator and manager; the authors emphasized that he kept the grove well maintained and pruned.

Freezes, a perennial problem for both grove workers and grove owners, are one reason for the need for constant adaptation in the citrus industry. James Taylor, one Merritt Island grove owner who was affected by a devastating freeze in 1895, had worked hard to start a five-acre grove in a mosquito-infested area only to see it frozen two years later. With the $1,800 he had invested in the trees gone, he mortgaged his land to finance more land and new trees. He did not see a profit until ten years later.[29]

One option for preventing freezes that was used primarily in the 1960s was dubbed "firing the grove." It involved warming the groves with a heat source. In an oral history with three current and former Brevard County citrus pickers, the men talked about this method of protection, emphasizing the urgency and scale of the workforce needed to ensure a safe crop. For freezes that lasted for days, workers attended to groves all night and then went back the next day to repeat the operation. The men remembered using wood in the early days, placing small campfires down the middle of the grove and replenishing the fires all night.[30]

Another riskier option was spraying the trees with a coating of water that froze and released heat to protect the bark from splitting. This water was a resource that residents in expanding neighborhoods nearby also coveted, and one picker, Alfonso Wilson, remembered a politician who advised him that "orange trees don't vote, and people do" when discussing water rights and the future of orange groves near urbanizing areas. Perhaps thinking beyond battles over water to the general landscape of Brevard in the 1960s, Wilson lamented that "people decided that it is more profitable to raise houses than it is to raise trees."[31]

NASA Moves In

For citrus workers and the orange industry in Brevard County, the Kennedy Space Center brought drastic changes. As Peter Redfield notes in his anthropological study of the French space agency's Guiana Space Center, "when one is seeking to leave the globe, wasteland becomes valuable, and underdevelopment can appear a virtue."[32] Of the three countries that were the first to launch their own satellites, the United States was alone in launching from a location in its mainland borders.[33] Because of this, it is easier to identify the French or Soviet installments in French Guiana or Kazakhstan as extensions of imperialism and by comparison celebrate the success of the American space program as a home-grown effort. Upon closer inspection, however, the parallels between the stories behind the centers are subtle but undeniable, particularly when one looks at NASA's use of eminent domain to build the Kennedy Space Center and the economic upheaval caused by a dramatic influx of the new technically skilled labor force.

Previously, the area around Cape Canaveral had gone through a series of changes beginning in 1938 with the US military's construction of defense facilities on the coast.[34] They were located on a barrier island called the Canaveral Peninsula, where the land juts into the Atlantic and connects to Merritt Island in the north and the Banana River separates the two landmasses to the west. Taking advantage of existing defensive infrastructure, the navy opened a station there during World War II. In 1949, the air force took charge of the facility, renaming it the Patrick Air Force Base. To the north on the promontory, the air force also constructed the Joint Long Range Proving Ground, situating launchpads along the beach and establishing it as the northernmost point of the Atlantic Missile Range. After NASA was established in 1958 and increased its testing, the limited facilities at the Cape struggled to accommodate both its needs and those of the military, resulting in a bottleneck of civilian and military rockets. By the end of the decade, so many launch towers had been constructed that one account described the scene as resembling "a Gulf Coast oil field." But even with the flurry of rocketry, by the time the Soviet Union launched Yuri Gagarin in April, it was clear that the United States needed to do something spectacular if they wanted to stay in the space race. A month later, President Kennedy announced the decision to go to the moon, setting in motion steps that would quickly expand NASA's scope.

Despite NASA's presence in the area, Cape Canaveral was not the only launch site considered for Project Apollo. Even as infrastructure around the Cape expanded during the 1950s, the location remained underdeveloped and the existing technical workforce was not sufficient to do work on the scale that would be required to land a man on the moon.[35] While the infrastructure for smaller rocket launches was a useful starting point, it was nowhere near large enough to support the launches proposed. Although the method of reaching the moon had not yet been determined, the liquid-fuel Nova rocket, which would have been used for a direct ascent mission, was too large to safely launch from the existing pads.[36] Even the Saturn V rockets proposed for use in the earth-orbit and lunar-orbit rendezvous methods took rocketry to a level that had not been seen at the Cape. In addition, the area faced weather challenges, including oppressive humidity and salty air with the potential to corrode metal over time.[37] Frequent yet unpredictable weather such as lightning, rain, and strong winds also created a less-than-advantageous environment for rockets that would sit in the open on launch pads.

NASA officials understood the importance of quickly choosing a launch site, since the construction of adequate facilities was considered a pacing factor in getting to the moon before the end of the decade.[38] In 1961, a NASA and Department of Defense joint report narrowed the search down to seven potential sites: Cape Canaveral, Mayaguana (an island in the Bahamas), Cumberland Island (in Georgia), Brownsville (in Texas), White Sands Missile Range (in New Mexico), Christmas Island (in the Republic of Kiritimati), and Hawaii.[39] Each was located in southern locations and almost all had a water buffer to the east to take advantage of the earth's rotational velocity at launch. Cape Canaveral was not the most geographically strategic choice, but Kiritimati, the closest option to the equator, and the other islands considered presented even greater logistical and construction difficulties.[40] The Brownsville location posed potential flyover risks for the southeastern United States, and White Sands, which would have likely cost the least to develop and operate, lacked access to water transportation.[41] The final alternative, Cumberland Island, had similar advantages to those of Cape Canaveral, but while the site was accessible via deep water transport, it did not have the communications network that had already been developed in Florida.

More by process of elimination than by merit, the report favored Cape Canaveral as the launch site, although several factors strengthened the case for the Cape, including the growing and "dynamic" surrounding area of

Titusville-Cocoa-Melbourne. The report exaggerated the strength of rocket personnel, saying that "practically the entire local area population [was] missile oriented" and said that NASA could anticipate a "minimum of public relations type problems due to missile hazards and inconveniences."[42] Already in the early hypothetical planning stages of expansion on the Cape, the gap between newcomers and those residents who already resided was evident, as the report brushed aside the concerns of the non-missile local labor force. Although the authors were reportedly "keenly aware" of the strain on Brevard County the expansion of NASA's had caused, the Cape was the clear front-runner in the report, and NASA had little time for lengthy deliberations about the concerns of locals as the deadline for a successful moon landing loomed.[43]

The final selection of Cape Canaveral was based primarily on financial advantages and access to existing infrastructure, including the crucial Atlantic Missile Range tracking system, a series of support stations running southeast from the Cape.[44] On August 24, after NASA announced plans to acquire the land on Merritt Island, the *New York Times* cited the potential concerns of residents of the area, reporting that some worried that rockets would "menace cities and towns for miles around" and that the "noise alone would damage windows and other structures, not to mention what it would do to people's nerves."[45] NASA, however, assured reporters that the risk of noise and blast damage would be "neutralized" by the ten miles that separated launch sites from populated areas, explaining that a 20-million-pound thrust rocket would generate approximately 120 decibels of noise, "well below" the danger level of 135 decibels.

Within a week, NASA asked Congress for the funds to purchase approximately 80,000 acres (324 square kilometers) of land northwest of the Cape Canaveral launch area. This was an additional $60 million on top of the initial $1.78 billion NASA budget approved in July. In his statement before the Senate Committee on Aeronautics and Space Sciences, NASA administrator James Webb argued that the land was urgently needed and that construction must begin immediately if NASA was to accomplish its goals in the timeframe the president had set for a lunar landing.[46] During the hearing on the budget amendment, Webb and his deputy Hugh Dryden answered questions, including some on the topic of housing facilities and citrus groves. Florida senator Spessard Holland said that he had not heard any complaints from homeowners on Merritt Island, who were "sorry to lose their homes, but . . . feel that if this is the preferred site and something to be done in the Nation's interest, they will have to accommodate themselves to it."[47] Whether he had directly heard this opinion from homeowners or if their silence had merely led him to this conclusion is uncertain, but it is clear that he differentiated the voices of homeowners, most of whom were laborers, from the voices of larger-scale citrus grove owners. Holland described the nearly 500 acres of "premium acreage" for Indian River fruit that would be lost. While he acknowledged that it was not a large area, the value of the fruit in both quantity and quality merited consideration from NASA. The senator argued for an option of ease-ment for owners and developers, who would otherwise lose not only their land but also the investment of years of growing and cultivating the trees. This solution would have allowed growers to maintain ownership of their groves but be required to evacuate during launches. It seems evident, however, that citrus groves had not been at the forefront of NASA's analysis when compar-ing launch sites. Webb skirted around the issue by saying that it would "not be wise" to make a statement about property at that time and Holland pushed back, again reiterating that although there was no objection to the taking it-self, the residents of the area who would be "vitally affected" deserved to know the effect NASA's presence would have on their livelihoods and homes.

Over the next year, as NASA put off making a decision about exactly which land would be affected, citrus growers grew more concerned as the Army Corps of Engineers began to appraise their land. In late 1961, NASA leaders promised goodwill and a quick decision to the Florida Citrus Mutual. But at an April 1962 Senate hearing, Holland again questioned NASA on behalf of his constituents, who had expressed "great anxiety" about the uncertainty surrounding the future of their property. Holland estimated that the affected citrus groves totaled around 2,000 acres owned by "good people [who were] entitled to a decision on this a long time ago."[48] NASA representatives at the hearing shifted blame for the delay to the Army Corps of Engineers and claimed that the corps had received no recent complaints from landowners. Holland pressed once more, asserting that he had personally passed along complaints to Administrator Webb from grove owners who needed to make decisions about spraying, fertilizing, and working the groves.

At the time that NASA acquired North Merritt Island, approximately sev-enteen communities, settlements, and towns with around 400 residents called the area home. Despite what Spessard Holland had told NASA when he advo-cated for citrus grove owners in Congress, not all of the former residents of the acquired land were happy to turn their property over to the government, even for the chance to contribute to a groundbreaking national project.[49] David Allan Taylor, a former resident who started working in the orange business around the age of six or seven, described the shock to older landowners spe-cifically. He said that his mother and father lived for less than a year after their land was taken.[50] For Taylor and others, the move was a devastating blow not only to their businesses but also to their memories and sense of place.

Many locals worried specifically about the level of transparency associated with the Army Corps of Engineers' efforts. Residents who refused the government's offers on the basis of "rumors" about neighbors getting better prices found themselves faced with condemnation proceedings, and when residents fought their low payments, almost all eventually lost in court. Some cases took nearly two decades to resolve.[51] NASA later purchased an additional 15,000 acres to the north in order to comply with studies that recommended a larger buffer radius for Saturn rocket launches.[52] When this buffer land was deemed too dangerous for habitation, the government created the Merritt Island National Wildlife Refuge. Although NASA framed this as a measure to assuage protests about its land acquisitions, it is unclear whether turning the once-inhabited land into a wildlife reserve placated local concerns about government overreach.

One instructive example is the case of Theodis Ray, an African American man who had grown up in Allenhurst, a Merritt Island freedom colony founded after the Civil War. From a young age, Ray worked as a fisherman around the area where NASA's giant Vehicle Assembly Building now towers over the landscape, packing and hauling fish with his father that they would sell in Titusville.[53] Ray's memories of the acquisition were based on his own experiences. "When the government came in and bought all this land, it wiped us out. I mean, you had these fishermen, that was their livelihood. Most of them didn't have education and all they did was fish." According to Ray, many orange groves in the area were owned by Black growers, a rarity during a time when few African Americans in Florida owned land. For the displaced African American grove workers and fishermen like Ray, employment options were limited, and many scrambled to claim the competitive janitorial and construction positions opening at NASA.[54] In Ray's case, the truck that he drove for a construction crew passed through very familiar territory as he worked to help erase the physical memory of his old community for the benefit of the space program.

NASA's final acquisition targeted a grove area of 3,206 acres containing 185,000 trees. Eventually, the two sides agreed on a deal that gave original grove owners in the buffer zone the ability to lease the land until 1968 with an option after that to renew the lease for five more years.[55] NASA's official 1972 history of the space center notes that the groves were "leased to their former owners who cared for the trees and harvested fruit. In return for this privilege, they paid annual fees to the U.S. Treasury."[56] This "privilege" was not what citrus growers originally envisioned. The prospect of easement proposed back in the September 1961 appropriations hearing had been abandoned for a system in which NASA leased land back to its former owners. After this, not only did growers and pickers have to contend with longer commutes, they were also not allowed to store their equipment near the groves.[57]

In a 2016 interview, Cocoa resident Frank Sullivan, whose family was heavily involved in the Brevard County citrus industry, described NASA's leasing program with distaste.[58] Compared to larger groves around the state, those on Merritt Island tended to be smaller properties that were owned and worked by families, which made this arrangement even more devastating. According to Sullivan, the leasing system was only intended to last three or five years, after which NASA consolidated smaller plots into 250-acre blocks that could then be leased out to the highest bidder.[59] This marginalized and created bitterness among smaller growers, who had no chance to compete with bigger citrus operations. To make matters worse, the Department of the Interior soon issued updated guidelines for areas that were part of the wildlife refuge that banned the cultivation of nonnative species, including oranges.[60]

Other small-scale grove owners on the island chose to sell their land outright at the low amounts offered and then faced the challenge of trying to relocate nearby, where the cost of land had skyrocketed with the growing demand. One woman recounted how she received only $244 per acre for her land from the Army Corps of Engineers and was shocked to see that a year later, land on South Merritt Island was selling for as high as $3,000 per acre.[61] Another man shared his experience with the lawsuits brought by families who objected to the low offers, relating how his family and others fought for more money in the courts. But even though his family won their case, the man noted wryly that by the time they had paid the lawyers, the amount that they received was about the same as the total originally offered.[62]

In general, NASA's process for choosing an Apollo launch site reflected a disregard for the local. Questions of colonial overreach do not appear to have been a topic of conversation as NASA made its decision, nor is it a theme that is widely explored in the well-known literature about the space program. NASA's decision to use Merritt Island came down to issues removed from local concerns, and in the end, tracking capabilities, cost, and resource redundancy meant more than the future of citrus groves or fishing docks. Eminent domain is a routine procedure for government projects that require land and there are organized systems in place for implementing it. In this case, it was the Jacksonville District US Army Corps of Engineers. But even in a history of this corps that details NASA's 1963 expansion, eminent domain is never mentioned.[63] The widespread use of this governmental prerogative that is written into the Constitution often makes it easy to overlook the lives that it affects, especially when those lives are underdocumented on lines of class or race.[64]

The Changing Face of America's Spaceport

As the Kennedy Space Center changed the demographics of Brevard County, older residents adjusted to the rapidly growing community and newcomers struggled to find a place for themselves. From 1950 to 1960, in large part due to the expansion of military and NASA operations on Cape Canaveral, Brevard became the fastest-growing county in the country. The population increase during this period (371.1 percent) dwarfed the average growth in Florida (78.7 percent) and the rate of population growth in the country as a whole (18.5 percent).[65] This area, which had supported 23,653 people in 1950, had ballooned to a population of 111,435 by 1960 and more than doubled to 230,006 by 1969.[66] In many areas of society in the county, tensions between the two groups could be seen in memories of shared activities and in general expressions of community values. In venues such as churches, schools, and community groups, attitudes toward social responsibility clashed as newcomers participated less or created separate and exclusionary groups related to professional connections.

In a 1966 report compiled by the Institute for Social Research at Florida State University, researchers highlighted population growth, education, government reach, community involvement, satisfaction, and perceptions of power structure. Noticeably missing from the studies are the voices of African American families in the area, who were excluded from the institute's surveys due to limited representation in the space industry and the low number of African American migrants to the area during the time of the study.[67] Although the population of Brevard County had boomed during the previous decade, the African American population remained steady, which meant that the group decreased significantly as a percentage of the population, from 25 percent in 1950 to only 11 percent by 1960.[68] Because of the absence of African American representation in the report, it is difficult to gauge nonwhite perspectives on issues that concerned all community members, including general satisfaction and involvement in community groups and local government.

The rapidly expanding education system in Brevard county is evidence of how the demographics of the area shifted during this time. Between 1951 and 1963 the average daily attendance in Brevard schools increased by 857.8 percent, from 4,163 students to 39,873, and the number of classrooms increased from only 117 to nearly 1,500.[69] Engineer Sam Beddingfield, who moved to the area with his family in September 1959, described the incredible growth and transformation of the Titusville school system, noting that his daughter was shuffled around to three different schools before he put in a complaint.[70] According to Beddingfield, at one point, the population was expanding so rapidly that they needed a new classroom every other day to keep up.

Participation in community organizations also showed the divide between "newcomers" and "old-timers." Original residents placed a high value on institutions such as churches, volunteer organizations, and school groups, newcomers found a stronger common bond in work relationships and activities.[71] There was further segmentation among NASA personnel, who rarely socialized outside circles in their own divisions.[72] While hobby groups related to specific interests such as bowling and music grew proportionally to the population, more established economic and service groups such as the Rotary Club and the Chamber of Commerce did not experience comparable growth, suggesting that the newer residents had different community priorities.[73]

For the engineers who worked on the Cape, most of whom were men, work hours were long and often included weekends. This left little time for participation in community affairs, which helps explain the appearance of apathy toward the broader community. Although their wives were more active in traditional women's community groups such as women's clubs and garden clubs, they also seemed to prefer profession-based groups with the wives of other NASA personnel. Sam Beddingfield remembered the heavy work schedules for engineers, lamenting the few days that he was able to spend with his family and commenting on the large number of divorces during Project Apollo.[74] A *Time* article from 1969 tied these difficulties to the nature of the engineering profession, which the magazine claimed valued the "technical world" over community.[75] In contrast to this fast-paced and stressful work environment, middle-class members of the community who were not employed on the Cape had predictable work schedules that allowed for regular meeting hours.[76]

NASA personnel over the years of Project Apollo had more in common with citrus workers than one might think. Both groups had long, unpredictable work hours that were dictated by factors largely out of their control and neither group participated in traditional middle-class activities of the time. For grove workers, race and class played heavily into these exclusions, as well as the migratory nature of a large proportion of grove work. Engineers, sensing career fluctuations and potential moves, were often insecure about their futures. Similarly, because of the expanding space-related industry and the increasing marginalization of citrus in the area's economy, grove laborers experienced uncertainty about job security. Many children of grove workers and the owners of smaller groves were removed from traditional school systems because of the seasonal need for extra labor. Although the children of engineers had secure educations, they experienced high levels of stress attributed to family tensions. One doctor at the time claimed that those stress levels led to a high number of children with ulcers in the area.[77]

As the population expanded, housing communities sprang up along the coast. Those developments settled largely along occupational and company lines.[78] In October 1963, Titusville's first high-rise apartment complex obtained a permit to build, and from 1950 to 1969, the number of housing units in the county grew from just over 9,000 to nearly 80,000.[79] Communities that had been largely self-sufficient made way for suburbs where vehicles became necessary, and the number of vehicles registered in the area surged from 4,163 to 61,824. While these numbers reflect the expanding population, this cannot be fully explained by population growth alone. While the population grew by approximately nine times, the number of cars in the county grew by fifteen times, reflecting the wealth of newcomers.[80] The land in previously undesirable areas, in many cases where African Americans lived, was taken for housing development. In Cocoa, the Brevard County Housing Authority seized land from predominantly Black neighborhoods in order to build "white-only" housing and former residents were pushed to the outskirts of the city in what one reporter called a "snake swamp."[81] Many of those who were evicted found themselves living far from their places of employment but unable to qualify for public housing without leaving their jobs. Although NASA had created jobs for these workers away from the citrus groves, in doing so, they set in motion the unintended consequence of increasing the population's wealth gap, often along racial lines.

One way locals took advantage of this growth was through support for a robust tourism industry centered on the space industry. In 1961, a *New York Times* article titled "Canaveral Boom in Missiles, Tourism" detailed the expected influx of tourism with NASA's acquisition of the additional land on Merritt Island.[82] According to the article, the area could look forward to a "flood of businessmen" looking to participate in the tourist industry by owning hotels and other attractions. Interestingly, alongside the obvious tourist attraction of rocket launches, the second biggest draw for prospective visitors the article highlighted was pristine fishing opportunities. As had been the case for tourists in the nineteenth century, the coast was seen as a place of wilderness and escape from city life. The only difference was that instead of encountering homesteaders on Merritt Island, fishing tours could now troll for trout within sight of launchpads that would send people to the moon.

Because the local economy and the tax base depended on high-paying technical jobs, the area realize the volatility of their position at the end of the 1960s, when NASA funding plateaued and then declined, taking thousands of jobs with it. In an NPR interview in 1971 at the tail end of Project Apollo, one local reported that "with each moonshot, the contractors in the area of the cape have laid off . . . [an average of] around 500 [workers]."[83] As engineers and technicians made an exodus from the area, retailers, construction workers, and others saw a drop in sales and business. In the context of this downturn, NASA leaders were challenged to maintain an "amicable" relationship with locals.[84] Brevard County attempted to stabilize the economy by attracting industry outside the aerospace sector, for example by encouraging seniors to move to the area to purchase surplus housing, and by promoting tourism.[85] Although the area experienced another boom in the 1970s with the arrival of the Space Shuttle program, the uncertain future of aerospace technology meant that Brevard County was perpetually at the mercy of political and administrative choices.

Although the newcomers and old residents existed in a codependent environment, for many non-NASA residents, the Kennedy Space Center and its operations existed on the periphery of their lives even as it was shaping the community in very significant ways. In many interviews with "old-timers" who grew up in the area, the space program was secondary to other concerns. During one 1994 interview, native Brevard County resident Kitty Bates described her experience growing up in the region but never mentioned NASA or the space center by name. Several times, she alluded to a more peaceful time in the region, referring to the activity in Titusville and on the Indian River almost disdainfully and lamenting that the "paradise" she once knew was gone.[86]

Even Kathryn Bouie, who had family with land on Merritt Island in the area around Clifton and Allenhurst and whose husband was a maintenance worker for Pan American Airways, was not preoccupied with the space center's presence in her life. Instead, she focused on the turpentine industry, church, college, and raising a family.[87] In another interview, Bouie's cousin, Sandra McMillan, provided more insight into the family's history in Allenhurst. She described one aunt named Eugenie who owned and worked her own orange groves on the island, budding stock and making a living from shipping her fruit to New York in addition to working at the Allenhurst Hotel as a cook.[88] During a time when membership in the NAACP was often an invitation for white backlash and Jim Crow made it difficult for African Americans to vote, Eugenie was a proud voter and a member of the civil rights organization.[89] Perhaps at least in part, this confidence and determination to exercise her rights can be tied to the nature of the independent community found in Allenhurst, which gave African Americans the ability to own land and cultivate business. As Theodis Ray, the Merritt Island fisherman turned NASA employee, emphasized, living in Allenhurst meant that he and his community were independent and free to some extent, but they understood the limitations of their situation.[90] In Ray's account, the erasure of Allenhurst and the settlements on Merritt Island in both a physical sense and in public memory represented a disregard for the work of the fishing and citrus workers that had called it home.

In contrast with the interviews done with "old-timers," those done with NASA personnel and contractors focused extensively on their work on the Cape rather than their lives in the community. Even JoAnn Morgan, who grew up in nearby Titusville and worked her first launch as an assistant engineer for the army at age 17 before going on to work on Apollo, rarely mentioned the outside community.[91] Morgan's experience would have no doubt intersected with the education system and community groups and she would have seen the effect the space boom had on the Cape, but those issues were not central to her narrative. Walter Kapryan, the longtime director of flight operations at Kennedy Space Center, spelled out the distance between the NASA community and the broader public more explicitly. When he was asked about his perspective on contemporary events like the Vietnam War, Kapryan replied that his life almost exclusively revolved around his family and the space program during those years. NASA personnel "ate together, [h]ad dinner together and partied together" in the closed world on Merritt Island.

During the transition, the citrus industry in Brevard County did not disappear, but it did change. By the 1960s, newspapers that had prominently advertised fertilizer or citrus buds in the early 1900s had few references to the fruit that remained one of the area's largest and most well-established products. In a 1970 Congressional hearing on NASA's future on the Cape, a local newspaper publisher described with pride the shift in industry, boasting that NASA had transformed the "23,000 orange growers of 1950 [into a] sophisticated population with over a quarter million persons engaged in some of the most complex technology known to man."[92] This exaggerated description both ignores the grove workers who had continued with their work and undervalues the complexity of knowledge that goes into growing citrus.

One such worker, John Moorer, a longtime citrus picker, described the tough choice many pickers made after Kennedy Space Center was built.

> When the Space Center came in . . . forty or fifty percent of the grove workers went over and got jobs . . . with benefits and better pay. . . . But there were some that couldn't go over there and get a job, some of them was too old, and some of them didn't even want that kind of job because see they done did this kind of work so long, and they've got addicted to it.[93]

Moorer, who opted to keep working in the groves, felt the pinch of economic fluctuations along with the rest of the community. People who worked in the citrus industry were forced to adapt to new housing developments, the ongoing menace of freezes, and incoming neighbors who increasingly valued engineering prowess over agricultural efforts.

Backing by the national government lent credibility and power to the technology of NASA, creating structures that placed larger sociopolitical factors over local considerations. In his 1969 work *Second-Order Consequences*, Raymond Bauer theorized about the unforeseen complexity of technology's impact on society and asked whether progress can be defined by technical milestones. He singled out the early space program as a discretionary and symbolic phenomenon that has a "relatively narrow over-all public impact."[94] While that claim can certainly be challenged by evidence of the value of NASA's technology for commercial spinoffs as well as broader scientific implications, few people have felt NASA's impact as directly as residents of Brevard County.[95] As the population rapidly grew and changed, the consequence was community displacement and disruption of existing structures, for good or ill. For some residents, the space center brought welcome new job opportunities and the chance to participate in a unique national effort. But for others, NASA existed on the sidelines of their lives, acting as a sometimes unnamed but significant catalyst for loss of land and livelihood.

Conclusion

In *Space in the Tropics*, Redfield asked if it ultimately matters where things happen and what it means when different things happen in the same place.[96] In the case of Merritt Island and the Kennedy Space Center, the interconnected yet distinctly separate histories of the national and the local are representative of many places in the United States affected by actions such as eminent domain. As Stuart Simms notes in his essay in this volume, as NASA was beginning proceedings to acquire land in Florida, it was making a similar effort to secure land for the Mississippi Test Operations (now Stennis Space Center). These case studies stand out because of the space program's mythological place in American memory, but there are countless cases of government land acquisitions where residents were displaced to make way for highways, stadiums, parks, and other projects deemed to be best for the public good. Brevard County citrus showcases one example of an agricultural technology with a long history of use and adaptation. It matters that orange groves and rockets both existed on Merritt Island because it brings into higher contrast their surprising similarities, contributions, and places in history.

Despite the apparent degrees of separation between NASA and local technologies like those associated with orange growing, neither existed in a vacuum. For orange growers and other preexisting industries, the relationship between themselves and the space industry was clearer, as they found themselves increasingly pushed to the periphery in both a geographic and cultural sense. Although citrus industry technology was sidelined in most discussions of the space center, when it is examined through a use-centered lens, it included operations such as grafting and freeze mitigation that boast extraordinary durability. As David Edgerton has noted, "In use-centred history technologies do not only appear, they also disappear and reappear, and mix and match across the centuries."[97] From the early use of grafting on Merritt Island that helped save the Florida citrus industry to the emergence of the world-famous Indian River Fruit and the explosion in popularity of orange juice and concentrate in the mid-twentieth century, citrus farming presents a clear case of use-centered technology. While its development can be written in a linear fashion, this history also represents cyclical patterns. Techniques like grafting, which have been in use worldwide for thousands of years, experience spans of disuse, rediscovery, and repurposing. Threats like freezes precipitated innovation and new ideas to protect trees, but they also prompted inventiveness based on existing methods.

Since 2005, the citrus industry in Florida and across the country has been besieged by a new threat: a bacterial disease called Huanglongbing (HLB), which causes damage to citrus products called greening. The disease, which is transferred between trees and groves by an invasive insect called the Asian citrus psyllid, causes small, malformed fruit and the eventual loss of the affected trees.[98] Although there is no cure, grafting offers hope for citrus growers, who have watched as their crops are decimated by HLB. In much the same way that Douglas Dummett grafted one kind of orange onto the more resistant rootstock of another, scientists in California have discovered ways to breed the affected citrus with HLB-resistant species and graft the resulting buds onto trees in the field to create a more resilient plant.[99] On Merritt Island in 2017, a two-acre USDA test grove located just south of the Kennedy Space Center contained five varieties of citrus trees grafted onto ten different root stocks, an experiment aimed at addressing HLB.[100] Brevard County has been hit particularly hard by citrus greening. From 2008 to 2017, production there fell by 87 percent. For the owner of the test grove, a local resident whose family has been in the citrus business for five generations, the experiment represents an opportunity to curb the decline of citrus in the area.

In her history of disruptive Canadian state-run megaprojects, Joy Parr wrote that "no place is merely local."[101] Similarly, no place is only national, even places like the Kennedy Space Center. By erasing the pre-NASA significance of citrus growing, a linear story of rocketry, engineering, spaceflight, and moon landings can too easily become the only story for Merritt Island. A more holistic account of the region's history includes not only the impact of the widely documented space program, but also the importance of the citrus industry, its workers, and their technologies before and after the arrival of NASA. The changes on Cape Canaveral cannot be viewed through a single lens, since many developments that were advantageous, such as improved infrastructure and job opportunities, came with the price tag of less desirable changes such as upheaval of community memory. Interviews with residents document the frustration that came with dispossession of land, while others illustrate the emergence of local pride in being host to the nation's launch site. The relationships between nationally backed technological forces and the local community are complex and unbalanced. Despite this, local memory and technology should be studied, not to create an argument for the moral superiority of local experience or to discount the achievements of national projects, but instead to recognize the value of both.

Notes

1 "Debrief and Goodnight," Apollo 16 Lunar Surface Journal, https://www.hq.nasa.gov/alsj/a16/a16.debrief1.html.

2 Andrew Chaikin, *A Man on the Moon: The Voyages of the Apollo Astronauts* (New York: Viking, 1994), 475.

3 The land that is referred to as Merritt Island is actually a peninsula that extends south from the mainland and is bordered to the west by the Indian River and to the east by the Mosquito Lagoon estuary. In this chapter, I will refer to the landmass as an island, as that is what is used locally. See Susan Parker, *Canaveral National Seashore: Historic Resource Study*, ed. Robert Blythe (Titusville, FL: National Park Service, 2008), 1.

4 Charles Benson and William Barnaby Faherty, *Moonport: A History of Apollo Launch Facilities and Operations* (Washington, DC: National Aeronautics and Space Administration, Scientific and Technical Information Division, 1978).

5 Jerrell H. Shofner, Jim Ball, and Vera Zimmerman, *History of Brevard County* (Melbourne, FL: Brevard County Historical Commission, 1995), 18.

6 Shofner et al., *History of Brevard County*, 25.

7 Robert I. Davidsson, *Indian River: A History of the Ais Indians in Spanish Florida* (West Palm Beach, FL: n.p., 2001), 137.

8 Kenneth Lipartito and Orville I. Butler, *A History of the Kennedy Space Center* (Gainesville: University Press of Florida, 2007), 29.

9 Lipartito and Butler, *A History of the Kennedy Space Center*, 29.

10 Merritt Precinct 14, Canaveral Precinct 15, and Haulover Precinct 16, Schedule No. 1—Population, Brevard County, Florida, June 5, 1900, manuscript census, National Archives and Records Administration.

11 Transcript of Marian Grant, interview with Nancy Yasecko, February 6, 1994, 13–14, Brevard County Historical Commission, https://www.brevardfl.gov/docs/default-source/historical-commission-docs/not-508-oral-history/marian-grant-interview-transcript.pdf?sfvrsn=3de11a28_2.

12 Richard Paul and Steven Moss, *We Could Not Fail: The First African Americans in the Space Program* (Austin: University of Texas Press, 2015), 35.

13 Scott D. Hussey, "The Sunshine State's Golden Fruit: Florida and the Orange, 1930–1960" (MA thesis, University of South Florida, 2010), 46.

14 Carita Doggett Corse, "The History of Citrus in Florida," typescript, Federal Writers' Project, n.d., 3, Special Collections Research Center, University of Kentucky Libraries, https://exploreuk.uky.edu/catalog/xt7nk9314n7q#page/1/mode/1up.

15 Louis W. Ziegler and Herbert S. Wolfe, *Citrus Growing in Florida* (Gainesville: University of Florida Press, 1961), 64; Hussey, "The Sunshine State's Golden Fruit," 10.

16 Ken Mudge, Jules Janick, Steven Scofield, Eliezer E. Goldschmidt, "A History of Grafting," *Horticultural Reviews* 35 (January 2009): 462.

17 John McPhee, *Oranges* (London: Daunt Books, 2016), 61; Alissa Hamilton, *Squeezed: What You Don't Know about Orange Juice* (New Haven, CT: Yale University Press, 2009), 5.

18 McPhee, *Oranges*, 64.

19 "East Florida," *Daily Kennebec Journal*, March 27, 1873, https://chroniclingamerica.loc.gov/lccn/sn82014248/1873-03-27/ed-1/seq-1/; "Indian River Farms Zu Vero, Florida," *Der tägliche Demokrat* (Davenport, Iowa), November 11, 1917, https://chroniclingamerica.loc.gov/lccn/sn84027107/1917-11-11/ed-1/seq-6/.

20 McPhee, *Oranges*, 71.

21 McPhee, *Oranges*, 66.

22 McPhee, *Oranges*, 64.

23 McPhee, *Oranges*, 40.

24 *Florida Star* (Titusville), October 8, 1909.

25 *Florida Star* (Titusville), December 10, 1909.

26 *Florida Star* (Titusville), August 20, 1909.

27 Department of Commerce, Bureau of the Census, Wilson Precinct 14, Haulover Precinct 12, and Shiloh Precinct 12, Population Schedule, Brevard County, FL, April 24, 1940, manuscript census, National Archives and Records Administration.

28 C. A. Bass, "Historical Sketch of the D.D. Dummit Grove at Allenhurst, Which Is Supposed to Be the Oldest Grove in Florida," *Florida State Horticultural Society*, January 4, 1926.

29 Roz Foster, "Explore Your History: Lost Communities of North Merritt Island: Shiloh," *Journal of the Brevard County Historical Commission* 14, no. 1 (2015): 22.

30 Coleman Mitchell, Alfonso Wilson, and John Moorer, "Firing the Groves," interview with Roz Foster, YouTube video, October 5, 2004, Brevard County Historical Commission, https://youtu.be/PlEg5IvGZ2Y.

31 Mitchell, Wilson, and Moorer, "Firing the Groves."

32 Peter Redfield, *Space in the Tropics: From Convicts to Rockets in French Guiana* (Berkeley: University of California Press, 2000), 125.

33 For a discussion of the Soviet space program and its decision to launch from rural Kazakhstan, where the nearest town was described as "a couple of two-story houses for the railwaymen, a couple of dozen small mud-plastered houses, and the tents of geologists prospecting for oil," see Asif Siddiqi, *Challenge to Apollo: The Soviet Union and the Space Race* (Washington, DC: NASA History Division, 2000), 133–135, quote on 135.

34 Lipartito and Butler, *A History of the Kennedy Space Center*, 30.

35 Charles Murray and Catherine Bly Cox, *Apollo* (New York: Simon & Schuster, 1989), 87.

36 Benson and Faherty, *Moonport*, 116.

37 Murray and Cox, *Apollo*, 87.

38 US Congress, House, Committee on Science and Astronautics, *1963 NASA Authorization: Hearings before the Committee on Science and Astronautics, Eighty-Seventh Congress, Second Session, on H.R. 10100 (Superseded by H.R. 11737), February 27, 8, April 12, and May 3, 1962*, Part I (Washington, DC: GPO, 1962), 623.

39 Benson and Faherty, *Moonport*, 126.

40 The UK and the US used Kiritimati, an atoll located in the South Pacific, for nuclear testing in the 1950s and 1960s.

41 Although the lack of overflight hazard was given as a strength for the Cape Canaveral site, launching rockets from Florida with a southeastern trajectory was not without overflight risk. An early account of launches from the Cape described how residents of Caribbean islands from the Bahamas to the Lesser Antilles collected debris from missiles that had fallen on their land—missiles that the account claims were "harmless if certain safety precautions were followed." William Roy Shelton, *Countdown: The Story of Cape Canaveral* (Boston: Little, Brown, 1960), 28.

42 NASA and Department of Defense, "Joint Report on Facilities and Resources Required at Launch Site to Support NASA Manned Lunar Landing Program," July 31, 1961, 19, NASA Headquarters, Washington, DC.

43 Benson and Faherty, *Moonport*, 127.

44 Benson and Faherty, *Moonport*, 129.

45 Richard Witkin, "Cape Canaveral Rocket Base to Be Expanded 5 Times in Size," *New York Times*, August 25, 1961.

46 US Congress, Senate, Committee on Aeronautical and Space Sciences, *Amending the NASA Authorization for the Fiscal Year 1962: Hearing before the Committee on Aeronautical and Space Sciences, Eighty-Seventh Congress, First Session, on S. 2481, a Bill to Amend the National Aeronautics and Space Administration Authorization Act for the Fiscal Year 1962, September 1, 1961* (Washington, DC: GPO, 1961), 3.

47 US Congress, Senate, Committee on Aeronautical and Space Sciences, *Amending the NASA Authorization for the Fiscal Year 1962*, 17.

48 US Congress, Senate, Committee on Appropriations, *Second Supplemental Appropriation Bill For 1962: Hearings before the Committee on Appropriations, Eighty-Seventh Congress, Second Session, on Apr. 4–6, 1962* (Washington, DC: GPO, 1962), 155.

49 Foster, "Explore Your History," 20.

50 US Fish and Wildlife Service, "The Merritt Island Adventure," November 17, 2014, 25:05, YouTube video, https://youtu.be/22IlBi3cR2I.

51 Lipartito and Butler, *A History of the Kennedy Space Center*, 87.

52 Lipartito and Butler, *A History of the Kennedy Space Center*, 99.

53 Paul and Moss, *We Could Not Fail*, 35.

54 Paul and Moss, *We Could Not Fail*, 37.

55 Benson and Faherty, *Moonport*, 144.

56 NASA, *Kennedy Space Center Story* (Kennedy Space Center, FL: NASA, 1970), 5.

57 McPhee, *Oranges*, 71.

58 Frank Sullivan, interview with Griffin Bixler and Michael Boonstra, YouTube video, July 21, 2016, Brevard County Historical Commission, https://youtu.be/D2SdBSG3ix4.

59 It is worth noting that this differs slightly from the details Frank Sullivan gave in his oral history interview, and it is possible that some plots of land were treated differently depending on whether they were within the bounds of the wildlife refuge or if they were part of the land NASA controlled directly.

60 Frank Sullivan interview.

61 Benson and Faherty, *Moonport*, 143.

62 US Fish and Wildlife Service, "The Merritt Island Adventure."

63 George E. Buker, *Sun, Sand and Water: A History of the Jacksonville District U.S. Army Corps of Engineers, 1821–1975* (Jacksonville, FL: US Army Corps of Engineers, 1981).

64 For more on race and eminent domain in Florida, see N. D. B. Connolly, *A World More Concrete: Real Estate and the Remaking of Jim Crow South Florida* (Chicago: University of Chicago Press, 2016).

65 Annie Mary Hartsfield, Mary Alice Griffin, and Charles M. Grigg, *NASA Impact on Brevard County: Summary Report* (Tallahassee: Institute for Social Research, Florida State University, 1966), 10.

66 Richard L. Forstall, ed., *Population of States and Counties of the United States: 1790 to 1990* (Washington, DC: US Dept. of Commerce, Bureau of the Census, Population Division, 1996), 30.

67 Hartsfield et al., *NASA Impact on Brevard County*, 100.

68 Benson and Faherty, *Moonport*, 377.

69 Hartsfield et al., *NASA Impact on Brevard County*, 51.

70 Sam Beddingfield, interview with Nancy Yasecko, YouTube video, January 15, 1994, Brevard County Historical Commission, https://youtu.be/5-IHqqZknEA.

71 Hartsfield et al., *NASA Impact on Brevard County*, 19.

72 Benson and Faherty, *Moonport*, 379.

73 Hartsfield et al., *NASA Impact on Brevard County*, 86.

74 Beddingfield interview.

75 Benson and Faherty, *Moonport*, 380.

76 Hartsfield et al., *NASA Impact on Brevard County*, 88.

77 Benson and Faherty, *Moonport*, 380.

78 Hartsfield et al., *NASA Impact on Brevard County*, 87.

79 "Consolidation, Growth Highlighted Area News in '63," *Star-Advocate* (Titusville, FL), January 1, 1964; NASA, *Kennedy Space Center Story*, 267.

80 NASA, *Kennedy Space Center Story*, 267.

81 Paul and Moss, *We Could Not Fail*, 44.

82 C. E. Wright, "Canaveral Boom in Missiles, Tourism," *New York Times*, November 5, 1961.

83 Mike Waters, Phil Miller, and Doug Terry, report on *All Things Considered*, National Public Radio, July 28, 1971.

84 NASA, *Kennedy Space Center Story*, 270.

85 Shofner et al., *History of Brevard County*, 212.

86 Kitty Bates, interview by Nancy Yasecko, YouTube video, February 5, 1994, Brevard County Historical Commission, https://www.youtube.com/watch?v=r92OnJdcUXw.

87 In addition to the orange industry in Brevard County, the turpentine camp in Mims was an important source of employment for many African Americans. Turpentine is made by harvesting and boiling the resin from pine trees and is primarily used for tar, oil, and medicinal purposes. In Brevard and elsewhere in Florida, turpentine was often harvested by forced convict labor. Kathryn Bouie, interview with Roz Foster, May 17, 1996, transcript, Brevard County Historical Commission, Cocoa, FL.

88 Sandra McMillan, interview with Roz Foster, August 27, 2004, YouTube video, Brevard County Historical Commission, https://www.youtube.com/watch?v=LgI-2SnbmO8.

89 In 1951, Harry T. Moore, the president of the state chapter of the NAACP, was assassinated in Mims when the KKK bombed his house on Christmas night in a blast that also killed his wife, Harriette. Moore's house was surrounded by a young orange grove that he had cultivated and hoped to live off in his retirement. For some, Moore's murder represents the beginning of the civil rights movement in Florida. For more on Moore, see Ben Green, *Before His Time: The Untold Story of Harry T. Moore, America's First Civil Rights Martyr* (New York: The Free Press, 1999).

90 Paul and Moss, *We Could Not Fail*, 40.

91 "JoAnn Morgan: Instrumentation Controller, Apollo Launch Control," in Billy Watkins, *Apollo Moon Missions: The Unsung Heroes* (Westport, CT: Praeger Publishers, 2006), 91–103; David Kamp, "Tracking Down JoAnn Morgan, a Semi-Hidden Figure of U.S. Space History," *Vanity Fair*, December 10, 2018, https://www.vanityfair.com/hollywood/2018/12/joann-morgan-nasa-apollo-11-interview.

92 US Congress, House, Subcommittee on Manned Spaceflight, *Future Utilization of The Kennedy Space Center, Cape Kennedy, Florida: Hearings before the Subcommittee on Manned Spaceflight, Hearings, Ninety-First Congress, Second Session, Apr. 10, 1970* (Washington, DC: GPO, 1970), 28.

93 Mitchell, Wilson, and Moorer, "Firing the Groves," 25.

94 Raymond S. Bauer, *Second Order Consequences: A Methodological Essay on the Impact of Technology* (Cambridge, MA: The MIT Press, 1969), 21, 98.

95 See the NASA Spinoff website, https://spinoff.nasa.gov/.

96 Redfield, *Space in the Tropics*, xiv.

97 David Edgerton, *The Shock of the Old: Technology and Global History since 1900* (Oxford: Oxford University Press, 2011), xii.

98 "Huanglongbing (HLB or Citrus Greening)," Center for Invasive Species Research, UC Riverside, n.d., https://cisr.ucr.edu/invasive-species/huanglongbing-hlb-or-citrus-greening.

99 "Citrus Clonal Protection Program," n.d., https://ccpp.ucr.edu/.

100 Dave Berman and Wayne Price, "As Brevard's Citrus Industry Declines, Growers Are Feeling the Squeeze," *Florida Today*, November 10, 2017, https://www.floridatoday.com/story/news/local/2017/11/10/brevard-county-florida-citrus-growers-feel-the-squeeze-oranges-grapefruits/840813001/.

101 Joy Parr, *Sensing Changes: Technologies, Environments, and the Everyday, 1953–2003* (Vancouver: UBC Press, 2010), 3.

6

Local Aerospaces

Cape Canaveral, Florida, and Baikonur, Kazakhstan

ARSLAN JUMANIYAZOV

For many American observers, the Russian ritual of watching the 1969 Soviet classic *White Sun of the Desert* (*Beloe Solntse Pustyni*) before flying into space from the Baikonur Cosmodrome in Kazakhstan is a strange practice. "The film has nothing to do with Space," says Eric Berger, senior space editor for the culture and technology website Ars Technica. Berger did note that there might be parallels between the symbolism of Sukhov, the main protagonist, wandering in the "empty dunes of the desert" and the cosmonauts and astronauts spending extended time "in the vast expanses of space." Both were away from home, family, and their social environment.[1] Those who are unaware of Soviet space history struggle to understand this ritual and to understand generally what Baikonur means to the peoples of Russia and Kazakhstan. The tendency in the historical literature is to see the building of spaceports and the development of space exploration in Russia as uncomplicated points of comparison to the same processes in the United States.

Cape Canaveral, Florida, and Baikonur, Kazakhstan, are two important names in the history of the Space Age. Points of comparison between history's two most famous launch sites are understandable and sometimes useful, but they can also be misleading. Cape Canaveral and Baikonur developed under starkly different political and economic systems and had dramatically different effects for the local populations. *White Sun of the Desert* may hold some clues. It is a classic Soviet "eastern," the Russian version of an American western. The film is filled with colonialist images of the Indigenous peoples of Central Asia and centers on Soviet-Russian conquests during the Russian Civil War (1918–1921). It conveys messages about colonization, violence, cultural assimilation, and crude Soviet propaganda that match well with the unique history of the development of the Baikonur spaceport.

Building on the well-known history of Cape Canaveral, this study seeks to illuminate the uniqueness of the development of the Baikonur spaceport and its effects on the local population through a comparison of the two sites. I outline the differences between these two historic spaceports in order to orient readers to the critical junctures that took their development along very different historical "plotlines." In contrasting the history of Baikonur with that of Cape Canaveral, I present a global context for understanding Russian and American development better. I will dissect some of the sociocultural context for the construction of the two facilities and will consider how the different historical plotlines have come to shape modern space policies.

These two launch facilities continue to have cultural significance as unique symbols of each nation's approach to space exploration. This piece builds on the works of historians such as Walter McDougall, who produced the pioneering comparative study of the space race between the US and the USSR during the Cold War. McDougall argued that the rise of technocracy in the US was largely a misguided reaction to a perceived challenge posed by Soviet success in space technology.[2] Von Hardesty has also written with care about this competition, as for example his comparisons of the two spaceports.[3] This chapter is not the first attempt to compare the expansion of American and Russian technological power at the expense of the Indigenous peoples. Steven Sabol and Kate Brown have insightfully compared how Americans and Russians carried the torch of western civilization into the American West and Central Asia, respectively.[4] In her innovative comparison of Billings, Montana, and Karaganda, Kazakhstan, Kate Brown concluded that the two small cities were "nearly the same place." Both cities exploited sparsely populated regions for their resources. Brown argued that Americans and Russians differed little in their treatment of Native Americans and Kazakhs. The violent deployment of European civilization and science against what actors from both governments viewed as "primitive" and "savage" Indigenous populations led to "destructive results" in both societies.[5]

Following Brown's approach, one could argue that Cape Canaveral and Baikonur were indeed "nearly the same place." Both sites were crafted in inhospitable environments as modern infrastructures. One was built on land that featured orange groves, marshlands, and seashores, where the first colonial surveyors encountered alligators, rattlesnakes, and fierce salt-marsh mosquitoes. The other was built at the site of a small railroad stop and a deserted sand quarry amid the salt marshes in the Kazakh steppes of the Kyzylkum Desert. There, builders and early residents from the Soviet metropole encountered wild dogs, rabid ground squirrels, and scorpions. They lived and worked under harsh conditions: temperatures that rose to 50 degrees Celsius (122 F) in the summer and dropped to -25 Celsius (-13 F) in winter; fierce winds that whipped sand into homes and into people's hair and faces. Most of the Russian newcomers came from central Russia's forests and steppes and struggled with acclimation. To help attract new residents, Russians built a huge park that they named Soldatski Park (Soldier's Park).[6] Both Cape Canaveral and Baikonur were built as close to the equator as possible in order to take advantage there of the greatest rotation of the earth (nearly 1,000 miles an hour) as an added boost for launching rockets eastward. Both sites were chosen with advantages for space launches in mind: Cape Canaveral site was built for launching ballistic missiles and rockets close to the Atlantic Ocean and Baikonur was built close to the prairies, forests, and tundra of Central Asia and Siberia.

NASA considered several other sites before choosing Cape Canaveral: Christmas Island in the Pacific; the southernmost point of Big Island in Hawaii; Brownsville, Texas; the Bahamas; and Cumberland Island, Georgia. The Cape offered a safe and reliable trajectory into the Atlantic, using tracking and control stations at the Bahamas, Brazil, and the Dominican Republic. The USSR also considered several other sites: Mari Autonomous Soviet Socialist Republic, Dagestan Autonomous Soviet Socialist Republic, a site near Stavropol, and a site near Astrakhan, but Soviet leaders chose Kazakhstan for its isolation and line of radio-control tracking stations all the way to Kamchatka (the target site for ICBM launches).[7]

Besides the hardships of the climates of Cape Canaveral and Baikonur, each site had its fair share of attractions. Florida had palm trees and tropical fruits as well as its beaches and salt-water fishing. Russians loved to fish in the Syr Darya, hunt on the steppes, and experience the colors of the steppe tulips in bloom or the taste of watermelons and garden fruits in summer. Both were in less developed regions of their countries: the American South and the Soviet "East," or what average Russians also called their "south." Both grew to become the main military and civilian launch centers of a Cold War superpower. Both could boast more triumphs than any other space launch sites in the world. The R series of ballistic missiles was tested and launched at Baikonur: the R-5, the great R-7 Semyorka, and the R-16 and R-36. Baikonur is also the site where the UR series, the Proton ICBM, the Energia rocket, and the Buran shuttle were developed and launched. All of these were built by lead designers Sergei Korolev, Valentin Glushko, Mikhail Yangel, and Vladimir Chelomey.

At Cape Canaveral, the Jupiter, Thor, Polaris, and Atlas ballistic missiles, the Titan and Saturn series, and the Space Shuttle and Space X were developed and launched.

In terms of space-launch infrastructure, Cape Canaveral in the 1960s and 1970s was what Silicon Valley is today to the world of technology. Similarly, Baikonur attracted the finest minds of Soviet military and space engineering. Employees at both sites were connected to vast research and development networks. These centers developed technologies that quickly became known by their acronyms, for example, LOC (Launch Operation Center) and NIP (Nauchno-izmeritel'nyi punkt; Scientific Tracking Point). Both were also connected to or defined by worlds of military secrecy.

Cape Canaveral and Baikonur had similar tragedies. The Apollo 1 and *Challenger* disasters were matched by Baikonur's Soviet R-16 disaster, the Soyuz 1 and Soyuz 11 disasters, and the four N-1 moon rocket failures. The accelerated pace of research and development driven by the space race in these research centers also exacted a high price from the engineers. They often suffered from heart attacks and other physical ailments and broken families.[8] Both Cape Canaveral and Baikonur experienced economic downturns, too: Cape Canaveral in the 1970s with layoffs and downsizing after the end of Project Apollo; Baikonur through the Soviet Union's bankruptcy and collapse. Lately, they have both attracted renewed investment. Port Canaveral has become one of the world's busiest cruise ports and the area has attracted new manufacturing helped by the state of Florida's space-industry initiatives. Space X and Blue Origins have competed for use of the updated launchpads at Cape Canaveral. Baikonur has received massive US aid and French investment as well as support from the Kazakhstan government, which intends to modernize its infrastructure. Some Kazakh officials go so far as to see Baikonur as the "Cradle of World Cosmonautics."[9]

The similarities between Cape Canaveral and Baikonur are certainly impressive. They had parallel development trajectories and similar goals during the Cold War rivalry between the two superpowers. They were similarly constructed and performed the same functions. Both had space vehicle launchpads and operation control centers, they had similar support services, and both were built close to railroad lines and access roads. NASA astronauts and engineers and US government sources spoke of Baikonur as the Soviet "Cape."[10] However, although they were in these rather formal ways the same *place* geographically and politically, Cape Canaveral and Baikonur were not the same *plot*. They were part of radically different political ideologies and historical plotlines.

Cape Canaveral was part of the American free-market, capitalist system, although it benefited from the power of the US government to transform and integrate the nation's economy and society to meet Cold War goals. When the Army Corps of Engineers set out to develop Merritt Island for NASA, they had to purchase property from landowners through legal means; they bought about 15,000 different tracts of land that covered over 100,000 acres. The corps and NASA had to contend with upset residents (who were none too happy with what some called a "northern" invasion), labor and union disputes, the creation of a wildlife refuge (by 1963), and a complex series of local relations.[11] Cape Canaveral is not just the spaceport at Kennedy Space Center on Merritt Island; it also hosts the Joint Long-Range Proving Ground and a US air force base. Astronauts and engineers remember them both simply as "the Cape." Both the base and the center were surrounded by growing communities in Brevard County, places like Cocoa Beach and Melbourne. These places experienced a "space boom" in the 1960s, as NASA-related in-migration created a significant population increase. Employees of Martin-Marietta, North American Aviation, Boeing, General Dynamics, Lockheed, McDonnell-Douglas, companies that contracted with NASA, also contributed to the population boom. The populations of some communities rose by 371 percent. As Kenneth Lipartito wrote, "The influx of people, money, and new industry from the lunar program was expected to lead Florida from rural underdevelopment to the industrial and space ages in a single great leap."[12]

Baikonur also served multiple functions, including one of deception. The real name of this area was Tyuratam, a railroad settlement on the Moscow-to-Tashkent railroad line built under Russian tsars. The real Baikonur was some 300 kilometers away to the northeast. The Soviet state renamed Tyuratam as Baikonur to fool American intelligence analysts after Yuri Gagarin's flight to Earth orbit in 1961. Until 1991, Baikonur was first and foremost a military base. One veteran called it a "small state in the desert" and a "unique military emirate."[13] At first, it was called Scientific-Research Firing and Testing Base #5 (Nauchno-Issledovetel'skii i Ispytatel'nyi Polygon) of the Ministry of Defense (or NIIP-5 MO). It consisted of at least 2,900 square kilometers in the Kyzylorda region of Kazakhstan and 4,900 square kilometers in the Karaganda region.[14] The military officers, engineers, rocketeers, and cosmonauts who worked there usually called it simply the Polygon. Many also called it the southern Polygon in contrast to its sister launch site, the northern Polygon in Plesetsk, which was located on the Arctic circle near Arkhangelsk.

I translate Polygon as "base." The word literally meant a military base, a secured geometric form upon the earth—by one measure 85 kilometers north

to south and 125 kilometers east to west—overlaying a grid of smaller geometric forms and lines. Baikonur's launchpads and control and support structures were framed with 1,200 kilometers of trolley lines and 500 kilometers of
railroad with 26 stations.[15] When the Presidium of the Communist Party and
the Council of Ministers of the USSR created Baikonur in a series of decrees
in early 1955, they did so with complete authority and with no regard for republican sovereignty or local autonomy. They simply expropriated the lands
to build the Polygon.[16]

Russians take great pride in Baikonur, which is now called the Baikonur
Cosmodrome in honor of Yuri Gagarin's achievement as the first human to
fly in space. It was a place of utopian vision and achievement. It was a "city of
the sun" (following Italian philosopher Tommaso Campanella's utopian essay
"The City of the Sun"), a place of high technology, good order, happy living,
and the joys of spaceflight. It was the place where the serious testing and improvement of space launchers and of their crews, happened, surrounded by
the sleepy villages and towns of backward Kazakhstan.[17] Or as the poet Robert
Rozhdestvenskii said: "On hearing the word Baikonur, the planet looks to the
sky" ("*Pri slove Baikonura, planeta smotri vverkh!*").[18]

Judging by newspaper and magazine articles and popular books on the
topic, Russians today associated Baikonur with the highest points of the Soviet years, including the launch of Sputnik in 1957 and especially Gagarin's
flight into orbit in 1961. Those who helped build and support Baikonur commonly refer to the center as an "oasis" and a "site of an advanced civilization"
in the desert. It was a modern "wonder of the world," a Russian version of
the great Egyptian pyramids.[19] The best and the brightest minds of Russia, its
"wonder workers" (*chudo-liudi*), built a "modern European-style city in the
middle of an empty desert."[20]

These boasts, however, came at the expense of the Indigenous Kazakh
populations, who were ostracized through caste-based racism and colonialist
narratives in order to justify the Soviet Union's capture and use of the area. As
one memoirist wrote, in a spirit of benevolence by the Soviet Union, before
Tyuratam became Baikonur, it was a "sun-burned, god-forsaken place on the
Kazakh steppes."[21] It was a "hell-hole" that was "at the edge of the universe,"
amid destitute Kazakhs "living on who knows what."[22] The Russian literature on Baikonur is filled with references to the backward "locals," or in the
worst case as *chiurki* (a racist and derogatory term meaning "black Turks"),
people who had been given the gifts of advanced science and technology.[23]
One memoirist called the Kazakhs mindless and "stumpy," just like the brutal
natural scenes around them. Another praised the Russians for building site #

10 at Baikonur, what became the city of Leninsk, because it allowed Kazakhs to leave their steppe yurts and mud homes for apartments with electricity and running water in cities with schools, and grocery stores.[24]

Kazakhs were part of the Polygon. They were literally at its base. They hosted some of the families of Soviet military leaders and personnel in the early days of Baikonur's construction in their modest homes near the train station. They were the security guards at its checkpoints, fences, and gates. They became service workers as janitors and street workers. They were also human points of dramatic contrast in many memoirs: props in the desert along with tumbleweeds and camels, signs of the backwardness surrounding Baikonur, the most modern of cities.

In these ways, Baikonur belongs to another paradigm: the Global South, what we used to call the Third World. In some ways, Baikonur is less similar to Cape Canaveral that it is to the French spaceport at Kourou, French Guiana, located in a former French colony in South America that included the infamous Devil's Island penal colony. It is also similar to the Brazilian air force's spaceport, which is located on former *quilombo* ("escaped slave") territory in the city of Alcântara on the Atlantic coast. Both sites were built to launch satellites and both expropriated land and displaced Indigenous populations in the interests of the state.[25] Baikonur belongs here, with these studies of imperialism and inequality, these distressing instances of the oppression and exploitation of local peoples and interests.

Baikonur, for example, was not the only Polygon in Kazakhstan. The Soviet military used Sary-Shagan Polygon for shorter-range ballistic missile tests. The larger Semipalatinsk nuclear test site brought formidable ruin and pollution to the ecology and humans of the nearby region. But the term polygon can also apply to any number of forced and violent Soviet-Russian impositions in Kazakhstan, such as the territorial delimitation of 1924 that created the country out of a larger and more Indigenous Turkestan, the Latinization and Russification campaigns that transformed the language and literature, the execution and killing fields of the Stalinist Terror (called polygons), the major sites of the Gulag in Kazakhstan, the areas of the forced famine in the 1930s, and the constant in-migration of Russian settlers, culminating in the Virgin Lands of the 1950s and 1960s. In these ways, all of Kazakhstan was one great polygon, one vast field for the great Soviet experiment between 1917 and 1991. The Polygon—the "Base"—may even be a valid metaphor for the whole Soviet experience. It is a figure of speech for the all-powerful military-industrial complex that ruled the country, that amassed and distributed its wealth and resources (natural and human), that imposed a "barracks socialism" on its people. This was all polygon.[26]

The Soviets learned just how different Baikonur was from its Cape Canaveral counterpart during the first Soviet-American joint spaceflight enterprise: Apollo-Soyuz in 1975. It was apparent to them that the cosmodrome was not up to western standards, so they rebuilt and improved local infrastructure such as the Lux Hotel for visiting NASA brass and the participating astronauts. They also attempted to deceive American visitors by putting identical black jumpsuits on soldiers civilian clothes on officers and transforming the commandant of Baikonur into someone who appeared to be "a respectable cosmodrome director."[27] Instead of genuine tours of the facilities, the Americans were provided with formal meals and feasts against a backdrop of local "color." NASA astronauts and engineers, dressed in Kazakh robes, ate lamb and ram's head, with oriental carpets and steppe ponies as props. American astronaut Thomas Stafford, though, was not fooled when he visited Baikonur. He saw the truth about the "poverty" and backward Kazakh "culture" all around, which "seemed to have changed little in hundreds of years." In contrast, Soviet teams and cosmonauts who visited the US and Cape Canaveral for Apollo-Soyuz were visibly impressed by the realities of American wealth and prosperity, realities that were very much at odds with the lies of Soviet propaganda. They were also surprised to see that tourists had free rein to "wander all over the Cape."[28]

Cape Canaveral and Baikonur met again in 1993, after the breakup of the Soviet Union, when US government and congressional delegations went to Baikonur to save it, along with the orbiting space station Mir. The new Russian Space Agency and Baikonur received nearly half a billion dollars in US aid. (Soon after that, Russia began leasing the site from a newly independent Kazakhstan.) Baikonur became an extension of America's spaceport infrastructure. NASA paid about $70 million for each of its astronauts to leave from Baikonur on the Russian Soyuz to ferry to the Mir space station and later to the International Space Station (after the Space Shuttle was retired). American visitors remember it as a city and space center in ruins, like something from the Third World. Russian memoirists made the same point: Baikonur was in a steep decline well before Kazakhstan "received" it, a decline brought on by the economic stagnation of the Soviet bloc and the Soviet state and social system after 1975.[29]

Kazakhstan saw things differently. At the collapse of the USSR, the Russian space program finally promoted several of the first Kazakh cosmonauts, Toktar Aubakirov and Talgat Musabayev. Kazakhstan president Nursultan Nazarbayev negotiated two lease agreements with Russia (with Boris Yeltsin in 1994 and Vladimir Putin in 2005) in which Russia paid Kazakhstan only $115 million a year for its use of Baikonur, an agreement that will be in force until 2050. Nazarbayev defined Baikonur as the science and technology center of Kazakhstan's spaceflight future, a centerpiece of what he has called "the Kazakhstan way."[30] Since the closure of the Space Shuttle program in 2011, more Americans have visited and experienced Baikonur. It has become more of a Kazakh city, at least in terms of de jure sovereignty and population. But it remains very much something of the wreck that it had already become in the late 1980s and early 1990s. Astronaut Scott Kelly recalls it as "a strange collection of ugly concrete buildings . . . with mounds of rusting, disused machinery piled everywhere. Packs of wild dogs and camels scrounge in the shadows of aerospace equipment. It's a desolate and brutal place . . . [filled with] dilapidated Soviet-era apartment complexes, huge rusted satellite dishes communicating with Russian spacecraft, mounds of garbage strewn about."[31]

The Russian-language literature on Baikonur is vast. But it is a very specific genre. Books and memoirs (I've counted at least four dozen so far) remember it as a peculiarly Russian city. It was one of the very few so-called administrative cities of the Soviet Union, essential capitals or borderland cities of supreme military significance. None of them were subject to any authority other than the Presidium or Politburo of the Communist Party and the highest state committees under the party's direct control and authority. There was no fiction or pretense of federalism or local sovereignty for administrative cities. They included Moscow, Leningrad, Sevastopol, and Baikonur. The last two are interesting. Sevastopol and Baikonur were both sites of military bases at the sensitive peripheries and borderlands of the Soviet and Russian states. Sevastopol was the center of the Black Sea fleet and a heroic city of World War II. In 2014, Vladimir Putin's regime invaded Crimea and seized control of Sevastopol. Baikonur was luckier. Yet both places attest to the irredentism at play in Russian foreign policies after the collapse of the Soviet Union. Putin's government and his leading supporters are not reconciled to the loss of its satellite states or to the loss of empire and security.

Veterans of the Baikonur Polygon and its cosmodrome, in fact, were some of the first elite and powerful voices to point out the militant and unforgiving irredentism of the Russian state. To Russian leaders, the loss of Baikonur represented a double blow: the loss of the empire and its non-Russian peripheries and the loss of its superpower status in terms of the high-technology machinery of a space launch center. In the words of one manifesto, Baikonur may have become the "territory of Kazakhstan," but to Russian leaders, it would always remain the "sacred property of Russia." To them, Baikonur was really Russian land, a symbol worthy of a true "superpower" and the "defense of the motherland [Rodina] and development of fatherland science and technology."[32]

These kinds of chauvinist images take us back to the characters and story-line of the ritual movie *White Sun of the Desert*. It is, no doubt, a personable and humorous story about overcoming hardship and a lovely kind of fable about spaceflight. But it also privileges the Russian characters at the expense of the local peoples. We need to turn to other sources for the truer voices of the Kazakhs, as for example the 2012 film *Baikonur* by the German director Veit Helmer. The movie offers a realistic portrayal of the Baikonur cosmodrome, one that aligns with Scott Kelly's description. It spotlights the dangers of the Russian Soyuz rocket and Proton rocket firings, the metal fragments from these rockets that fell back on the earth, along and the spread of toxic pollution from heptyl fuel on the steppe. The story also describes the main character, Gagarin, as a man who rejects a false and shallow understanding of Russian spaceflight in order to discover his real identity as "Iskender." Iskender's true personality is founded on the folk wisdom and culture of his native land, not on modern western technology or propaganda. He learns to love his Kazakh homeland more than Russian spaceflight.[33]

Notes

1 Eric Berger, "I Was Bored, So I Watched the Movie That Astronauts Must View Before Launch," Ars Technica, April 8, 2020, https://arstechnica.com/science/2020/04/i-was-bored-so-i-watched-the-movie-that-astronauts-must-view-before-launch/.

2 Walter McDougall, *The Heavens and the Earth: A Political History of the Space Age* (New York: Basic Books, 1985). For the Soviet and Russian angles, I have relied on Anatoly Zak's posts to RussianSpaceWeb.com.

3 Von Hardesty with Gene Eisman, *Epic Rivalry: Inside the Soviet and American Space* (Washington, DC: National Geographic, 2007), 138–139.

4 Steven Sabol, *The Touch of Civilization: Comparing American and Russian Internal Colonization* (Boulder: University of Press of Colorado, 2017); Kate Brown, "Gridded Lives: Why Kazakhstan and Montana Are Nearly the Same Place," *American Historical Review* 106, no. 1 (2001): 17–48.

5 Brown, "Gridded Lives," 30–32. Also see John Collins and Carole McGranahan, eds., *Ethnographies of U.S. Empire* (Durham, NC: Duke University Press, 2018).

6 Jacques Villain, "A Brief History of Baikonur," *Acta Astranautica* 38, no. 2 (1996): 132–133.

7 To be fair, we ought to be comparing Baikonur to the White Sands Proving Ground or Edwards Air Force Base. These launch sites in the deserts of the American Southwest provide the best comparison with the firing ranges in Russia's central Asian deserts.

8 Mike Gray, *Angle of Attack: Harrison Storms and the Race to the Moon* (New York: Norton, 1992); Iu. V. Ivanchenko and K. V. Gerchik, eds., *Baikonur: Pamiat' serdtsa. Sbornik podgotovlen sovetom veteranov kosmodroma Baikonura* (Moscow: Terra, 2001).

9 Yelena Levkovich, "Baikonur Modernization Projects Brought Closer to Fruition at Space Days in Nur-Sultan," *Astana Times* (Kazakhstan), November 18, 2019, https://

astanatimes.com/2019/11/baikonur-modernisation-projects-brought-closer-to
-fruition-at-space-days-in-nur-sultan/. See also "Forum 'Space Days in Kazakhstan:
Baikonur—the Cradle of World Cosmonautics' Kicks Off in Kazakhstan," November
12, 2019, FOR.kg, https://www.for.kg/news-622487-en.html.

Further research on the topic would require a variety of approaches: comparing
the populations and demographics of Cape Canaveral and Baikonur (within the centers and in the regions, especially in terms of in-migration and minority populations);
weighing the regional investments in higher education, especially NASA's "Supporting
University Program" in Florida and the work of the Moscow Aviation Institute at Baikonur; and comparing the US War on Poverty as it related to the space program and
the USSR's development programs in Central Asia for the same.

10 For the use of this term, see United States Congress, Senate, Committee on Aeronauti-
cal and Space Sciences, *NASA Authorization for Fiscal Year 1973: Hearings before the
Committee on Aeronautical and Space Sciences, Ninety-Second Congress, Second Ses-
sion, on S. 3094, a Bill to Authorize Appropriations to the National Aeronautics and
Space Administration for Research and Development, Construction of Facilities, and for
Other Purposes* (Washington, DC: GPO, 1972), 879; Congressional Research Service,
Soviet Space Programs: 1976–1980 (Washington, DC: GPO, 1982); Committee on
Commerce, Science, and Transportation, 99th Congress, 1st Session, US Senate, PRT
98–235, Part 3 (May 1985): 1052.

11 Kenneth Lipartito and Orville Butler, *A History of Kennedy Space Center* (Gainesville:
University Press of Florida, 2007), 27–29, 87–88, 96–99; Charles Benson and William
Faherty, *Gateway to the Moon: Building the Kennedy Space Center Launch Complex*
(Gainesville: University Press of Florida, 2001), 87–107; William Faherty, *Florida's
Space Coast: The Impact of NASA on the Sunshine State* (Gainesville: University Press
of Florida, 2002).

12 Lipartito and Butler, *A History of Kennedy Space Center*, 96.

13 Vitaly Leonidovich Katayev, *A Memoir of the Missile Age: One Man's Journey* (Palo
Alto, CA: Hoover Institution Press, 2015), 57.

14 Mike Gruntman, "From Tyuratam Missile Range to Baikonur Cosmodrome," *Acta As-
tronautica*, December 2018, 2.

15 V. L. Men'shikov, *Baikonur: Moia bol' i liubov'* (Moscow: Garant, 1994), 230.

16 I. M. Baturin, ed. *Sovetskaia kosmicheskaia initsiativa v gosudatsvennykh dokumentakh*
(Moscow: RTSoft, 2008), 43–47.

17 B. E. Chertok, *Hot Days of the Cold War*, vol. 3 of *Rockets and People* (Washington,
DC: GPO and NASA, 2005), 348–353; Viktor Pelevin, "Code of the World," *Franfurter
Allegemeine Zeiting*, February 28, 2001.

18 M. I. Kuznetskii and I. V. Strazheva, eds., *Baikonur: Chudo XX veka. Vospominaniia
veteranov Baikonura* (Moscow: Sovremennyi pisatel', 1995), 5.

19 Men'shikov, *Baikonur*, 118, 141, 210–211.

20 Sovet Veteranov Kosmodroma Baikonur, *Proryv v kosmos: Ocherki ob ispytateliakh,
spetsialistakh, i stroiteliakh kosmodroma Baikonura* (Moscow: TOO VELES, 1994),
112; Ivanchenko, *Baikonur*, 139–141, 217, 226–227.

21 Ivanchenko, *Baikonur*, 69.

22 Chertok, *Rockets and People*, 1:311–315.

23 Men'shikov, *Baikonur*, 39.

24 Katayev, *A Memoir of the Missile Age*, 39; Kuznetskii, *Baikonur*, 126.

25 Peter Redfield, *Space in the Tropics: From Convicts to Rockets in French Guiana* (Berkeley: University of California Press, 2000); Sean Mitchell, *Constellations of Inequality: Space, Race, and Utopia in Brazil* (Chicago: University of Chicago Press, 2017).

26 For context, I have relied on Ann Laura Stoler, *Imperial Debris: On Ruins and Ruination* (Durham, NC: Duke University Press, 2013).

27 Ivanchenko, *Baikonur*, 286.

28 Thomas Stafford and Michael Cassutt, *We Have Capture: Tom Stafford and the Space Race* (Washington, DC: Smithsonian Institution Press, 2002), 180–184.

29 Committee on Science, Space, and Technology, *Oversight Visit, Baikonur Cosmodrome: Chairman's Report of the Committee on Science, Space, and Technology, House of Representatives, One Hundred Third Congress, Second Session* (Washington, DC: GPO, 1994). Men'shikov, *Baikonur*, 33.

30 See Nursultan Nazarbayev, *The Kazakhstan Way* (London: Stacey International, 2008); Pamela Blackmon, *In the Shadow of Russia: Reform in Kazakhstan and Uzbekistan* (East Lansing: Michigan State University Press, 2011).

31 Scott Kelly, *Endurance: My Year in Space* (New York: Crown Books, 2018).

32 Sovet Veteranov, *Proryv v kosmos*, 6, 112.

33 The movie was inspired by Chingiz Aitmatov, *The Day Lasts More than a Hundred Years* (Bloomington: Indiana University Press, 1988), a classic science fiction tale. It is a story of steppe herders set against the backdrop of a nearby cosmodrome and cosmic soap operas in outer space. Aitmatov told the original fable about trusting memory and culture over dreams and propaganda.

7

"Fire in the Bucket"

The Impact of Mississippi Test Facility on Local Residents

STUART SIMMS

"I'll never forget Gainesville and I don't believe anyone else who's lived here, even for a little while, will either," Mrs. Louise Jones observed on the loss of her home in the longleaf pine forests of Mississippi in November of 1961. She held the family dog, Chico, and her three children close after hearing the announcement that she and her family would be displaced to accommodate the rocket test site coming to Hancock County, Mississippi.[1] Mrs. Jones lived within the boundaries of one of NASA new facilities. The Jones's story is not an uncommon one. It is one of acknowledgment and reconciliation but also one of sacrifice and sadness, removal and frustration, and "progress." In 1961, the American space program came to Hancock County and adjacent areas with a *hard* landing.

What is now called Stennis Space Center went by several names before administrators settled on giving it the name of an ultra-conservative Democratic senator from Mississippi, John C. Stennis. Originally called Mississippi Test Operation, it later became Mississippi Test Facility, and still later the National Space Technology Laboratories. The various names reflect contestation on multiple levels—the space race between the United States and the Soviet Union, the dynamics inside the newly created federal space agency, disagreements between notable political figures, and standoffs between "outsiders" and residents of Hancock County. While Mississippi Test Facility was built at a moment when NASA had captivated the American imagination (and the American treasury), it swept away local communities that had been evolving for centuries—communities that, for many, represented an entire world of existence.

The new facility also erased evidence of a complex past of myths and legends, trials and tribulations, exploitation and freedom. This chapter looks at four nodes in the story of Hancock County and Mississippi Test Facility: 1) the rich, complex history of the county before the space center's development; 2) the methods NASA used to sweep into Hancock County and sweep residents out; 3) former residents' deep ties to local communities and homesites decades after they were removed; 4) and how the official narrative associated with the test facility overlooks the stories of former residents.

Mississippi Test Facility was arguably one of NASA's most contentious US sites. After Apollo 11 landed on the moon, Mississippi Test Facility faced hard times. Mothballing emerged as a new worry for local outlier communities that had become increasingly dependent on the jobs this Cold War industry had brought to south Mississippi. By consolidating control over a vast amount of land in Hancock County, NASA had created a place marked by a framework of Cold War development that included displacement of communities, increased "scientific" management of the land, enclosure of land, and local protest. These factors make the Hancock County site part of the larger pattern of Cold War development in the Deep South. They also invite a larger conversation about the site's impact on the people who lived there who were removed and silenced in the wake of the facility's development. These local people depended on an intricate commons system for sustenance, versions of which had operated in that place for centuries. This system faced constant challenges in the twentieth century until NASA's incursions finally displaced it in the 1960s. NASA created a new kind of space in Hancock County, but it did not do so in a vacuum. NASA replaced the unique flavor of local life with a space that was framed according to paradigms of Cold War development in the United States.

Cathedral of Forests—"The Thorn Before the Rose"

A series of events shaped Hancock County into a modern hub of rocket engine testing during the space race.[2] Similar transformations took place across the United States as the federal government and large corporations subjugated and transformed landscapes and exploited the people living there for economic gain and national security. The story is modern, but it began well before the first rumbles of rocket engine testing at Mississippi Test Facility in 1966. Folklore and oral tradition in Hancock County reveal a history of longstanding, interwoven communities around the site.

In the latter half of the nineteenth century, Hancock County began transforming from a coastal frontier to a thriving lumber hub. A gradual rise in population and increased interest in southern longleaf pine fueled the exploitation of the vast coastal pine forests throughout Hancock County (and along

Figure 7.1. H. Weston Lumber Company mill. Pre-NASA, date unknown. Courtesy of NASA.

much of the Gulf Coast). The Poitevent & Favre Lumber Company, the H. Weston Lumber Company, and others dominated the landscape of Hancock County after the Civil War (fig. 7.1). These companies used the wealth they accumulated from longleaf pine and their political power to build a series of railroads and other infrastructure improvements around their massive processing plants and shipment operations. River and rail access made Hancock County desirable for these large mills.[3] As the nation moved into the Progressive Era, Hancock County remained tied to the longleaf pine as one of its primary economic drivers.

However, as timber gave way to steel and concrete, the county's economic needs shifted. In the process, the lumber companies' stranglehold over the local communities became even more apparent. The industry was hard hit during the Great Depression, which led to mass unemployment and challenging times in the local communities that depended so heavily on the lumber industry. Once the companies left the county, residents worked in small lumber-adjacent industries or other small-scale industries such as illicit moonshine distilling and wood pulp production or moved away to find employment elsewhere. A new face also changed Mississippi politics. A conservative judge, John C. Stennis, replaced Senator Theodore Bilbo in a special election following Bilbo's death in 1947. Stennis used Cold War fearmongering and jingoism, his connections to boosters interested in regional development, and a burgeoning national interest in the space program to gain momentum for a project in his home state. This shifting political landscape made it easier for Hancock County residents to accept the prospect of a new rocket facility's inception, one that threatened their homes.

A young national space agency was desperate for a place to test its rocket engines. The year 1957 tumultuous for space exploration: the launch of the Soviet Sputnik satellite, the rhetoric around what President Eisenhower described as a "missile gap," and the onset of the space race. As competition grew, America looked to the moon as a site where it could secure a symbolic victory over communism. German rocket pioneer Wernher von Braun needed to test new iterations of American rockets like the massive Nova rocket, which required engines capable of producing thrust even more powerful than that of the Saturn V launch vehicle. By the 1960s, Huntsville, Alabama, the site of von Braun's Marshall Space Flight Center, was too densely populated for such testing because of the massive influx of facilities and workers dedicated to the Saturn V. Scientists had determined that subsequent rocket testing at Marshall Space Flight Center would place the surrounding city in danger because of the overwhelming noise pollution and vibrations of the F-1 and J-2 engines that personnel at the center had designed to send the Saturn V to the moon. Von Braun knew he needed a vast, "uninhabited" space to test his engines, one that was close enough to the rest of his facilities that he could personally oversee testing and coordinate transportation and delivery of components. Senator Stennis, who had become an influential member of both the Senate's Appropriations Committee and its Committee on the Armed Services, accepted NASA's proposal that it use a piece of his state to build the new testing site. The first choice of NASA's site selection committee was Hancock County plus adjacent pieces of Pearl River County and St. Tammany Parish in Louisiana became the first choice for the site selection committee.

Similar instances occurred across the country around the same time as the federal government sought host sites for the nuclear arms complex, nuclear power generation facilities, and other military installations. These places all have similar stories of contention, removal, and remaking of place and identity for residents like Louise Jones during post–World War II industrialization (fig. 7.2). One of the most colorful of these folk stories during the displacement of the local communities involved Cora E. Davis, affectionately called

Figure 7.2. Postmistress Lollie B. Wright lowers the American flag at the Logtown Post Office for the last time, 1963. Courtesy of NASA.

"Aunt Blue" by many residents. Aunt Blue had lived in Gainesville all her life and intended to remain there until the day she died. Unfortunately for her, NASA claimed her land and forced her out. To express her disdain for the displacement, she stayed on the front porch as movers moved her house from Gainesville to the town of Nicholson (fig. 7.3). Like other folk stories about displacement, the story of Aunt Blue lives on as a testament to the reluctance of people to comply with the forced moves.[4] Such stories were familiar to many Hancock County residents.[5]

By November 1961, after announcements made by Stennis and local boosters, the most prominent outsider influence in the county shifted from the lumber companies to America's new space program. The agency's arrival involved mass meetings, forced removals, and battles with the landscape.[6] Senator Stennis framed the sacrifices local residents were making so the facility could be built as an opportunity to "take part in greatness."[7] Initial engine tests began at the space center in April 1966 and continued for three years, leading to the US landing on the moon in July 1969. However, a series of dis-

Figure 7.3. Cora E. Davis ("Aunt Blue") sitting on her front porch, date unknown. Courtesy of NASA.

putes arose during this period as local people stepped forward to seek compensation for damages and to accuse the test center of wrongdoing during engine tests. NASA claimed little to no responsibility for these issues and continued to push ahead with its mandate of beating the Soviets to a successful moon landing.

NASA's acquisition of the land it needed was a complicated process. At the center of the site, NASA purchased 13,500 acres from landowners. This land, which NASA referred to as the operational area of the site, was also known as the "fee area."[8] The remaining land area of the center, totaling around 125,000 acres, served as an acoustical buffer zone. Although NASA did not purchase the land for the buffer zone, it placed restrictive easements on the land that prevented human habitation and the construction of specific structures but allowed residents to continue growing crops, timber, and maintaining livestock (fig. 7.4). Although owners of land in the buffer zone were limited in how they could use their property, they were not exempt from increasing property taxes. By mid-1962, disputes about property in the fee area had ended and residents who lived on property in the buffer zone had until 1964 to vacate.[9] As NASA moved in, crews demolished or removed homes inside the boundaries of the fee area. On April 23, 1966, the facility tested an engine for the first time, a Saturn V S-II-T stage (fig. 7.5).[10]

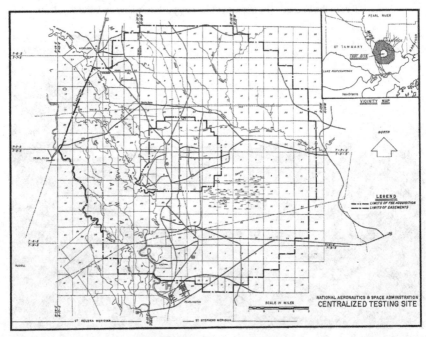

Figure 7.4. An Army Corps of Engineers map indicating the borders of the acoustical buffer zone and the fee areas for land acquisition, 1961. Courtesy of NASA.

Figure 7.5. First engine test at Mississippi Test Facility, April 23, 1966. Courtesy of NASA.

One Small Step...

Less than a month after Apollo 11 landed on the moon on July 20, 1969, a massive natural threat loomed in the Atlantic Ocean. Hurricane Camille made landfall on August 17. As NASA was celebrating its biggest success to date, the people of the Gulf Coast were suffering as the result of a historic storm.

Before Hurricane Katrina, Hurricane Camille lingered in the imagination of Gulf Coast residents as one of the most powerful storms to hit the shoreline.[11] Hundreds of lives and billions of dollars were lost to the hurricane or to the destructive tide it created. As devastation rocked the Gulf Coast, one small hub of activity spearheaded recovery efforts and acted as a haven for locals—Mississippi Test Facility. Less than a decade before, NASA had removed people and communities from their land. Following Camille, however, Mississippi Test Facility provided essential relief and support for a community in extreme distress.[12]

In oral histories about this period, the importance of the facility's role as a refuge is prominent. As the Gulf Coast braced for an impending natural disaster, residents came to the NASA facility seeking refuge. Over 1,000 people sought safety in the offices and cramped spaces of Mississippi Test Facility's central complex. After the storm, Mississippi Test Facility led the way in the recovery effort by lending vehicles and labor to clean up debris and search for people. It also served as a command center for cleanup efforts.

As the facility began helping people recover, an order issued by NASA officials caused rumors to spread like wildfire. Officials in Huntsville ordered the discontinuation of engine testing for Project Apollo, which had been the only mission of the facility since its inception. Mississippi Test Facility faced budget cuts and growing tensions with residents who had leased them land in the buffer zone when rumors of decreased testing grew.[13] As site director Jackson Balch and Senator Stennis worked with NASA and government officials to advocate for the site's survival, local people used the opportunity to call for change.

Newspaper headlines alluding to reductions in aerospace industry spending and Saturn V testing fueled local tensions. Headlines such as "Mississippi Test Facility May Be Erased" and "NASA To Abandon Facility" sparked the hopes of local residents that they could regain access to their land and restore their communities.[14] Their voices come out in letters they wrote to the political face of the space center: John Stennis. However, residents' pleas fell on unsympathetic ears. The facility's managers and the region's politicians had no interest in granting the requests of local people. This conflicted with the view of many residents. They had done their part: they had sacrificed their land at Stennis's request, and now that the sacrifice was no longer necessary, they wanted their land back. From late 1969 to 1976, dozens of people wrote to the senator about land issues with the facility. They asked for such things as the right to visit cemeteries, access to roads and water, and changes to easement restrictions and they expressed concerns over illicit activity in the buffer zone. These people were aware that the facility's future was far from certain and wanted to use that uncertainty to spark a conversation aimed at restitution.

Like Cora Davis, these residents exemplify local resistance to the site's enclosure of the buffer zone. For example, Mr. A. K. Northrop of Pass Christian, Mississippi, a local farmer, wrote to Senator Stennis on August 27, 1969, requesting permission to build a new home on the property he owned in the buffer zone. In the weeks before his letter to Stennis, Northrop had lost his home to Hurricane Camille. He criticized both the restrictive building easements placed on the buffer zone and ever-climbing tax rates for property around the facility. He wrote, "[I] cannot even build a cowshed much less a home. . . . Taxes have quadrupled since my land was seized." Northrop accused the government of land speculation. "I was paid about $17.00 net per acre for the easement in spite of the fact that I paid $110 per acre for 44 acres that I bought approximately ninety days before the government confiscated same." He also alluded to landholders' growing reluctance to continue with the space agency's current system for dealing with residents and property owners, especially after Hurricane Camille. He said of the space venture in south Mississippi, "I am getting a bit gun shy." Northrop's statement highlights the growing doubts many local landowners had following reductions in space funding and rumors of site closure.[15] Senator Stennis, NASA administrator Tom Paine, and legislative affairs officer H. Dale Grubb maintained that the center should remain on "standby status" to guarantee sufficient land in the buffer zone for tests in the coming years.[16]

David H. Stockstill, a teacher at the Charles B. Murphy Elementary School, was another frustrated Mississippi Test Facility neighbor. In October 1970, Mr. Stockstill asked for access to the highway on the site because of hardships in his morning commute. Before the site was built, Stockstill had a direct route to his school. But after NASA enclosed the land around its site, Stockstill was forced to take a different route. He wrote, "This route is much longer; the roads are in poor condition, and there is greater risk to my personal property and safety." He described an elaborate carpooling arrangement with neighboring teachers that alternated which car was used to minimize damage to vehicles because of the poor road conditions. He wrote to Stennis as a last resort because of the reluctance of site security officials to consider the change in access: "I accepted this teaching position on the strength of a written promise from NASA security that necessary passes would be issued to me for the use of Road 'A' through MTF." Site manager Jackson Balch responded to this claim by stating that the only way to accommodate this request would be to give passes to others as "equally deserving." NASA officials cited security risks and the continued need for limited access to protect the critical feature of the facility, now deemed a "national asset."[17] This continued refusal of access fueled narratives of distrust.

In July 1971, Mrs. L. C. Cuevas from Pearlington, Mississippi, registered her objections to the facility's land management practices. She wrote to Senator Stennis about the graves of family members inside the buffer zone. Her family's graves were initially located in the Gainesville Cemetery, on land where the Mississippi Test Facility was built. NASA had moved them to the buffer zone and they had begun to sink. Cuevas said that when she asked for access to the highway on the site check on the graves, she was treated "very cold" by NASA officials. As a result, she had to travel from Pearlington to Slidell up the Pearl River rather than along Highway 607, the main highway into the facility. "Its not right," Cuevas wrote.[18] Unfortunately for Cuevas, Stennis had no new information and stated that the buffer zone would not change because NASA wanted to protect the space for further testing and other activities. Stennis also noted the distance to Pearlington via Interstate 10 was five miles longer and took about the same time as the route Cuevas currently used.[19]

In the mid-1970s, after Mississippi Test Facility was renamed the National Space Technology Laboratories, the rhetoric of Hancock County residents became even more heated. Augustus Elmer Jr., a Pass Christian resident, wrote to Stennis in 1975 in the hope that the senator could help him regain access to fishing grounds on land he owned that was restricted by easements. Elmer's comments revealed that some landowners saw NASA use of that land as abuse. "As it stands now it is being operated as a private club," he wrote. Elmer's letter described the position he faced as a landowner. He could not access his land, yet he had to watch as facility employees and their guests enjoyed fishing on property he once controlled and for which he still paid taxes.[20]

Newspapers did the facility no favors as tensions flared with local communities. In 1976, Bill Gaudet of the Biloxi Sun-Herald wrote an exposé on the buffer zone and its impact on Hancock County residents and on the property of the landholders in the easement zone. "Buffer Zone Evolving into Dumping Site" described the zone as a "no man's land and a breeding place for crime." According to Gaudet, since 1972, there had been four murders in the buffer zone, more than thirty cars had been stripped of moveable parts and left to rust, campers had trashed campsites, multiple arrests had been made for firearms and drug violations, and the land had become a hideaway for escaped convicts. These issues plus a reduction in security patrols and funding led local people to push for protection and commemoration of the site. Roy Baxter, a marina owner in Pearlington, described how he provided an armed escort and transportation to and from the site to women who visited family graves in Logtown, within the buffer zone.[21] Gaudet's account of the buffer zone leaned on local knowledge and the misgivings outspoken opponents such as L. C. Cuevas, David Stockstill, and Augustus Elmer Jr. expressed about the enclosure of the buffer zone. Despite local critiques, however, the facility maintained its operational status and evolved, incorporating multiple federal agencies into the site to quell rumors that the site would soon be closed. By 1988, the facility had been renamed the John C. Stennis Space Center and had separated from Marshall Space Flight Center, its parent facility.

One Giant Leap . . .

The hardships of dispossession linger in local memory. In 2015, for example, the *Biloxi Sun-Herald* interviewed the Moran family in their former family home in Kiln, Mississippi, not too far from the outer boundary of the buffer zone. The Morans—father Sonny Jr., mother Mary Louise, and daughter Kathy Hendry—talked about their lives before NASA came to town.

Mary Louise recalled that when the family heard that the federal government had decided to displace them, she was beside herself. "Every day when she gathered eggs from the old home place, Mary Louise Moran cried," the article said. "'I cried and cried,' she said. 'We all cried when we left.'" Kathy Hendry remembered playing in the house as a child and teared up during fond memories of living inside the buffer zone.[22] This emotional response from multiple family members to the removal suggests a deep, powerful connection to the land and the community that most displaced people expressed in their discussions of displacement. Usually, the families of local people had lived and worked on these lands for generations. Families often described themselves as having deep cultural connections to it.

Such connections to land resonate with other Cold War–era stories of displacement in the service of national goals. For example, in "Sending the Flood Upriver: Impersonal Change and Personal Stories in the Savannah River Valley," Robert P. Shapard describes the development of the Clarks Hill Dam project in parts of Lincoln County, Georgia, and McCormick County, South Carolina. Shapard interviewed a resident in one of the communities that the building of the dam flooded, Joe Holloway Jr., who saw the federal government's action of taking the land of local people for the reservoir as wrongdoing. The impact that event had not only when it happened but also years later echoes the stories Hancock County residents told. Together, these stories suggest displacement as a paradigm of federal power during the Cold War. By examining people's connections to places on the land such as fishing holes, gravesites, and home sites, historians can highlight counternarratives to the traditional, government-approved stories of "progress" and "development" that pervade discussions of government projects. These studies bring significant issues to the fore. As the federal government continues to tackle the questions of development for the "greater good," it must also weigh issues such as the rights of landowners and local communities and how to navigate that discussion justly.[23]

Similarly, in Jake Kosek's study of the impact of the Los Alamos National Laboratory, he interviewed a resident from one of the outlier communities close to Los Alamos whom he calls "Paula Montoya."[24] She described the problems she experienced while living on the fringes of the facility and the sparse resources in surrounding communities. Montoya compared living near the facility to living with a "bad boyfriend." She described the initial allure of inviting the facility to the area and felt that she and others had been victimized by displacements that were presented to them as "progress" and "positive" federal development. Montoya mentioned significant increases in crime and the degradation of communities in the valley.

Montoya's brother, Ernesto, lost his life in a heroin overdose. She feels that Ernesto's death was not simply about the night he overdosed. Instead, she understood it as unfolding over an extended period and developing from his lack of access to a good job or economic standing, followed by desperation about an inability to pay mounting bills.[25] She knew the immediate negative impacts of the Los Alamos facility on the land, but she also was aware of the harm the facility's presence did to local knowledge of the land that demonstrated a profound connection to the community. She was able to tie her experiences to place and could also fully articulate the intricacies of living on the edges of a "progress" that the federal government defined solely in terms of national security needs.

Hancock County's Cold War story of the local consequences of a government development project echoes these other stories. Though they would not likely have used the same terminology, Stockstill, Cuevas, and Elmer would certainly have found resonance with Montoya's "bad boyfriend" analogy. All three accused site managers of failing to live up to written NASA promises and Stennis's guarantees. The Moran family expressed the same sense of profound loss as Joe Holloway Jr. Together, narratives from local communities suggest the outlines of a more authentic narrative of Cold War development in America that considers not only gains but also profound losses.

Conclusion

The space that became Stennis Space Center exemplified how the cost of the moon landing depends on who you ask. While the benefits convinced policymakers that the center should be maintained, some people living on the fringes of the facility's buffer zone expressed different opinions on the matter. While it led the way to the moon, NASA perpetuated a set of paradigms typical of federal developments during the Cold War. Rural displacement and land enclosure proved to be two major factors. Although the federal government justified these developments by referring to Cold War politics and national pride, some residents quickly lost faith as they faced the complex realities of development. They saw a shrinking program and a national agency that lost its way following initial successes and investments. Examining such discussions of Stennis Space Center helps us recognize similar problems in the national narrative. We still do not know much about the Mississippi Test Facility's remaking of Hancock County. For instance, the racial dimensions of this story remain buried almost entirely.[26] Nor has that facility yet received the general historical attention it deserves. Key historiographical works such as Bruce Schulman's *From Cotton Belt to Sunbelt* barely register Mississippi Test Facility. This oversight must be corrected. Discussions of Cold War America demand interdisciplinary exploration, and few places exist that represent that need better than Stennis Space Center and the ongoing relationship it shares with local communities.

This story also troubles the overarching, complicated concept of "progress" that is associated with the space race and NASA's inception. Although Project Apollo and subsequent scientific discoveries by NASA at Stennis Space Center brought positive changes, the human costs of the project need to be acknowledged appropriately. As the clock continues to count down for the leases granted to the federal government by landholders in the buffer zone, a question remains: What will happen to the buffer zone and to the families that still own that land? The consequences of NASA's impact on what is (and what *was*) Hancock County continue to plague both local people and the agency. Moving forward, these problems will persist. For people like the Morans, the past seems like a rose before the thorn. Although residents like Mrs. Louise Jones are long gone, the people still tied to the buffer zone continually consider their collective fate as owners of nonexistent homesteads. These conversations suggest that residents continue to feel connections to places where their families lived—places removed by "progress."

Notes

1 Winfred Moncrief, "How Rocket Test Site Folks Feel," *Hattiesburg American*, November 9, 1961. The phrase "fire in the bucket" refers to an exclamation an employee at Stennis Space Center made on the morning of April 29, 1966. That was the day of the first test of a rocket engine in NASA's history. When the engine ignited and flames lit the early morning sky, someone at the center shouted "We have fire in the bucket!" Paul Foreman, "'Fire in the Bucket'—Stennis Begins Rocket Engine Testing," April 11, 2008, https://www.nasa.gov/centers/stennis/news/newsreleases/2008/CLT-08-042.html.

2 "Watch for Speculators," *Rural Electric News*, December 1961. In a speech delivered to residents announcing the removal, Senator John C. Stennis said, "There is always the thorn before the rose." Quoted in Mack R. Herring, *Way Station to Space: A History of the John C. Stennis Space Center* (Washington, DC: National Aeronautics and Space Administration, NASA History Office, Office of Policy and Plans, 1997), 25.

3 Graham Calloway, "Henry Weston and the H. Weston Lumber Company," 2010, unpublished document.

4 Herring, *Way Station to Space*, 36.

5 I have drawn from several works among the vast historiography of these trends: Kate Brown's discussion of nuclear communities and families in *Plutopia: Nuclear Families, Atomic Cities, and the Great Soviet and American Plutonium Disasters* (Oxford: Oxford University Press, 2015); Kari Frederickson's study of the Savannah River Plant in Aiken, South Carolina, *Cold War Dixie: Militarization and Modernization in the American South* (Athens: University of Georgia Press, 2013); Robert P. Shapard's essay on the Clarks Hill Dam project in 1954, "Sending the Flood Upriver: Impersonal Change and Personal Stories in the Savannah River Valley," in Kathryn Newfont and Debbie Lee's anthology *The Land Speaks: New Voices at the Intersection of Oral and Environmental History* (Oxford: Oxford University Press, 2017); Andrew Needham's discussion of the Four Corners Power Plant in *Power Lines: Phoenix and the Making of the Modern Southwest* (Princeton: Princeton University Press, 2014); Carol Rainey's discussion of relationships between local communities and plants in the Ohio River nuclear complex in *One Hundred Miles from Home—Nuclear Contamination in the Communities of the Ohio River Valley: Mound, Paducah, Piketon, Fernald, Maxey Flats, and Jefferson Proving Ground* (Cincinnati, OH: Little Miami Press, 2008); Caroline Peyton's "Kentucky's 'Atomic Graveyard': Maxey Flats and Environmental Inequality in Rural America," *Register of the Kentucky Historical Society* 115, no. 2 (2017): 223–263; and Jake Kosek's analysis of the local communities surrounding Los Alamos National Laboratory in *Understories: The Political Life of Forests in Northern New Mexico* (Durham, NC: Duke University Press, 2006).

6 For more on the battles with natural factors of the landscape such as mosquitos, snakes,

and alligators and the mobilization of the federal government, see Cindy Donze Manto, *Stennis Space Center* (Chicago: Arcadia Publishing, 2018). Other interesting environmental history works include the militarization of responses to natural phenomena such as flooding and insects; see Richard M. Mizell Jr., *Backwater Blues: The Mississippi Flood of 1927 in the African American Imagination* (Minneapolis: University of Minnesota Press, 2014).

7 Mack R. Herring, *Way Station to Space: A History of the John C. Stennis Space Center* (Washington, DC: NASA History Office, 1997), 26.

8 Paul Foerman [*sic*], "Stennis Officials Issue Buffer Zone Reminder," October 21, 2008, https://www.nasa.gov/centers/stennis/news/newsreleases/2008/HEC-08-158.html.

9 *Information Concerning Land Acquisition Program for National Aeronautics & Space Administration Centralized Testing Site Mississippi & Louisiana* (Mobile, AL: US Army Corps of Engineers, 1961).

10 Herring, *Way Station to Space*, 113–114.

11 Eric Jeansonne, "Hurricane Camille Remembered 50 Years Later," *KLAB*, August 12, 2019, https://www.kalb.com/content/news/Hurricane-Camille-remembered-50-years-later-537633351.html. According to many, Camille remains a benchmark for storm damage and intensity for older generations.

12 Herring, *Way Station to Space*, 149–160.

13 Herring, *Way Station to Space*, 149–160.

14 "Mississippi Test Facility May Be Erased," *Times-Picayune*, September 28, 1969; "NASA To Abandon Facility," *West Point Times-Leader*, September 29, 1969.

15 A. K. Northrop, Pass Christian, Mississippi, to John C. Stennis, August 27, 1969; Thomas O. Paine to John C. Stennis, October 8, 1969; H. Dale Grubb to John C. Stennis, July 29, 1970; John C. Stennis to A. K. Northrop, August 15, 1970. All in series 25, box 7, folder 31, John C. Stennis Collection, Congressional and Political Research Center, Mississippi State University Libraries.

16 The Mississippi Test Facility's standby status ensured the facility's claims on property following the Apollo missions, allowing them to maintain the buffer zone for future testing.

17 David H. Stockstill, Picayune, Mississippi, to W. H. Dearing, October 20, 1970; David H. Stockstill, Picayune, Mississippi, to John C. Stennis, December 12, 1970; Jackson Balch to John C. Stennis, January 15, 1971. All in series 25, box 8, folder 13, John C. Stennis Collection.

18 Unless otherwise noted, I have retained the spelling and grammar of the original letters in order to preserve the language of residents.

19 L. C. Cuevas, Pearlington, Mississippi, to John C. Stennis, July 20, 1971; and John C. Stennis to L. C. Cuevas, August 13, 1971. Both in Series 25, box 8, folder 12, John C. Stennis Collection.

20 Augustus Elmer Jr., Pass Christian, Mississippi, to John C. Stennis, March 29, 1975, series 25, box 9, folder 24, John C. Stennis Collection.

21 Bill Gaudet, "Buffer Zone Evolving into Dumping Site," *Biloxi Sun-Herald*, September 6, 1976.

22 Anita Lee, "Stennis Buffer Zone Split This Home in Two. But Kiln Family Never Let Go of 'The Old House,'" *Biloxi Sun-Herald*, October 17, 2015.

23 Shapard, "Sending the Flood Upriver," 137–154.
24 Montoya's name was changed in Kosek's story to protect her identity.
25 Kosek, *Understories*, 228–275.
26 Alexandra Hedrick, "The Possum Walk Before the Buffer Zone," *Picayune Item*, May 16, 2014.

8

Moonshots and Moonshine

Project Apollo and the Appalachian Regional Commission

MAX CAMPBELL

On May 25, 1961, President John F. Kennedy made his second address to Congress after a turbulent first four months in office. This address, commonly known as his "moonshot" speech, aimed to restore confidence in his leadership after the failed Bay of Pigs invasion, tensions between East and West Berlin, and the rapidly increasing US military presence in Southeast Asia. Kennedy set some lofty goals. He began his speech by proclaiming, "These are extraordinary times. And we face an extraordinary challenge." He famously declared, "I believe that this nation should commit itself to achieving the goal, before this decade is out, of landing a man on the Moon and returning him safely to the Earth." These are the words dedicated to the new Apollo program that many Americans who were alive at the time remember. Yet seven sections earlier in the address, he announced an entirely different challenge related to economic disparity in the Appalachian region: "Large-scale unemployment during a period of prosperity would be intolerable. I am therefore transmitting to the Congress a new Manpower Development and Training program, to train or retrain several hundred thousand workers, particularly in those areas where we have seen chronic unemployment."[1] Appalachia would not be left behind in a spacefaring nation.

With his words, Kennedy launched two massive new government programs: Project Apollo and the Appalachian Regional Commission. Each of these "extraordinary challenges" built upon his early campaign promises and required massive federal funding, committees of leaders, and positive media attention to get off the ground. The tragedy of an assassin's bullet prevented Kennedy from witnessing the culmination of the goals he set forth on that early summer day. Yet each of these challenges was a mark of his administration's sense of confidence and purpose: to take the nation higher in space, reach the moon, and surpass the Soviet Union and simultaneously to raise a region out of the depths of poverty and inequity. America needed to close both gaps. From the earliest years of Kennedy's administration, spaceflight was compatible with social justice.[2]

Kennedy's New Frontier

The president's attention to Apollo and Appalachia was no surprise. He had already threaded them into his presidential campaign in the preceding years. They were hallmarks of his political life. As a senator from Massachusetts, Kennedy had witnessed the Soviet Union's rise to space supremacy as a member of the Senate Foreign Relations Committee, and he argued publicly that Moscow was surpassing the United States in launch vehicle technology, creating a missile gap. "Now it is time to take longer strides—time for a great new American enterprise—time for this nation to take a clearly leading role in space achievement, which in many ways may hold the key to our future on Earth."[3]

As a rising political star, Kennedy chose to use President Eisenhower's rather apathetic response to Sputnik 1 as an opportunity to gain momentum for his presidential campaign in 1960. In a speech entitled "Can We Compete with the Russians?" delivered on October 18, 1957, at the University of Florida in Gainesville, Kennedy asked the audience, "Can we compete with a nation that has absolute power to determine the number of scientists it shall produce and the projects on which they shall work?" He continued with many more rhetorical questions about America falling behind, reminding the audience about the presumed bomber gap, according to which the Soviet Union had more bomber aircraft at the ready than the United States. Kennedy stated, "and now we are in the age of Sputnik." He called it a paradigm shift in the Cold War, not because of "crude imitation or the result of espionage and stolen secrets," but because "the Soviet Union has available for this kind of work more engineers and scientists than we presently have in any capacity in this country." Alluding to the science gap, Kennedy advised Americans to vote the Republicans out of office. "I would not want our future course in the cold war determined by either those with their eye on the nuclear stockpile, glibly assert that we have nothing to fear, or those who, with their eye on the Russian Moon, cry out frantically that all is lost. Let us be calm, let us be realistic, let us be determined."[4]

Candidate Kennedy used similar language at other speeches around the country and incorporated this rhetoric into his 1960 presidential campaign.

One focus of Kennedy's campaign included the perceived missile gap between the United States and the Soviet Union. Journalists such as Joseph Alsop interpreted leaked classified evidence and published articles in the media, beginning with the *New York Times*, arguing that there was a lopsided missile surplus in the Soviet Union's favor while the United States lagged behind. Kennedy recited Alsop's claims even though they were not accurate.[5] In 1960, Kennedy argued on the Senate floor that it was "indisputable" that the United States was deficient in several areas, including ballistic missiles.[6]

During his presidential campaign, Kennedy witnessed firsthand the catastrophic effects of unemployment during his visits to the Appalachian region. Areas such as West Virginia and the Cumberland Plateau of southeastern Kentucky had experienced rapid unemployment and labor migration out of the region in the 1950s and 1960s. By 1955, 50 percent of coal miners had lost their jobs.[7] Campaigning in Huntington, West Virginia, Kennedy made multiple stops, and on April 20, 1960, he gave a speech titled "Aid to Depressed Areas" in which he spoke of the importance of bringing jobs to the area and improving living conditions. "To attract new industry in this day and age, a city must have adequate water, streets, schools and other community facilities."[8] Kennedy emphasized the importance of training unemployed workers in a new skill and said that he would help pass an area redevelopment bill to elevate West Virginia to the status of its more affluent neighbors, such as the coastal areas of Virginia and Maryland. Returning to Huntington on May 6, 1960, Senator Kennedy remarked, "Today's government is not doing the people of West Virginia—and the people of our other depressed areas—the things that must be done, the things you people cannot do for themselves." He added, "Only the help, the understanding and the leadership of national government can do the job."[9] This was Kennedy's fifth time visiting the area leading up to the May 10th Democratic primary vote, in which Kennedy won West Virginia over Hubert Humphrey. He also won the state in the general election. Like his slogans about the missile gap, West Virginia helped Kennedy win the presidency.

Spaceflight and regional poverty were both signature aspects of Kennedy's campaign platform. He highlighted these key initiatives in his acceptance speech at the Democratic National Convention on July 15, 1960, nine years and a day before the Apollo 11 mission launched. "We stand today on the edge of a New Frontier—the frontier of the 1960s—a frontier of unknown opportunities and perils—a frontier of unfulfilled hopes and threats." One of those frontiers was "the uncharted areas of science and space." Another was an appeal to "the unemployed miners and textile workers," though it received only a paltry round of applause.[10]

These issues prefigured new frontiers within the country too, the geographic footprints of the new Apollo and Appalachia programs. Both encompassed regions that stretched thousands of miles across the country. The Apollo program relied on state-of-the-art aerospace development and launch facilities in Florida, Maryland, Alabama, Louisiana, New York, Texas, and California. These included a vertical "footprint" too: a global satellite communication network to track and communicate with a spacecraft that was like a vast "overhead cable thousands of miles in the air to which users all over the world may make connections by means of invisible radio beams."[11] Appalachia was far more modest: the region encompassed the mountainous areas of Pennsylvania, Maryland, Ohio, West Virginia, Kentucky, Virginia, Tennessee, North Carolina, Georgia, and Alabama. However, transportation of goods, services, and people was a major concern for Appalachia since most of the region was difficult to navigate. Revitalizing Appalachia called for the creation of roads, airports, and bridges in areas that were inaccessible and would now be open to new industry, business, and communications.

Appalachia in Need

In the 1962 book *Night Comes to the Cumberlands: A Biography of a Depressed Era*, Harry Caudill described the conditions and plights of the region in great detail. The book is a sort of call to arms. Caudill, a lawyer and a member of the Kentucky legislature, argued that the Cumberland Plateau country of eastern Kentucky had become an "anchor dragging behind the rest of America." Caudill summarized the Kentucky Appalachian as the "illiterate son of illiterate ancestors, cast loose in an immense wilderness without basic mechanical or agricultural skills, without the refining, comforting and disciplining influence of an organized religious order, in a vast land wholly unrestrained by social organization or effective laws, compelled to acquire skills quickly in order to survive, and with a Stone Age savage as his principal teacher."[12] His brutal description of the Cumberland Plateau residents garnered national attention and revealed to the country a region that suffered from lack of basic health care, electricity, and telephones, and a dependence on subsistence farming and hunting. While Caudill's description was extremely blunt, it came from a place of genuine concern. Caudill's family had been in the region in 1792. He spent much of his life after World War II as a state legislator touring the region and experiencing the plight of its residents firsthand. One pinnacle moment came when he visited a rural schoolhouse and watched children sing "America the Beautiful" as rain poured through the leaking roof.[13] This event, among others, convinced him to do research to find the reasons why this part of the country had lagged so far behind. Historically, land speculation, the lumber industry, and coal mining had exploited the people and land of the Appalachian region. Speculators and financiers in the nineteenth and twentieth centuries had made massive profits by purchasing the property of mountain residents without disclosing the lumber or coal resources on their land. Many were illiterate and signed the contracts with an X.[14]

Caudill especially blamed the speculative lumber and coal industries for the region's problems. Overlogging and overmining had permanently altered and destroyed the Appalachian region. The extraction of coal, he wrote, "leaves a legacy of foul streams, hideous slag heaps and polluted air. It peoples this transformed land with blind and crippled men and with widows and orphans. It is an extractive industry which takes all away and restores nothing. It mars but never beautifies. It corrupts but never purifies."[15] As a result, by the 1950s, one in every three Appalachian residents was impoverished, the per capita income was 23 percent lower than that of the average American, and roughly two million people had left the region in search of better jobs. As they moved north for jobs, cities met them with prejudice. As David Walls and John B. Stephenson wrote in *Appalachia in the Sixties*, "Even the better educated viewed the region as little more than a setting for moonshining, feuds and Little Abner."[16]

Because organized churches and schools were poorly established during the eighteenth and nineteenth centuries and few mountaineers were fully literate, superstitions and traditions often filled those gaps. Above all, mountaineers considered the moon to be full of supernatural influences. Caudill wrote, "The phases of the Moon were carefully noted and, to a considerable degree, life was regulated by its growth and decline. Some crops could be planted only when the Moon was new, while others must be withheld from the soil until an old Moon arrived. Meat would spoil even in the coldest weather if the animal was slain when the Moon was new." The moon even dictated the future of babies born in the Appalachian region: "The frontier wife was likely to hope that her child would be born when the moon was a tiny crescent. Such good timing was thought to ensure strength and a good mind in the baby."[17] A coincidence of this superstition would later become apparent during one of the tensest moments of the Apollo 11 mission.

Apollo in Progress

Apollo and Appalachia were inheritances from the late Eisenhower years. But in the 1960s they were spearheaded by federal agencies that headed classic large-scale programs for solving America's problems and had the stamp of presidential executive authority.[18] By the time Kennedy was inaugurated on January 20, 1961, NASA had already selected its first seven astronauts to fly in the Mercury program. The first mission, which launched on May 5, 1961, made astronaut Alan Shepard America's first person in space and answered Yuri Gagarin's orbital flight from a few weeks before. On May 21, 1961, after only one crewed spaceflight, Kennedy committed the nation to a lunar landing before the end of the 1960s. He was battling a stereotype of a once-great America that was losing ground to the Soviets. A few months before Kennedy made his address to Congress committing the United States to a moon landing, a working group at NASA had laid out the initial plans for achieving this goal. "A Plan for Manned Lunar Landing" was created by the Manned Lunar Working Group, headed by George Low. The plan was transmitted to the NASA associate administrator Robert Seamans on February 7, 1961. This was the first fully developed plan for how NASA proposed to send humans to the moon.[19] The vision of American human spaceflight had materialized into reality with the Mercury program. Using the Army's Redstone rockets led by Wernher von Braun in Huntsville, Alabama, the Mercury program sent six astronauts into space with each subsequent mission pushing the limits of human spaceflight further. The Manned Lunar Working Group report described the Mercury program as the "beginning of a series of programs of ever-increasing scope and complexity." The early Mercury missions tested human functions in microgravity such as range of motion, testing vision, eating, taking photographs, and performing basic scientific observations. Later missions included longer-duration missions that tested sleeping, worldwide tracking networks, recovery procedures, and communications. The report continued, "The next step after Project Mercury is a Project Apollo. The multimanned Apollo spacecraft will provide for the development and exploration of manned space flight technology in Earth orbit."[20]

Kennedy tasked NASA with completing a lunar landing before the end of the decade. Led by James E. Webb, who had been born in Tally Ho, North Carolina, at the foothills of the Appalachians, NASA also relied on the collaboration of federal and state agencies to fulfil Kennedy's wish. These include the national laboratories that previously had been under the authority of the National Advisory Committee for Aeronautics, the Federal Aviation Administration, the Department of Defense, and the state governments of Florida, Texas, Alabama, Louisiana, and Maryland, to name a few. Extensive federal-state initiatives could bring assistance to areas of the country in need, such as Appalachia. Webb's agenda was "to use science and technology, and now Apollo, to strengthen the United States educationally and economically."[21] Webb expressed this idea to Kennedy in a report jointly drafted with Secretary of Defense Robert S. McNamara titled "Recommendations for Our National Space Program: Changes, Policies, Goals." It detailed how Webb foresaw NASA accomplishing Kennedy's goal and recommended an "immediate initiation of an accelerated program of spacecraft development."[22]

During the Johnson administration, Webb also advocated for the Sustaining University Program, which aimed to increase partnerships between NASA and universities and the local communities surrounding them. Webb argued that a partnership with universities would benefit NASA because the universities would provide innovative research and facilities. One example Webb touted to Johnson was that Rice University had 3,800 acres of land available for potential new research facilities. The universities would receive research labs and prestige. Local communities would benefit economically from the increased interest and activity in the area.[23] Webb recognized the advantage of contracting with companies in different regions of the country and the positive effect their work would have in those regions.[24] In university towns across the South, including the Research Triangle Park region in North Carolina, the high profile and longevity of the Apollo program provided economic prosperity through prolonged government spending.[25] A similar benefit occurred with spending for Appalachian programs. A new and improved highway system across Appalachia increased access to the region, providing new opportunities for business and commerce. Appalachia served as a "national laboratory" for solving regional problems in other areas of the United States.[26]

Appalachian Regional Commission

On May 8, 1961, the eight members of the Conference of Appalachian Governors attended a luncheon at the White House to speak to the president about their efforts to save the region. A statement to the press from the president after the event spotlighted his commitment to address "the economic problems faced by the Appalachian region. Here many of the economic problems that face our nation find special emphasis." President Kennedy directed the newly formed Area Redevelopment Administration to focus on Appalachia, expand the retraining program in the region, ask the Defense Department to review its policy on assigning contracts in areas of substantial unemployment, and set up a special liaison between the Conference of Appalachian Governors and the Area Redevelopment Administration to facilitate suggestions by the governors for the federal government. The statement concluded with Kennedy saying, "The Appalachian Governors are to be complimented upon their resourcefulness in the treatment of unusual multi-State regional problems.

It is the first time an entire section of the Nation has been organized to develop an important regional program of this magnitude."[27] Two weeks later, during his May 25, 1961, "moonshot" special message to Congress, Kennedy reinforced the need for economic advancement in areas such as Appalachia. In his message, Kennedy committed the nation to large-scale federal- and state-run programs aimed at improving the lives of neglected regions of the country. Kennedy clearly felt that Apollo and Appalachia were policy goals of equal weight and composition. Both of these programs were regional-based, benefited local economies through focused government spending, and were urgently needed to support Kennedy's New Frontier.

On April 9, 1963, just seven months before he was assassinated, President Kennedy formed the President's Appalachian Regional Commission (PARC) to administer redevelopment in the region. This joint federal and state program called for "an immediate, or short-run, investment to provide basic facilities and programs not provided in the past but which are essential to the growth of the region and opportunity for its people."[28] There was a sense of urgency about this project, just as there was for Project Apollo. After President Kennedy's death, PARC chair Franklin D. Roosevelt Jr. submitted a plan for President Johnson's approval that by the eight Appalachian governors had already vetted. It would soon to be a part of the Great Society program. Several members of the president's cabinet and heads of federal agencies also approved the plan, including the secretary of defense and NASA head James Webb. Appalachia had turned into a national security issue. The commission divided the plan into three main parts: investment in social overhead, investment in human resources, and investment in economic resources. The plan stated that these three main parts must be balanced and structured to achieve the ultimate goal of "emergency development action though which the Appalachian people may fully join the progress of a growing America."[29]

The report defined Appalachia as a region of "gaps" that were holding all of America back. There was an employment gap. More than 7.1 percent of Appalachians were unemployed, compared to 5 percent of most other Americans. Rampant unemployment had created an income gap. About a third of Appalachians lived below the poverty line (at the time, a family income of less than $3,000 a year). And, directly related to the aerospace dreams of the United States, there was an education gap. Fewer than a third of Appalachians over 25 had a high school diploma, more than 11 percent of people had not finished grade school, and only 5 percent of people in the region were college graduates. The solution was not to impose artificial solutions or throw money at the region but to identify potential hubs of regional and economic growth and support them and the locals around them, drawing rural folk into the modern world. One of the models for these hubs was Huntsville, Alabama, home to the University of Alabama-Huntsville and NASA's new Marshall Space Flight Center.[30]

Media Coverage of Apollo and Appalachia

The glamor of the space program only added to John F. Kennedy's persona as a media president. He used live television to broadcast the 64 press conferences he gave, an average of one every 16 days.[31] Unlike previous administrations, Kennedy recognized the power of directly interacting with the American public through these televised conferences. At the first live conference on January 25, 1961, a reporter asked Kennedy if he thought there was any risk involved with the president speaking live, to which Kennedy replied that "this system has the advantage of providing more direct communication."[32] Magazines and newspapers also recognized the effect of Kennedy's direct communication with the public. *Life* magazine featured Kennedy on the cover seven times during his presidency. Other magazines such as *Time*, *Look*, *Ebony*, *Newsweek*, and the *Saturday Evening Post* also featured Kennedy on their covers. Newspapers and magazines also featured some of his policy issues in prominent ways. A study of *New York Times* articles reveals that Project Apollo garnered much more attention than efforts in Appalachia. On May 2, 1961, front-page stories, in adjacent columns, described the beginnings of the two major federal government programs. One column detailed the announcement of the first Mercury flight that day, which gave President Kennedy the confidence to commit the nation to Project Apollo. The other summarized an event that had happened the day before, May 1, in Washington DC, the passing of the Area Redevelopment Act, which later coalesced into the Appalachian Regional Commission. The paper reported both items as rather tenuous. The Mercury article quoted NASA administrator James E. Webb, who described it as "one of the many milestones we must pass."[33] The author focused on how poorly this American attempt compared to the Soviet's successful orbit one month earlier. The adjacent column recounted the scene at the Oval Office, where Kennedy signed the Area Redevelopment Act, a significant public relations photo opportunity. Kennedy proudly referred to this moment in his young presidency, saying that there was "no piece of legislation that has been passed which gives me greater satisfaction to sign" (fig. 8.1). The bill, which was enacted to last four years, created "a $451 million program in an effort to aid in alleviating the suffering in distressed areas hit by unemployment and underemployment."[34] An Oval Office ceremony of this magnitude,

Figure 8.1. President Kennedy signing the Area Redevelopment Act in the Oval Office on May 1, 1961. Courtesy of Abbie Rowe, National Park Service and John F. Kennedy Presidential Library and Museum.

with dozens of members of Congress, governors, the vice-president, members of the cabinet, and members of the newly created administration, spotlights the significance Kennedy placed on the signing of this bill. Less than a week after this ceremony, Kennedy entertained a similar sized crowd, but this time for the first American in space.

This kind of shared coverage did not remain the rule. For the next several weeks, the pages of the *New York Times* were obsessed with Alan Shepard and his flight, largely ignoring the Area Redevelopment Act. Starting on May 4, 1961, Shepard appeared on the front page on five consecutive days surrounding his Mercury flight. Articles spoke about America's success with the mission, compared the US space program favorably with that of the Soviet Union, and even included an article titled "Symbol of the Nation: Shepard Restores the Capitol's Faith in Virtues of the American People." Photos of the ceremony at the Rose Garden at which President Kennedy awarded Shepard the NASA Distinguished Service Medal, photos of Shepard's motorcade down Pennsylvania Avenue, and a complete transcript of Shepard's press conference all made the front page (fig. 8.2). These articles touted Shepard as a hero, proclaimed him a national icon, and lauded the US space program as successful. Little attention was paid to the Council of Appalachian Governors, which was facing poverty, hunger, and unemployment in their own backyards. Only a few articles appeared on new aid to the region. They were buried deep in the paper, framed by advertisements for orange juice and clothing, cars and cameras. The governor of Kentucky, Bert Combs, was asked by the *New York Times* about how much money the governors required for their region and he replied, "We want to get all that we can, because we believe it will be an investment for America."[35] Much like Webb's reasoning that the space program's benefits extended far beyond the aerospace industry, Combs saw that direct investment into the Appalachian region had benefits that reached farther than just his state. Ultimately, the program was an investment for the entire United States.

Media coverage of Kennedy's May 25, 1961, Special Message to Congress on Urgent National Needs, the moonshot speech, prioritized the high-flying Apollo mission over groundbreaking Appalachia initiatives. The *New York Times* headlined President Kennedy's request for "An Urgent Effort to Land on the Moon" in full capital letters. The potential of the United States to beat the Soviet Union was front-page news despite Kennedy's focus on more pressing issues. The paper covered other topics Kennedy highlighted in his speech, including a civil defense initiative for the construction of fallout shelters and more funds for military spending in defense against Communism. Most articles made no mention of the Appalachian region, unemployment, or poverty despite Kennedy's prioritization of "Economic and Social Progress at Home." Only certain aspects of Kennedy's speech warranted a front page, above-the-fold article. The *Times* mentioned Appalachia, but only in the final two sentences of the very last paragraph of a three-column article.[36] It was clear where the media, and perhaps national public opinion, placed its ultimate priorities. Apollo was front and center; Appalachia remained an afterthought.

The *Times* reported criticisms of both programs, but not surprisingly, Project Apollo received far less in the early 1960s. Journalists spotlighted the newness and suddenness of spaceflight, noting that "scientists and a public at large [are] still uncertain and debating over the objectives, value and urgency of man's exploring space." They reminded readers just how "modest" and "belated" Shepard's flight was in comparison with Gagarin's. Shepard's flight was "magnified beyond its intrinsic importance. Perhaps inevitably, it was cast as America's answer to the Soviet feat, which it was not."[37] Articles like this that criticized the space program were not common during NASA's infancy. Most of them praised and celebrated the federally funded missions.

Figure 8.2. President Kennedy awards Alan Shepard NASA's Distinguished Service Medal in the Rose Garden on May 8, 1961. Courtesy of Abbie Rowe, National Park Service and John F. Kennedy Presidential Library and Museum.

Appalachian redevelopment programs were not so fortunate. Criticism and scrutiny abounded. Opponents of the programs called on participants to voluntarily pull out and "undertake development on their own." William L. Batt Jr. of the Department of Commerce defended the program. Yes, he argued, there were areas not yet receiving funds. Batt also noted that funds had been allocated in order to prioritize projects and direct aid in the most efficient way possible. "This is a new program," stated Batt. "Every project we undertake now sets a precedent. We take a long hard look at everything." He added, "Every community that wants help first has to develop its own 'overall economic development plan,' go to its state development officials and then come to us. We don't go to them." Finally, he said, "Critics say this is red tape. But we don't want to invest Federal or local money on projects that turn out to be useless."[38] On one crucial topic, however, critics did scrutinize the excessive speed by which both of these federal programs progressed. With Apollo, it was a seemingly impossible game of catch-up to the Soviet Union; with the

Area Redevelopment Act, it was either too much help to some areas, too little help to others, or not fast enough help overall.[39]

In 1961 and 1962, *Life* magazine put astronauts, space travel, or X-15 rockets on eighteen of their magazine covers. Most notably, *Life* had an exclusive contract with the Mercury 7 astronauts to interview and share their lives with the American public. *Life* featured the seven astronauts in one issue and the astronauts' wives in the next. The articles, with titles such as, "The Astronauts— Ready to Make History," and "I Know It Can Be Done and I Want To Do It," which was about Alan Shepard, showed both color and black-and-white photos of the astronauts training. Each astronaut was featured in an article that described their likes, dislikes, relationship status, background information, military records, and hobbies in a sort of baseball card fashion. Photographs also featured the astronaut families and their tidy middle-class homes.[40]

In contrast, *Life* published only one article about the Appalachian redevelopment programs, a 1964 piece titled "The Valley of Poverty."[41] The featured article contained several detailed full-page photographs and smaller articles that documented the region's plight. The differences between the photos of the astronauts and those of Appalachia are striking. One photo shows Alan Shepard relaxing with his family on his patio answering fan mail. Another article and series of photos shows John Glenn's entrance into politics. Another shows an immaculate-looking Rene Carpenter, wife of astronaut Scott Carpenter, counseling their daughter Kristen to mind her manners. The Appalachian photos are dramatically different. One photo shows Mrs. Delphi Mobley in her shack cradling her daughter, who was suffering from measles. Other Appalachian photos depict children scavenging on railroad tracks for dropped coal to heat their homes or newspapers lining walls of shacks for added insulation. The message to American readers could not have been clearer. These photos exposed our national subconscious: the hopeful Apollo and the desperate Appalachia. The stark contrasting vision of American life offered two choices: take the moonshot or fall back into moonshine.

Divergent Legacies

The legacies of these two programs reverberate into the present. President Kennedy's 1961 address rallied the nation around them, one to accelerate the exploration of space, to regain national prestige after losing the space race to the Soviets, and the other to heal a broken region whose people were suffering daily and required immediate assistance. The government spent a significant amount of money on each: about $25 billion for Apollo and at least $15 billion for the initial Appalachia programs.[42] America responded with federal and state leadership to carry out both program goals, But with a difference. Apollo was canceled in victory after achieving six lunar landings, while the Appalachian Regional Commission still exists today, its mission not yet completely fulfilled, although it made significant strides through the 1960s and 1970s. By 2015, the region still lagged behind the rest of the nation, but the poverty rate had been cut almost in half, the unemployment rate had been lowered, and the high school graduation rate had been increased, as have the number of degrees in higher education.[43] As of 2016, the Appalachian Development Highway System, which began in 1965, has increased the amount of roads into the region by 88 percent.[44] The number of counties that the ARC labels as distressed fell from 216 in 1960 to just 78 in 1980. As of 2018, the number of counties was 84.[45] Some of the current initiatives by the ARC include combating the opioid epidemic, increasing access to high-speed broadband, and fighting COVID-19.[46] With federal support, Appalachians closed some of the gaps just as new problems presented themselves.

In American popular culture today, we celebrate the Apollo anniversaries widely on the internet. Monographs and biographies recall the achievements and struggles of the Apollo program.[47] Theaters bring in large crowds for blockbuster movies like *Apollo 13* and *First Man*. Historians and sound engineers have re-created the audio of the Apollo missions so we can follow along and emulate them in real time.[48] We remember the Appalachian initiative with books like J. D. Vance's *Hillbilly Elegy* and Steven Stoll's *Ramp Hollow: The Ordeal of Appalachia*, works that remind readers of past and present failures in the region.[49] Movies like *Deliverance* and television shows like *The Beverly Hillbillies* and *Hee Haw* reinforce the worst stereotypes, while comedians Larry the Cable Guy and Jeff Foxworthy embrace them. Regardless of the rather superficial and divergent interpretations, there are grounds that help bridge the two programs in retrospect.

It is important to remember how Americans came together in the space program, people of all backgrounds from both the North and the South. Norman Mailer once described his experience watching the launch of Apollo 11 in these terms: "It was hardly just middle-class America here tonight, rather every echo of hard trade-union beer-binge paunch-gut-and-muscle, and lean whippy redneck honky-tonk clans out to bird-watch in the morning with redeye in the shot glass. . . . One felt the whole south stirring on this night."[50] Homer Hickam's memoir *Rocket Boys*, which was later made into the feature film *October Sky*, told the story of the burgeoning US space program of the 1950s *and* of his beloved and broken Appalachia. A young boy and his friends growing up in Coalwood, West Virginia, inspired by Sputnik, begin to build amateur rockets and win a science fair contest, an experience that ultimately led Hickam to a career at NASA. One of their rockets just happened to be propelled by what the boys called zincoshine, a combination of zinc, sulfur and what Hickam described that "fiery liquid . . . one hundred percent pure two-hundred proof alcohol . . . the best corn likker in the county"—moonshine.[51]

There is also the story of Charlie Duke, born at the edge of the Appalachians, in Charlotte, North Carolina. Serving as capsule communicator (capcom) during the Apollo 11 lunar landing in his unmistakable Appalachian accent, Duke delivered one of the most famous lines during the entire mission. "I was so excited, I couldn't get out 'Tranquility Base.' It came out sort of like 'Twangquility,' you know. And so it was, 'Roger, Houston. Twangquility Base here. . . . We've got a bunch of guys about to turn blue. But we're breathing again."[52] Duke later served as lunar module pilot for Apollo 16, becoming the tenth person to walk on the moon. Prior to the mission, Duke specifically requested to have the southern staple of grits on board.[53] Perhaps the Appalachian folklore about the moon is true: Duke was born on October 3, 1935, under a crescent moon.

Notes

1 John F. Kennedy, Special Message to Congress on Urgent National Needs, May 25, 1961, Archives, John F. Kennedy Presidential Library and Museum, https://www.jfklibrary.org/asset-viewer/archives/JFKPOF/034/JFKPOF-034-030.

2 For recent studies of these questions, see Neil Maher, *Apollo in the Age of Aquarius* (Cambridge: Harvard University Press, 2017); and Matthew Tribbe, *No Requiem for the Space Age: The Apollo Moon Landings and American Culture* (New York: Oxford University Press, 2014).

3 Robert Dallek, *An Unfinished Life: John F. Kennedy 1917–1963* (Boston: Little, Brown, 2013), 224; Kennedy, "Special Message to Congress on Urgent National Needs."

4 John F. Kennedy, Speech at Blue Key Banquet, University of Florida, Gainesville, Florida, October 18, 1957, Archives, John F. Kennedy Presidential Library and Museum, https://www.jfklibrary.org/archives/other-resources/john-f-kennedy-speeches/university-of-florida-19571018.

5 Christopher A. Preble, "'Who Ever Believed in the "Missile Gap"?' John F. Kennedy and the Politics of National Security," *Presidential Studies Quarterly* 33, no. 4 (2003): 801–826.

6 John F. Kennedy, "An Investment for Peace," speech on Senate floor, February 29, 1960, Archives, John F. Kennedy Presidential Library and Museum, https://www.jfklibrary.org/asset-viewer/archives/JFKSEN/0906/JFKSEN-0906-035.

7 Richard Drake, *A History of Appalachia* (Lexington: University Press of Kentucky, 2001), 207.

8 John F. Kennedy, "Aid to Depressed Areas," speech in Huntington, West Virginia, April 20, 1960, Archives, John F. Kennedy Presidential Library and Museum, https://www.jfklibrary.org/asset-viewer/archives/JFKSEN/0908/JFKSEN-0908-020.

9   John F. Kennedy, campaign speech at Huntington, West Virginia, May 6, 1960, Archives, John F. Kennedy Presidential Library and Museum, https://www.jfklibrary.org/asset-viewer/archives/JFKSEN/0909/JFKSEN-0909-013.

10 John F. Kennedy, Acceptance Speech at the 1960 Democratic National Convention, Los Angeles, California, July 15, 1960, Historic Speeches, John F. Kennedy Presidential Library and Museum, https://www.jfklibrary.org/learn/about-jfk/historic-speeches/acceptance-of-democratic-nomination-for-president.

11 Lyndon B. Johnson to John F. Kennedy, May 8, 1961, Box 30, Folder 19, President's Office Files 1961–1963, The John F. Kennedy Library, Boston, MA.

12 Harry M. Caudill, *Night Comes to the Cumberlands: A Biography of a Depressed Area* (Boston: Little, Brown, 2017), 12, 39.

13 John Cheves, "At 90, Anne Caudill Talking to New Generations about Her Family's Appalachian Experience," *Lexington Herald Leader*, April 26, 2014.

14 Caudill, *Night Comes to the Cumberlands*, 77. See also Carolyn Clay Turner, *John C. C. Mayo, Cumberland Capitalist* (Pikeville, KY: Pikeville College Press, 1983), 144.

15 Caudill, *Night Comes to the Cumberlands*, 10.

16 "ARC's History and Work in Appalachia," Appalachian Regional Commission, https://www.arc.gov/about/ARCHistory.asp; David Walls and John B. Stephenson, *Appalachia in the Sixties: Decade of Reawakening* (Lexington: University Press of Kentucky, 1972), 25.

17 Caudill, *Night Comes to the Cumberlands*, 33–37. Caudill included a whole chapter, "Moonshiners and Mayhem," on the war between moonshiners and the law during Prohibition. The terms moonshot and moonshine were originally used interchangeably to refer to moonlight. The term moonshine came to mean the luck of "a spirit that was illicitly produced outdoors, under the light of the Moon"; Kevin Kosar, *Moonshine: A Global History* (London: Reaktion Books, 2017), 14–17. See also Daniel S. Pierce, *Tar Heel Lightnin' How Secret Stills and Fast Cars Made North Carolina the Moonshine Capital of the World* (Chapel Hill: University of North Carolina Press, 2019).

18 See John Logsdon, *The Decision to Go to the Moon: The Apollo Project and the National Interest* (Cambridge: MIT Press, 1970); and Roger D. Launius and Howard E McCurdy, eds., *Spaceflight and the Myth of Presidential Leadership* (Urbana: University of Illinois Press, 1997).

19 John M. Logsdon and Roger D. Launius, *Exploring the Unknown: Selected Documents in the History of the U.S. Civil Space Program*, vol. 7 (Washington, DC: NASA, 1995), 458.

20 Logsdon and Launius, *Exploring the Unknown*, 459.

21 W. Henry Lambright, *Powering Apollo: James E. Webb of NASA* (Baltimore, MD: Johns Hopkins University Press, 1998), 99; James E. Webb, *Space Age Management: The Large-Scale Approach* (New York: McGraw-Hill, 1969).

22 Lambright, *Powering Apollo*, 2. See also the record of Lyndon Johnson's involvement in the folder Johnson, Lyndon B., 1961: January–May, John F. Kennedy Presidential Library and Museum, https://www.jfklibrary.org/asset-viewer/archives/JFKPOF/030/JFKPOF-030-019?image_identifier=JFKPOF-030-019-p0024.

23 Lambright, *Powering Apollo*, 99–100.

24 Walter A. McDougall, *The Heavens and the Earth: A Political History of the Space Age* (Baltimore, MD: Johns Hopkins University Press, 1997), 382.

25 Lambright, *Powering Apollo*, 137.

26 Appalachian Regional Commission, *Annual Report, 1969* (Washington, DC: n.p., 1969), 4.

27 "Statement by the President Following a Meeting with the Conference of Appalachian Governors, May 8, 1961" in *Public Papers of the Presidents of the United States: John F. Kennedy: Concerning the Public Messages, Speeches, and Statements of the President, January 10 to December 31, 1961* (Washington, DC: Government Printing Office, 1962), 365.

28 President's Appalachian Regional Commission, *Appalachia: A Report by the President's Appalachian Regional Commission, 1964* (Washington, DC: GPO, 1964), II.

29 President's Appalachian Regional Commission, *Appalachia*, III.

30 Walls and Stephenson, *Appalachia in the Sixties*, 35. See also Monique Laney, *German Rocketeers in the Heart of Dixie: Making Sense of the Nazi Past during the Civil Rights Era* (New Haven, CT: Yale University Press, 2015).

31 "John F. Kennedy and the Press," n.d., John F. Kennedy Presidential Library and Museum, https://www.jfklibrary.org/learn/about-jfk/jfk-in-history/john-f-kennedy-and -the-press.

32 John F. Kennedy, News Conference 1, January 25, 1961, John F. Kennedy Presidential Library and Museum, https://www.jfklibrary.org/archives/other-resources/john-f -kennedy-press-conferences/news-conference-1.

33 Richard Witkin, "U.S. Space Flight Scheduled Today," *New York Times*, May 2, 1961.

34 W. H. Lawrence, "President Signs Bill to Increase Needy-Area Jobs," *New York Times*, May 2, 1961; Legislative Summary: Economy & Finance, 1961, John F. Kennedy Presidential Library and Museum, https://www.jfklibrary.org/archives/other-resources/ legislative-summary/economy-finance.

35 "Kennedy Pledges Appalachian Aid: Confers with 8 Governors on Moves to Spur Jobs," *New York Times*, May 9. 1961. The same dynamic, with the spotlight on Shepard, appeared in the West Virginia newspapers. See UPI, "For Aid to Appalachian Region: Governors Draft Plans to Present to Kennedy," *Raleigh Register* (Beckley, WV), May 8, 1961.

36 W. H. Lawrence, "President to Ask an Urgent Effort to Land on Moon," *New York Times*, May 24. 1961; and W. H. Lawrence, "Kennedy Asks 1.8 Billion This Year to Accelerate Space Exploration," *New York Times*, May 26, 1961.

37 John Finney, "The Race in Space Still Has a Long Way to Go," *New York Times*, May 7, 1961.

38 "Hodges Criticizes Development Aid," *New York Times*, October 4, 1961; Peter Braestrup, "First Job Training Program Set Under '61 Area Development Act," *New York Times*, October 12, 1961; Peter Braestrup, "Needy-Area Head Defends Program," *New York Times*, October 16, 1961.

39 "Red Tape Delaying Jobless Retraining," *New York Times*, November 10, 1961, 21; "West Virginians Retrain Miners: State Program Shows Hope for U.S. Test Project," *New York Times*, December 9, 1961.

40 "Seven Brave Women Behind the Astronauts," *Life*, September 21, 1959, 142–163.

41 John Dominis, "The Valley of Poverty," *Life*, January 31, 1964, 54. See also Ben Cosgrove, "War on Poverty: Portraits from an Appalachian Battleground," https://www.life.com/history/war-on-poverty-appalachia-portraits-1964/.

42 For Apollo program funding, see Linda Neuman Ezell, *NASA Historical Data Book Volume II: Programs and Projects 1958–1968* (Washington D.C.: U.S. Government Printing Office, 1988), 121. For funding for Appalachia, see Ben A. Franklin, "Saving Appalachia: Was $15 Billion Well Spent?" *New York Times*, September 27, 1981.

43 Franklin, "Saving Appalachia." See also Center for Regional Economic Competitiveness and West Virginia University, *Appalachia Then and Now: Examining Changes to the Appalachian Region since 1965* (Washington, DC: Appalachian Regional Commission, 2015), 5–7, 9, https://www.arc.gov/assets/research_reports/AppalachiaThenAndNowCompiledReports.pdf.

44 Economic Development Research Group, Inc., "Appalachian Development Highway System Economic Analysis Study: Synthesis of Findings to Date," Economic Development Research Group, Boston, 2016, 4. https://www.arc.gov/report/appalachian-development-highway-system-economic-analysis-study-synthesis-of-findings-to-date/.

45 "Classifying Economic Distress in Appalachian Counties," Appalachian Regional Commission, October 1, 2020, https://www.arc.gov/classifying-economic-distress-in-appalachian-counties/.

46 Kelsey Souto, "Major Broadband Developments Announced," https://www.wsaz.com, October 14, 2020, WSAZ NewsChannel 3, https://www.wsaz.com/2020/10/14/major-broadband-developments-announced/.

47 See Douglas Brinkley, *American Moonshot: John F. Kennedy and the Great Space Race* (New York: Harper Collins, 2019); Jay Barbree, *Moon Shot: The Inside Story of America's Apollo Moon Landings* (New York: Open Road Integrated Media, 2011); James Donovan, *Shoot for the Moon: The Space Race and the Extraordinary Voyage of Apollo 11* (New York: Little, Brown and Company, 2019).

48 See Ben Feist's website Apollo in Real Time, https://apolloinrealtime.org/.

49 Note also the retrospective article about Caudill's book *50 Years of Night* (2014) in the *Lexington Herald Leader*. Caudill's wife and research partner recounted that "thousands flocked to Eastern Kentucky to take the 'poverty tour' or volunteer their services. And everyone wanted to meet Harry Caudill." See Cheves, "At 90, Anne Caudill Talking to New Generations."

50 Norman Mailer, *Of a Fire on the Moon* (New York: Random House, 1969), 59.

51 Homer Hickam, *Rocket Boys* (New York: Random House, 2000), 269–279.

52 Transcript of Charles M. Duke Jr., interview with Doug Ward, Johnson Space Center Oral History Project, March 12, 1999, page 12-24, https://historycollection.jsc.nasa.gov/JSCHistoryPortal/history/oral_histories/DukeCM/CMD_3-12-99.pdf. See also Eric M. Jones, "The First Lunar Landing," Apollo 11 Lunar Surface Journal, https://www.hq.nasa.gov/alsj/a11/a11.landing.html.

53 Eric M Jones, "Wake-Up for EVA-1," Apollo 16 Lunar Surface Journal, https://www.hq.nasa.gov/alsj/a16/a16.eva1wake.html.

9

Apollo and the Creation
of the Clear Lake Community

JENNIFER ROSS-NAZZAL

In the spring of 1961, President John F. Kennedy committed the nation to sending a man to the moon and returning him safely to Earth. "It will not be one man going to the Moon," he explained. "It will be an entire nation. For all of us must work to put him there." The plan, the president admitted, was a "heavy burden" and the United States must be willing to shoulder the load to achieve this goal.[1] Only a few months later, NASA announced plans to build a center in Houston. Employees and wives of the Space Task Group, the pioneering team that managed Project Mercury at the Langley Research Center in Hampton, Virginia, accepted the president's challenge, uprooted their families, and moved to the Gulf Coast of Texas, an area that had recently been ravaged by Hurricane Carla. Over the course of the decade, others from across the nation and even around the globe followed. Those who moved to Clear Lake were "excited about a new kind of frontier" and the creation of "a new hometown."[2]

This chapter explores the impact of Project Apollo and NASA's role in the creation of the Clear Lake community, the area surrounding Houston's Manned Spacecraft Center. In Houston and across the country, "men of science and industry, academics and engineers, politicians and dreamers, in short, people from all walks of life, united in an unlikely alliance to reach the Moon. It was a time like no other."[3] Those who lived in Clear Lake echoed this idea repeatedly, even those who did not work for the space agency. As scientist-astronaut Edward G. Gibson recalled, Clear Lake and the NASA area were "a certain little community that you just don't run into again. Most of the environments you live in, people are aloof, and that was not true here."[4] The thrill of working toward the president's goal served as a unifying force for the community.

In one sense, Clear Lake was similar to other suburbs across the United States that had "neither history, tradition nor established structure, no inherited customs, institutions, socially important families or big houses."[5] But at Clear Lake there was one important difference: the region was home to the astronauts, the celebrities of the Space Age. The connections between the space program and the region were conspicuous mainly because employees outnumbered locals. NASA was the dominant influence in the region. Interest in the space program was reflected in the overwhelming media spotlight the astronauts, their wives and families, and their neighborhoods experienced. Art in the area reflected the excitement of Project Apollo. Residents established neighborhood and community traditions around space missions. NASA's influence even extended to local churches.

In many ways, however, the residents of the neighborhoods surrounding the spacecraft center were not all that unique. Those who moved there were like other Americans, people moving into suburbs across the country created new communities and established their own traditions. Like citizens elsewhere, Clear Lake transplants found the quality of services and institutions in the area lacking and they came together to build churches and demand better schools.[6]

In 1960, the year before the decision to build a center in Texas, the area southeast of Houston now known as Clear Lake had only 6,500 people. The region was sparsely settled, and most of the homes that now surround NASA did not exist. Soon after NASA released its plan to build a center in Houston in September 1961, builders purchased land formerly tended by farmers and ranchers. Developers built homes, schools, and churches on the former Pearson 1776 Ranch, located across the street from NASA, now known as the city of Nassau Bay. Around the same time, the Del E. Webb Corporation and the Humble Oil and Refining Co. announced plans to build a new development called Clear Lake City next to the center (fig. 9.1). It was the first master planned community in Texas.[7] In only a few years, as aerospace firms relocated to the Houston area, the technical, scientific, and engineering population outnumbered locals who called the nearby small towns of Kemah, Seabrook, Webster, Friendswood, and League City home. By 1966, the population had swelled to 33,000, forever changing the landscape and the area's rural communities.[8]

The dream of working for the space program lured many to the Manned Spacecraft Center. Newcomers came from across the country. Few recall any locals in the new neighborhoods. People remember that everyone on their block was a newcomer to Texas. Only a handful of people knew any of their new neighbors. There was no one from the Welcome Wagon to share infor-

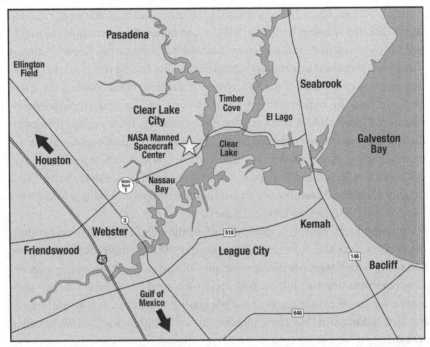

Figure 9.1. Map of the Clear Lake region in Harris County, Texas. Map created by Cynthia Bush.

mation about the area, so newcomers turned to their colleagues or members of their church for suggestions on locating a doctor, a hairdresser, and the best places to shop. With such a small population, options were limited. People leaned on each other and "learned to share," recalled Lockheed employee Bill Strait.[9]

Most civil servants and contractors who worked at NASA were men and women of faith who attended church. Churches played an important role in building fellowship and friendships. Astronauts, flight controllers, managers, engineers, scientists, secretaries, and technicians needed a spiritual home, and many NASA employees founded local churches. In addition, Webster Presbyterian Church, an established congregation, provided a ready-made community for newcomers, fostered fellowship among couples, and even arranged lectures and special events.[10]

The response of locals to the influx of new residents and the building of the NASA center is surprisingly difficult to track. NASA retiree William A. Larsen believed that Webster and League City residents thought that "we would land on the Moon, maybe do it a couple times more, the place would close up, everybody would go away, and then everything would come back to normal."

Houston had experienced this several times with Ellington Air Force Base. The facility opened when a need arose and then closed.[11] Some locals might have seen this as a new start. Some in the community might have wanted an opportunity to find a local job that paid higher wages. But not everyone was happy, as transplant Betty Ulrich explained in a letter to her mother. "Old timers in this area resent, naturally, that so many people are flocking in." They were not opposed to the money flowing into the area, but things were moving too quickly and "they are getting cantankerous."[12] But what could they do? The federal government had decided to build a NASA center in Clear Lake, and that ignited the explosion that led to a building boom and a population boom.

Clear Creek Independent School District initially struggled to meet the educational needs of the children of new residents. In the 1964–1965 school year, more than 1,000 new students arrived. When the school year ended in May, the district identified an additional 1,200 students that would be attending the following year, so schools had to be creative with limited space. In Clear Lake City, the local school was still being built and 110 students had to attend class in a recreation center until Clear Lake City Elementary opened. Classes were held in the lounge rooms of the center and the district brought in four trailers for additional classes. Although the temporary classrooms were air conditioned, not all of the district schools were.[13] Clyde Mason, the assistant superintendent of schools for the district, remarked that the new parents recognized the need for and demanded a better school district.[14]

Friendships blossomed quickly in the area. All sorts of get-togethers were common in the Apollo years: cocktail parties, impromptu block parties, coffee klatches, and teas. Community dances also brought people together. An advertisement for Clear Lake City in a local newspaper emphasized the feeling of neighborliness in the area: "Everyone's so friendly in this community, it's contagious. Newcomers aren't strangers here for long."[15] Families involved in Project Apollo held mission parties in their homes. At the end of a successful flight, they had splashdown parties to celebrate. The layout of neighborhoods also fostered friendships. In Nassau Bay, for instance, many of the backyards connected and allowed families to visit each other without going through the front door. Astronaut Buzz Aldrin's property adjoined the property of seven other neighbors, including that of fellow astronaut Charles A. Bassett. The gate between the Aldrin and Bassett backyards led to a solid friendship between the two families.[16]

As the men worked on Apollo, women built the community during those formative years. They served as "the telephoners, organizers, and arrangers of community life."[17] They pushed for better schools. Mothers shuffled kids to and from swim meets. They participated in the Parent Teacher Association, taught Sunday school, and volunteered at the library. They often bonded as they watched their children at activities such as the neighborhood swim meets. They joined social clubs that included the Lockheed Ladies and held fashion shows, played bridge, and hosted luncheons. They babysat for one another.

There were few cultural outlets like theaters in the area, so wives had to travel to nearby communities for entertainment or establish new spaces. Joan Aldrin, a mother of three and the wife of astronaut Buzz Aldrin, had studied theater in college and earned a master's degree in drama from Columbia University before meeting her husband. She joined the Pasadena Little Theater. In 1965, she helped found the Clear Creek Country Theatre, which continues to this day. Jane Conrad, a mother of four boys and the wife of astronaut Charles "Pete" Conrad, was also active in theater. She remembered doing "everything from cleaning the toilets to being president of the board."[18] Conrad starred in a few plays and helped establish an educational program for children about theater.[19]

Kids became friends in the usual activities: Boy Scouts, Girl Scouts, riding bikes, swimming, diving, and playing in the neighborhood. They participated in youth groups in local churches. Their parents, including some NASA fathers, played a role in these activities. Astronaut John Glenn headed a Boy Scout troop. Gerald P. Carr and his wife JoAnn headed up the youth group for high schoolers at Webster Presbyterian Church. When he was in orbit for Skylab IV, the kids in the group had a chance to talk with him and wish him a merry Christmas.[20]

In a young community with few public gathering places and little formal entertainment, NASA often filled that void. In 1965 and 1966, the center sponsored a variety show in their auditorium. Members of the Manned Spacecraft Center Charm Club served as usherettes for the 1965 program and NASA photographer Terry Slezak sang opera.[21] The club, which counted secretaries and wives of astronauts among its members, held numerous activities such as space-age fashion shows and Hawaiian luaus. The Employees Assistance Program planned dinners, dances, and picnics for employees and maintained a list of monthly activities for children.[22] The space center held annual picnics at Galveston County Park in League City.[23]

In 1966, to celebrate the first rendezvous in space, Clear Lake established the Festival of the Ten Days to Rendezvous (now known as the Lunar Rendezvous Festival). Numerous events aimed to include everyone in the fun, even man's best friend. Activities for families included a canoe race, a parade, and swim relays. Of special note was a water ballet arranged by astronaut wives Janet Armstrong and Beth Williams.[24]

The people working on Project Apollo were young, and the community reflected the demographics of the center. Many came to Texas with pregnant wives and young children. Others who moved there in support roles were willing to stake out new lives for themselves on the frontier of space exploration. Charter members of the Clear Lake United Methodist Church, which included NASA employees and contractors, remember their congregation as youthful, enthusiastic, energetic, dedicated, outgoing, and willing to tackle new projects.[25] Ed Gibson summed up the community of the time, "Everything was young, vibrant, make it happen, we're all in this together."[26]

Neighborhoods were close-knit. There was a sense among the young residents of the area that they could accomplish anything, and the community reflected NASA's can-do spirit. The newest residents of the Timber Cove neighborhood "moved in about the same time with children, pets, station wagons and a common cause," Jane Conrad wrote.[27] In those early years, Clear Lake citizens felt that the area was a supportive community.

Life reporter Dora Jane Hamblin wrote that residents of the Clear Lake suburbs described their community as "Togethersville." It was a "close-knit, protective society which has grown up around the front men of the Space Age. Tradesmen treat them with respect, plumbers plumb at cost, tourists gawk, and every astronaut and his family are at the instant disposal of any other astronaut and his family who may need them in in a small crisis or an overwhelming tragedy. Togethersville can be a warm and loving family or just a great gelatinous blob which closes around and smothers the individual."[28] In other words, Togethersville represented both the best and the worst of the suburbs.

James Schefter, a *Time-Life* reporter who lived in El Lago, invited Hamblin to a backyard barbecue in the neighborhood to "see the feeling that we have here that I don't think you'll find other places." Hamblin cringed at the homogeneity of the place and encouraged her hosts to recruit new residents. "You are too all alike," she said. El Lago resident Betty Geehan always remembered that comment and disagreed "because we liked our life the way it was." Young families purchased homes, raised children, and built a life in their new neighborhoods. Men supported their family while the women stayed at home. Everyone seemed to be at the same stage of life: people having children and raising kids. There were so many families moving in with young children that one street in El Lago had eighty kids.[29] For Geehan, it was a "utopia-type life."[30]

The neighborhoods looked like most other suburbs at the time. Most were solidly middle class. People of color were not part of Clear Lake communities.[31] But there were rumors of "unsuccessful attempts to probe the willingness of developers to let them come in."[32]

These neighborhoods, astronaut Eugene Cernan recalled, were some "of the most unique in America—all Program." Nassau Bay, Timber Cove, and El Lago drew the attention of tour buses and visitors from across the country who were eager to get a snapshot of an astronaut or an astronaut's home. Curious drivers would stop and oftentimes unknowingly asked astronauts or their family members for directions. Astronauts often got the last laugh by sending them in the opposite direction. "If we weren't in space suits," Cernan wrote, "no one recognized us."[33]

Wives, on the other hand, could not avoid the media spotlight when their husband was in orbit. Preparations for TV coverage began in earnest before a mission launched. The networks erected transmitter towers in the neighborhoods and a "tent city" outside astronauts' homes to broadcast the latest reports from the homesteads of America's space pioneers.[34] Traffic congestion during flights was a major issue, especially for residents of El Lago; there was only one way in and out of their neighborhood and the streets blocked off due all the media attention.[35]

Life reporters and their photographers sat with families in their homes to capture an inside story.[36] For example, during Gemini V, Miguel Acoca of *Life* reported on activities at the Conrad house. Any other wife might have been horrified by what occurred by the end of the mission. The flight had been a challenging eight days for Jane Conrad. Their kitten kept having accidents on the floor, and her youngest son had poured some of her perfume on a particularly stinky pile. Their maid called and said she could not come to work because she had been stabbed in a fight, and their dog was sick. On top of it all, in the middle of the flight, the air conditioner broke. Conrad temporarily escaped from the situation for a day. When she returned, she sat in front of the squawk box.[37] She needed to hear Pete's voice. Later she told her sister that it reminded her of being pregnant, when you wonder if the baby is still alive. "You know he's alive," she said, "but you sit in fear and wait for him to kick, and only then are you reassured. That's the way I felt. I rushed back home and turned on the squawk box and begged it to let me hear Pete's voice."[38] Ococa asked if he could include the story in the magazine. "I was horrified," she remembered. "Personal stories yes—but THAT personal?"[39]

The Manned Spacecraft Center community worked long hours to meet the deadline Kennedy had set for a lunar landing. Everyone at the center, from astronauts to secretaries, had "10 jobs to do and they had to have them done tomorrow, and they spent all of their waking hours doing it," Flight Director Christopher Kraft remembered. "Even when they were at home having dinner at night they were thinking about getting their job done."[40] Secretaries in the Astronaut Office often babysat for the astronaut families and updated manuals and checklists at home.[41] William Larsen, who moved to Houston in 1965, remembered working thirty-six hours on his first shift. Most men often worked on Saturdays, and even when they were at parties or other get-togethers, they talked about the program. No one thought that their work schedules were inconvenient. On the contrary, most "seemed to think it was worth the inconvenience and some of the contention that inevitably came out." Men gave up vacations and received promotions as incentives to "work more."[42] People poured their hearts and souls into the program. Buzz Aldrin called NASA employees "'workaholics,'" adding that "we had to be, even in our social life."[43]

Everyone living in the area felt a connection to the space program and people often highlighted their connections with its celebrities, the astronauts. The Clear Lake City Newsletter welcomed and listed the new astronauts who chose to buy homes in the neighborhood.[44] One local family, the Ulrichs, often highlighted their connections to the spacefarers, like the time their son George made friends with John Glenn's children. Glenn was a member of their church.[45] The Ulrichs were especially thrilled when NASA named Aldrin, who also belonged to their church, to the Apollo 11 crew, noting that he "will probably be the first man on the Moon!"[46]

One of the most distinguishing characteristics of the area was the unification behind the goal President Kennedy had set at the beginning of the decade. Astronaut Walter Cunningham summed it up best, "We never lost sight of one thing: *the mission always came first*. Where it counted, deep in our guts, nothing else existed. Everything else was parsley."[47] The center and its surrounding community had a unified sense of purpose—to achieve the goal of landing a man on the moon by the end of the decade. Everyone working for the space center had a desire to get a man to the moon, as Marlowe Cassetti, chief of the Guidance and Performance Branch, explained: "There was this intense feeling of teamwork" because NASA had "a common goal."[48] Employees, regardless of their role in the organization, recognized how being part of the Manned Space Center team would enable them to achieve the extraordinary.[49]

Employees were not the only ones committed to achieving the goal; wives and families were part of the effort. "Everybody had one goal in mind. That was to eventually send man to the Moon," Jan Evans, wife of astronaut Ronald E. Evans, remembered. "Everybody felt a part of this program."[50] The moon landing was "what we were all here for," JoAnn Carr said. Families were excited about the moon missions, and when NASA canceled Gerald Carr's Apollo flight, the couple was heartbroken.[51] Janet Armstrong, wife of astronaut Neil A. Armstrong, told historian Jim Hansen, "Our lives were dedicated to a cause, to try to reach the goal of putting a man on the Moon by the end of 1969. It was an all-out effort on everyone's agenda. It wasn't just our astronaut families that had put our lives on hold; thousands of families were in the same mode."[52] NASA was their focus and their lives. As one engineer's wife told Mary Wiegers of the *Washington Post*, the space agency has become "our god, our government, and our goal."[53] Wives and the local community supported the center "in every way they knew how," Jane Conrad recalled. "It was an exhilarating time and place to live."[54] There were plenty of exciting events and moments, but spaceflight—the launches, spacewalks, landings, and time apart from family—could also be scary and stressful. As Barbara Cernan, wife of moonwalker Gene Cernan, once said, "If you think going to the Moon is hard, try staying home."[55]

Mothers shopped for food, did the laundry, mowed the lawn, managed the family finances, and oversaw the construction of homes in Clear Lake. They provided stability for families, giving their husbands the space to focus on the moonshot. Moms provided routines—getting the kids ready for school, making dinner, and seeing that the chores were done. For some it was overwhelming.[56] "My wife's always complaining that she can't go on running the home by herself," astronaut Donald K. "Deke" Slayton told author Oriana Fallaci.[57] But this experience was not that unusual. The wives had grown up during World War II and remembered the absence of family members.[58] Plenty of women outside the space program had experience with long separations—military wives, for instance. Men fighting in Vietnam might be gone for months at a time. In Houston, men who worked in the petrochemical industry often worked long and irregular hours, leaving the wives to handle the day-to-day affairs of the household.

What was unusual was the media coverage, lack of privacy, and public commentary about the astronauts' wives and families and their neighborhoods. "Home was where the 'personal stories' were," as Michael Collins explained in his autobiography. "It was tough sledding indeed for *Life* or anyone else to bring a tear to the eye of these mechanical men, but at home—ah, that was different! Flip the armored beetle over and inspect the soft underbelly. What did little Sarah Jean think about Daddy's impending departure, temporary or permanent. How did Mom feel when Dad was up?"[59] The press camped outside the homes of crews in orbit. Capturing the mundane was their business: "pictures of the family, anything: the wife, kids, goldfish, parakeet."[60] During particularly dull moments, the press egged kids on to get a better picture or story. During Gemini 5, the Conrad boys played on the roof of their home and rode their bikes into the pool to impress reporters.[61]

Even with the media around, kids who lived in the area acted like children in other suburbs. They climbed trees and rode bikes. No one felt more special than other kids because his or her dad was an astronaut. For the children of Apollo, spaceflight was a part of life; it was normal, not necessarily eventful. Apollo children did not know another life. As Joan Robertson, daughter of an Apollo engineer explained, "I didn't think it was a big deal, because doesn't everybody's dad work at NASA? Doesn't everybody's dad deal with Apollo?"[62] Kirk Griffin, son of a flight director, thought that "everybody's dad wore short sleeve white shirts with ties and no sport coat and worked at NASA."[63] Their dads might have been astronauts or rocket scientists, but they were just dads to their kids.

When the fathers were home, they spent time with their children and their friends. Gene Cernan taught his daughter and her friend Amy Sue Bean to ride a horse. Astronaut Alan L. Bean built a playhouse from start to finish for his daughter.[64] Pete Conrad took his sons hunting and flying. He even took his oldest son with him to the center to fly in the simulators for Gemini and Apollo.[65] He taught Timber Cove kids how to sail, while his fellow astronaut and neighbor Virgil I. "Gus" Grissom taught them how to water ski and took them to baseball games.[66]

Local churches became involved in the Gemini and Apollo projects in a variety of ways. Pastors recognized that many members of their congregations played important roles before and during the missions and therefore needed their spiritual guidance. Ministers provided prayers during flights or communion to family members before liftoff.[67] Church members relayed warm wishes to the astronaut families. The Ulrichs, for instance, sent blessings to the Aldrins, praying for Buzz's "safety" and Joan's "serenity" as she handled the "ordeal of publicity" surrounding the first lunar landing flight.[68]

Some of the astronauts involved their churches in their missions. Frank Borman, commander of the Apollo 8 mission, served as a lay reader for St. Christopher Episcopal Church in League City. He was scheduled to give the reading on Christmas Eve in 1968, but he had to bow out because he would be circling the moon that evening if all went according to plan. Someone from the church joked that he should have to read from the spacecraft. That planted a seed, and Borman decided to send a prayer of peace, the subject of the sermon that evening, to his church. Rodney G. Rose, an engineer at the Manned Spacecraft Center, agreed to record the prayer and deliver it to St. Christopher's. Rose also selected the prayer that Borman would recite in orbit. (Rose called the prayer "'Experiment P-1,' for first prayer from space.")[69] While in orbit around the moon, Borman dedicated the prayer to "Rod Rose and the people at St. Christopher's, actually to people everywhere."[70] Rose delivered the audio recording of Borman's prayer and his reading from the book of Genesis to the church when he came off shift.

Perhaps the most prominent example occurred during the Apollo 11 mission, when Buzz Aldrin took communion on the moon. Aldrin, an elder at the Webster Presbyterian Church, asked Rev. Dean Woodruff to help him find a way to "give thanksgiving for all people" during the first lunar landing in 1969. Together they came up with the celebration of communion on the lunar surface, and every year since that historic day, the congregation celebrates lunar communion.[71]

Webster Presbyterian was not the only local church with ties to the space program; almost all of the churches in the area around the space center had someone in the pews who worked for a contractor or for NASA directly. An entire congregation became involved in celebrating the Apollo 12 mission. The celebration acknowledged the families and others working on the mission as well as the unique relationship between local churches and NASA. Like his Apollo 11 colleague, Al Bean wanted to find some way to celebrate his moon mission. He decided to fly a banner embroidered by another charter member of the church, Mary Keathley, to the surface of the moon. The banner, about 18 inches wide by 14 inches tall, contained religious symbols depicting the history of Christianity including the cross and flame, the symbol of the United Methodist Church.

Rev. John E. Fellers and many others in the Clear Lake United Methodist Church congregation went to Florida to witness the launch of Apollo 12. In a letter to Bean's parents, Fellers expressed his excitement about the flight. Both he and his wife could "hardly wait for Friday to arrive. Alan will never know how much it means to us to share this great moment with him."[72] For those unable to attend, the church held a brief service in Texas at the time of liftoff.[73]

That Sunday, the families of the crew returned home from Florida and attended church. The church decorated the altar with flowers in honor of the Apollo 12 crew. The day was especially notable because the Beans were among those who, only five years earlier, had helped found the church. Fellers delivered a sermon he called "Outer Space and Inner Resources." He spoke with pride about the good fortune of the church that Bean, "a friend, neighbor and fellow church member . . . was to be part of this crew" and that they were "personally involved."[74]

When Bean was released from quarantine, he presented the banner to the church and addressed the congregation from the pulpit. He was reluctant to do so but agreed because he believed he needed to repay a debt he owed to the church. "You visited Sue while I was on the mission. You had a special worship service and you supported our crew with your prayers. As a matter of fact," he admitted, "your prayers are in some part responsible for our successful mission."[75] There was some interest in commercially reproducing the banner, but the church declined. The purpose of embroidering a banner that would be carried to and from the surface of the moon was to create a one-of-a-kind memento for the Clear Lake congregation, not to make money for the Methodist denomination.[76] This was an opportunity for all members of the church to feel as if Bean had taken something to the moon for each of them, a way for everyone in the congregation to participate in the historic mission.

After the mission, Al Bean enrolled in classes with Ardis Shanks, an artist who also lived in Nassau Bay. He told Shanks he wanted to paint when he retired from space. "I'll be the only artist that's ever been in space that's painting space. I can make a living doing it." She hoped to teach him the basics, but during class he painted space themes, not the subjects she wanted him to capture.[77]

Space inspired other local artists with connections to NASA. Nassau Bay resident Helen M. Garriott, the wife of scientist-astronaut Owen K. Garriott, began creating space-themed pottery. In 1970, she built a pot that looked like the moon; the earth formed the handle of the lid. Garriott made thousands of the Earth Rise Pots (or Moon Pots, as some call them). The jars featured tiny footprints at Tranquility Base, the site of man's first steps on the moon, and rhinestones marked other landing sites. She also sold moon chips that could be worn as a necklace or on bolo ties.[78]

The neighborhoods around Clear Lake reflected the excitement of the Space Age in numerous ways. In Timber Cove, where most of the Mercury 7 astronauts chose to live, the community rallied behind the idea of creating a pool in the shape of a Mercury space capsule. Arthur E. Garrison, a NASA procurement officer and a resident of the neighborhood, proposed the idea. The steps on one end represented the retrorockets, and the pool was cast so that the sides appeared to be tapered, just like the spacecraft. When the Mercury Project ended, the pool opened and became a hub of activity for the neighborhood families.[79]

A Timber Cove tradition began in 1968, when NASA first sent men around the moon aboard Apollo 8. The three-man crew included astronaut and resident James A. "Jim" Lovell. On Christmas Eve, as the crew traveled on the dark side of the moon, neighbors put out luminaries around their yards and driveways. As day turned to dusk, homeowners lit the candles in celebration of the mission and in recognition of the men who were the first to see the moon from a different perspective and then sang Christmas carols. The tradition continues to this day.[80]

When missions ended, crews returned home to nearby Ellington Air Force Base, to the welcoming arms of families and friends.[81] Gemini crew returns were more personal, but the sizes of the crowds that greeted Apollo crews grew over the years as the excitement about the program increased. More than 2,000 people greeted the Apollo 8 crew at Ellington. The number of fans surprised Lovell, who said, "At two in the morning I expected to get in my old blue bomb and go home."[82] More than 5,000 came to welcome his Apollo 13 crew back to Earth.

Neighborhoods welcomed their heroes in different ways. Timber Cove residents greeted astronaut Fred W. Haise from Apollo 13 at the entrance of their neighborhood with flashlights and torchlights. They placed their hands on his car and walked him to the safety of his home. Haise, exhausted from his aborted moon trip, said very little other than, "I'm glad to be back."[83] On the last mission to the moon, the residents of El Lago lined the street that ran through the neighborhood with flags, all the way to the home of the Evanses. The pride in the program and the accomplishment was obvious to everyone, recalled Jan Evans. "There were people on horseback carrying flags, and any child that had a bicycle or tricycle had red, white, and blue decorated streamers and in their wheels."[84]

The uniqueness of this time and place is a thought echoed again and again by residents who moved to the area, even those who did not work for the space agency. Are they being nostalgic or was this the reality? At the time, there was a sense that Clear Lake was a unique community that had not been seen before. Some families who moved to Clear Lake believed that it was "a very special place" with a "close community feeling."[85] This belief remains strong for many who lived there. Jan Evans remembered that people in the area "developed pseudo families. Everybody took care of everybody else."[86] Jane Conrad agreed. "We felt like we were special. Because everybody was so close we were all like family."[87] Children who grew up in the area fondly remember their time here and say they have never found a neighborhood or place like Clear Lake in the 1960s.

Thinking about Project Apollo makes many people who lived through that period nostalgic. This was their generation's moment to triumph over the Soviet Union. They are proud of their accomplishment. Apollo was a strenuous effort that they lived through and participated in, and many residents of Clear Lake look back on the program with fond memories. The feelings harken back to a period in their lives when it seemed like everything was possible. Many historians have noted that Apollo was a unique project that cannot be replicated today. The same appears to have been true for the community that was created outside of the gates of the Manned Spacecraft Center.

Few felt that they made sacrifices to see the country land men on the moon. Jane Conrad said, "I didn't have to give up anything. I gained a lot of things."[88] Betty Burghduff said that she "was proud to be part of it in a secondary way."[89] Dotty Duke compared her Apollo experience with her family's experiences during World War II and said of the war years, "That was a sacrifice."[90]

In an era where the Vietnam War, feminism, civil rights, and the counterculture drove Americans apart, the space program appeared to pull Americans together, especially those who lived in the area surrounding the Manned Spacecraft Center. Everyone felt excitement about the program and felt that they were a part of it, even if only in a small way. The Apollo community was unique for three reasons: it was home to the astronauts, it received significant media scrutiny, and the ways that art, churches, and community traditions captured and reflected the excitement of the time. Reminiscing about living in Clear Lake reminds people not only of the program's accomplishments but also just how special that time was. Jane Conrad summed it up this way: Pete's moon mission was a "gift that elevated and enriched" her life.[91] Marilyn Lovell called it the "best time in [her] life."[92]

Notes

1 John F. Kennedy, Address to a Joint Session of Congress, May 25, 1961, JFK Presidential Library and Museum, https://www.jfklibrary.org/learn/about-jfk/historic-speeches/address-to-joint-session-of-congress-may-25-1961.

2 Transcript of Betty K. Burghduff and Richard D. Burghduff, interview with Jennifer Ross-Nazzal, June 13, 2019, 2, Johnson Space Center Oral History Project, https://historycollection.jsc.nasa.gov/JSCHistoryPortal/history/oral_histories/BurghduffRD_BK/BurghduffRD_BK_6-13-19.pdf. This chapter includes quotes and information culled from oral history interviews conducted for the Johnson Space Center Oral History Project, the University of Houston-Clear Lake, and the Presbyterian Church (U.S.A.) from 1998 to 2019. Interviewees include those who moved to the Clear Lake area in the 1960s and created the community around NASA—civil servants and contractors and their wives and children and other residents. When I combined information from the oral history interviews with other sources such as archival documents, magazine articles, memoirs, books, and newspaper articles, a common pattern emerged that formed the basis of this chapter. Bound together by their experiences and shared values and a desire to achieve President Kennedy's goal, the Clear Lake area united behind Apollo.

Some might question the lack of diversity or conflict in the text. To the extent that I could verify conflicts, I shared those details. Mainly, however, the trends point to a generally positive response to NASA's decision to build the Manned Spacecraft Center.

The Clear Lake population mirrored the folks who worked at NASA. As I note in the text, like most other suburbs at the time and in that socioeconomic range, residents were overwhelmingly white. Even today whites are the majority in this area. At that time, people of color tended to live farther south in working-class communities such as Dickinson and La Marque.

3 Rod Pyle, *Destination Moon: The Apollo Missions in the Astronauts' Own Words* (New York: Carlton Publishing Group, 2005), 13.

4 Transcript of Edward G. Gibson, interview with Carol Butler, December 1, 2000, 88, Johnson Space Center Oral History Project, https://historycollection.jsc.nasa.gov/ JSCHistoryPortal/history/oral_histories/GibsonEG/GibsonEG_12-01-00.pdf.

5 Rosalyn Baxandall and Elizabeth Ewen, *Picture Windows: How the Suburbs Happened* (New York: Basic Books, 2000), 145.

6 Marilyn Lovell, who was married to astronaut James A. Lovell, "was mortified" by the elementary school where she registered her children. School clothes included cutoffs and shoes were not required, so many students walked around barefoot. Instructors used corporal punishment. Robert Zimmerman, *Genesis: The Story of Apollo 8: The First Manned Flight to Another World* (New York: Four Walls Eight Windows, 1998), 82; Baxandall and Ewen, *Picture Windows*, 151–157.

7 Master-planned communities are designed to foster a sense of fellowship and friendship between residents. Neighborhoods in these areas feature a mix of housing in a variety of price ranges and offer residents access to numerous amenities, including pools, recreation centers, community-owned golf courses, and clubhouses. Shops and services are located just outside these self-contained spaces.

8 I. P. Halpern, *Some Major Impacts of the National Space Program*, vol. 5 (Menlo Park: Stanford Research Institute, 1968), 36.

9 Transcript of Bill Strait, interview with Rick A. Jones and Jennifer Ross-Nazzal, April 23, 2019,13, Johnson Space Center History Office, Houston, Texas (hereafter JSC History Office).

10 Strait interview.

11 Transcript of William A. Larsen, interview with Jennifer Ross-Nazzal, August 12, 2003, 9, Johnson Space Center Oral History Project.

12 Betty Ulrich to her mother, February 10, 1964, box 1, folder 4, Ulrich Collection, University of Houston-Clear Lake Archives, Houston, Texas (hereafter Ulrich Collection).

13 Betty Ulrich to her mother, September 16, 1964, box 1, folder 4, Ulrich Collection.

14 Earle G. Lass, "The Reach for the Moon," *The Rotarian: An International Magazine*, May 1966, 19. In 1960, the year before NASA's announcement, the district had 1,900 students. By 1966, that number increased to 6,700. Halpern, *Some Major Impacts of the National Space Program*, 36.

15 Clear Lake City advertisement, *Webster Spaceland Star*, March 26, 1964, box 1, folder 4, Ulrich Collection.

16 Edwin E. "Buzz" Aldrin with Wayne Warga, *Return to Earth* (New York: Random House, 1973), 156, 163.

17 Baxandall and Ewen, *Picture Windows*, 152.

18 Jane Conrad divorced Pete Conrad and later remarried, becoming Jane D. Dreyfus. I will refer to her as Jane Conrad in chapter text but will use her current name in the

notes. Transcript of Jane D. Dreyfus, interview with Jennifer Ross-Nazzal, November 12, 2019, 17, Johnson Space Center Oral History Project, https://historycollection.jsc .nasa.gov/JSCHistoryPortal/history/oral_histories/DreyfusJC/DreyfusJC_11-12-19 .pdf.

19 Sue Dauphin, *Houston by Stages: A History of Theatre in Houston* (Burnet, TX: Eakin Press, 1981), 164–165, 202, 205.

20 Transcript of JoAnn Carr, interview with Rick A. Jones and Jennifer Ross-Nazzal, April 24, 2019, JSC History Office.

21 Summer Variety Show Program, July 16, 17, 18, 1965, Clear Lake Community Collection, University of Houston-Clear Lake Archives, Houston, Texas; "MSC's 1966 Variety Show, Moonglow 1966, Goes Into Rehearsals," *Roundup*, October 28, 1966, 5.

22 "Charm Club's Style Show Featured 'Space Age' Dress for Women," *Roundup*, April 16, 1965, 2A; "MSC Charm Club Schedules Lectures on Self Improvement," *Roundup*, November 11, 1966, 6; "EAA Calendar of Activities," *Roundup*, April 16, 1965, 2A.

23 See, for instance, "MSC Employee Picnic (EP-3) Ticket Sales to Close Monday," *Roundup*, September 17, 1965, 7.

24 "Lunar Rendezvous Golden Jubilee, 1965–2015," souvenir booklet, JSC History Office.

25 Clear Lake United Methodist Church Remembrance Letters, 1984 and 2013, Clear Lake United Methodist Archives, Houston, Texas. In 1967, Procurement Officer Dave W. Lang reported that the average age of MSC employees was 35.1 years; see "The Impact of the Manned Spacecraft Center on the Houston-Gulf Coast Area," July 1967, 13, general reference, file cabinet 2, drawer 1, JSC History Collection, University of Houston-Clear Lake, Texas.

26 Edward D. Gibson interview.

27 Jane D. Dreyfus, unpublished memoir, chapter 4, private collection of Jane D. Dreyfus.

28 Dora Jane Hamblin, "The Fire and Fate Have Left 8 Widows," *Life*, January 26, 1968, 62.

29 Transcript of Gratia Kay Lousma, interview with Jennifer Ross-Nazzal, 17, July 10, 2019, Johnson Space Center Oral History Project, https://historycollection.jsc.nasa .gov/JSCHistoryPortal/history/oral_histories/LousmaGK/LousmaGK_7-10-19.pdf.

30 Transcript of Betty Geehan, interview with Rick A. Jones and Jennifer Ross-Nazzal, April 23, 2019, 6, JSC History Office.

31 Aldrin, *Return to Earth*, 155.

32 Robert E. Huldschiner, "Ministry in the Space Age," *The Lutheran*, 10, Clear Lake United Methodist Archives. Few African Americans lived in the area before NASA moved in. A school population study from 1961 found only twenty black families in the entire district. Wallace H. Strevell and Leo G. Mahoney, *School Population Study of Clear Creek Independent School District* (Houston: Gulf School Research Development Association, 1961), 30. At least one Japanese-Hawaiian family, the Kozumas, lived in Oakbrook; unsigned Ulrich letter, May 8, 1964, box 1, folder 4, Ulrich Collection.

33 Eugene Cernan with Don Davis, *The Last Man on the Moon* (New York: St. Martin's Press, 1999), 72.

34 Dreyfus, unpublished memoir.

35 George Walraven, Charlie Hill, and Glendora Hill, interview with Jennifer Ross-Nazzal and Rebecca Wright, August 21, 2003, tape recording, JSC History Office.

36 *Life* had an exclusive contract to capture the personal stories of astronauts. Other media outlets were forced to gather outside the home.

37 NASA installed squawk boxes inside the families' homes so they could follow the missions. The boxes transmitted the audio conversations between the crews and Mission Control.

38 Miguel Acoca, "Success—and Yet Another Conrad Splashdown," *Life*, September 10, 1965, 38.

39 Dreyfus, unpublished memoir.

40 *Managing the Moon Program: Lessons Learned from Project Apollo: Proceedings of an Oral History Workshop Conducted July 21, 1989* (Washington, DC: NASA, 1999), 43.

41 Transcript of Jamye Flowers Coplin, interview with Rebecca Wright, November 12, 2008, 12, Johnson Space Center Oral History Project, https://historycollection.jsc.nasa .gov/JSCHistoryPortal/history/oral_histories/CoplinJF/CoplinJF_11-12-08.pdf.

42 Larsen interview, 8.

43 Aldrin, *Return to Earth*, 175.

44 Clear Lake City Newsletter, January 24, 1964, box 1, folder 4, Ulrich Collection.

45 Christmas newsletter, 1963, box 1, folder 4, Ulrich Collection.

46 Letter, January 12, 1969, box 1, folder 4, Ulrich Collection.

47 Walter Cunningham, *The All-American Boys: The U.S. Space Program* (New York: ibooks, 2003), 86, italics in original.

48 Transcript of Marlowe D. Cassetti, interview with Carol Butler, December 21, 1998, page 12-30, Johnson Space Center Oral History Project, https://historycollection.jsc .nasa.gov/JSCHistoryPortal/history/oral_histories/CassettiMD/MDC_12-21-98.pdf.

49 Transcript of Harvey L. Hartman, interview with Sandra Johnson, May 7, 2002, Johnson Space Center Oral History Project, https://historycollection.jsc.nasa.gov/ JSCHistoryPortal/history/oral_histories/HartmanHL/HartmanHL_5-7-02.pdf.

50 Transcript of Janet M. Evans, interview with Jennifer Ross-Nazzal, August 7, 2003, 20, Johnson Space Center Oral History Project, https://historycollection.jsc.nasa.gov/ JSCHistoryPortal/history/oral_histories/EvansJM/EvansJM_8-7-03.pdf.

51 Transcript of JoAnn R. P. Carr, interview with Jennifer Ross-Nazzal, July 2, 2019, 39 (quote), 33 (heartbroken over canceled flight), Johnson Space Center Oral History Project, https://historycollection.jsc.nasa.gov/JSCHistoryPortal/history/oral _histories/CarrJ/CarrJ_7-2-19.pdf.

52 Jim Hansen, *First Man: The Life of Neil A. Armstrong* (New York: Simon & Schuster, 2005), 290.

53 Mary Wiegers, "Space Community: Emphasis on Excellence," *Washington Post*, July 13, 1969.

54 Dreyfus, unpublished memoir.

55 Cernan, *The Last Man on the Moon*, 254.

56 Walter Cunningham, *All-American Boys*, 74; Charlie Duke and Dotty Duke, *Moonwalker* (Nashville: Oliver-Nelson Books, 1990), 243–244.

57 Oriana Fallaci, *If the Sun Dies* (London: Collins, 1967), 93.

58 Dotty Duke's father, for instance, spent two years overseas during the war and sent only a few letters to the family. For more details, see transcript of Dotty Duke, inter-

view with Sandra Johnson, November 12, 2019, Johnson Space Center Oral History Project.

59 Michael Collins, *Carrying the Fire: An Astronaut's Journeys* (New York: Farrar, Straus and Giroux, 1974), 54.

60 Cunningham, *All-American Boys*, 175.

61 Jane Dreyfus interview; Dreyfus, unpublished memoir.

62 Transcript of Joan Robertson, interview with Jennifer Ross-Nazzal, May 30, 2019, 16, Johnson Space Center Oral History Project.

63 Transcript of Kirk Griffin, interview with Sandra Johnson, November 13, 2019, 11, Johnson Space Center Oral History Project.

64 Amy Sue Bean, "Eulogy for My Father," Moonwalker Daughter, accessed February 20, 2020, https://Moonwalkerdaughter.com/.

65 David Schlom and Matt Fidler, "Blue Dot 169: Growing Up Apollo: An Interview with Moonwalker's and Flight Controller Kids," North State Public Radio, https://www.mynspr.org/show/blue-dot/2020-01-03/blue-dot-169-growing-up-apollo-an-interview-with-moonwalkers-and-flight-controller-kids.

66 Transcript of James Kinzler, Kim Kinzler, and Doug Shows, interview with Jennifer Ross-Nazzal and Rebecca Wright, August 28, 2003, tape recording, JSC History Office.

67 Kendrick Oliver saw the churches in Clear Lake as a place where the men of NASA could "mitigate" the costs of meeting President Kennedy's goal—the long hours away from family spent working on the project. *To Touch the Face of God: The Sacred, the Profane, and the American Space Program, 1957–1975* (Baltimore, MD: Johns Hopkins University Press, 2013), 34–35.

68 Paul Ulrich and Betty Ulrich to Joan Aldrin, July 16, 1969, box 1, folder 4, Ulrich Collection.

69 Transcript of Rodney G. Rose, interview with Kevin Rusnak, November 8, 1999, page 12-30, Johnson Space Center Oral History Project, https://historycollection.jsc.nasa.gov/JSCHistoryPortal/history/oral_histories/RoseRG/RGR_11-8-99.pdf; Frank Borman, "Message to Earth," *Guideposts*, April 1969, 3, Apollo Series, box 071-31, JSC History Collection.

70 Apollo 8 Technical Air to Ground Transcription, December 1968, tape 50, page 7, Apollo Series, box 078-13, JSC History Collection.

71 "Presbyterian Minister Recalls His Secret Apollo Mission," Voice of America, November 2, 2009, https://www.voanews.com/a/a-13-2009-07-15-voa9-68747877/410330.html. Details about the communion service and plan can be found in a history of the Webster Presbyterian Church: Judith Haley Allton, Patricia M. Brackett, and Dana Ray, *The Little White Church on NASA Road 1: From Rice Farmers to Astronauts* (Webster, TX: Webster Presbyterian Church, 1993), 88–92.

72 John E. Fellers to Mr. and Mrs. Arnold H. Bean, November 10, 1969, Clear Lake United Methodist Archives.

73 Lee Holley, "Space Scoop," *Clear Lake City Spaceland News Citizen*, November 12, 1969, Clear Lake United Methodist Archives.

74 John E. Fellers, "Outer Space and Inner Resources," November 16, 1969, sermon, 1, JSC History Office.

75 *The Texas Methodist*, February 6, 1970, Clear Lake United Methodist Archives.

76 John E. Fellers to Stephen Rosskamm, December 10, 1969, Clear Lake United Methodist Archives.

77 Transcript of Ardis M. Shanks, interview with Jennifer Ross-Nazzal, July 10, 2019, 8, Johnson Space Center Oral History Project, https://historycollection.jsc.nasa.gov/JSCHistoryPortal/history/oral_histories/ShanksAM/ShanksA_7-10-19.pdf.

78 Bay Area Museum case featuring a Moon Pot, February 19, 2020, Houston, Texas.

79 Jack Kinzler, Sylvia Kinzler, Art Hinners, and Smitty Hinners, interview with Jennifer Ross-Nazzal and Rebecca Wright, August 7, 2003, tape recording, JSC History Office; Aleck Bond, Tassie Bond, and Bernie Goodwin, interview with Jennifer Ross-Nazzal and Rebecca Wright, August 21, 2003, tape recording, JSC History Office.

80 Rebecca Wright, "A Home for Heroes—Timber Cover," *Houston History* 6, no. 1 (2008): 50.

81 Three crews, those of Apollo 11, 12, and 14, greeted their families from the safety of the Mobile Quarantine Facility. The astronauts from these missions could not have any personal contact with their families until they were released from quarantine.

82 "Astrotrio Arrives Home," *Abilene Reporter-News*, December 30, 1968.

83 Interview with Jack Kinzler, Sylvia Kinzler, Art Hinners, and Smitty Hinners.

84 Janet M. Evans interview.

85 "Dear neighbor," letter, n.d., box 1, folder 4, Ulrich Collection; Dad to unknown, November 25, 1963, box 1, folder 4, Ulrich Collection.

86 Janet M. Evans interview.

87 Jane Conrad Dreyfus, interview with Shelly Kelly, July 18, 2009, Timber Cove Oral History Project, Timber Cove Collection (#2011-0008), University of Houston-Clear Lake Archives, Houston, Texas.

88 Conrad Dreyfus interview.

89 Interview with Betty K. Burghduff and Richard D. Burghduff.

90 Dotty Duke interview.

91 Dreyfus, unpublished memoir.

92 Lily Koppel, *The Astronaut Wives Club* (New York: Grand Central Publishing, 2014), 279.

Part 3

Leveraging Aerospace for Tourism

10

Gateway to the Stars

Carl McIntire, Space Tourism, and the Cape Canaveral Bible Conference

KARI EDWARDS

The whole emphasis upon creation and God's great and glorious power in making it is in the hearts of those who walk in Cape Canaveral.

Carl McIntire, "Hallelujah—Praise Ye the Lord," *Christian Beacon*, January 9, 1975, 1

A new morning dawned at the Cape Kennedy Hilton on February 1, 1971. As Florida sunshine poured through the large windows and bathed the impressive Spanish-style lobby in daylight, reminders of the wild night before littered the Pieces of Eight Lounge, the hotel's popular bar. Cheap glass ashtrays full of discarded cigarette butts sat on each of the tables and the faint smell of stale beer lingered in the air. A recklessly loud rock band had finished playing around 5:00 that morning while scores of partiers—both locals and out-of-towners—had danced and drunk without a care until the last possible moment, when they all knew those familiar doors would be closed to them for good. It was the night of January 31, a launch weekend, and Apollo 14 was on its way to the moon. But the Hilton crowd was not simply celebrating another NASA mission taking flight just down the road from them. "All seemed to be caught up in a kind of frenetic, frantic fever to live it up now as never before," wrote Mary Ann Hill, the amusements editor for Brevard County newspaper *Florida Today*. "Never mind the general recession or the cutbacks or the layoffs or all the other worrisome spooks haunting us lately. It was shot weekend and a time to celebrate and the last time to do it at the Hilton."[1]

The party was a raucous send-off for a storied Cape Canaveral gathering place that had played host to astronauts and distinguished journalists alongside trendy young locals looking for a good time from the moment it first opened in November 1967.[2] But on February 1, 1971, everything would change. The new owners had already auctioned off the last of the liquor from Pieces of Eight Lounge and had swiftly discarded the ashtrays. Only hours after the Cape Kennedy Hilton "roared out of existence with all the fanfare of a New Orleans Mardi Gras," Rev. Carl McIntire moved in.[3] Only a day after the "wake" for the Hilton, the hotel and its adjacent conference center hosted Christian guests from twelve states for the opening session of a fundamentalist Bible conference. The next night, attendees adjourned to the parking lot so they could view the launch of a NATO satellite from the nearby space center. Afterward, the entire crowd made an orderly return to the hotel's lobby to hear one of the featured guests, Dr. Fague Springman, sing "How Great Thou Art."[4]

Carl McIntire's presence in Cape Canaveral marked the beginning of a drastically new chapter in the history of the Florida town the US space program had built. By 1971, Cape Canaveral enjoyed international recognition as the home of Kennedy Space Center and the location of America's famous "moonport" that all of NASA's space missions were launched from. The town was synonymous with space exploration, with the wonders of modern technology and the promises of the future. Local residents wholeheartedly embraced the space industry as the key element in their community's livelihood—and the 1960s were indeed a good decade to enjoy a prosperous livelihood in Cape Canaveral. Aerospace jobs abounded throughout Brevard County as NASA raced headlong toward fulfilling the nation's goal of beating the Soviet Union to the moon. Those jobs brought families, more housing, new businesses, and a steadily skyrocketing tourism industry built around America's newfound obsession with all things space. But life in Cape Canaveral grew more precarious as Project Apollo neared its end. The space program suffered through a series of major cutbacks, which meant a loss of jobs and an exodus of aerospace professionals from the area.[5] Even though tourist dollars still trickled in during mission launches, Cape Canaveral—or Cape Kennedy, as it was often called during this period—was headed straight into a potentially irreversible decline by the time Apollo 14 left its shores and the Hilton hosted its final batch of revelers. "Motels and restaurants have closed, and the ones that have stayed open are busy crying the blues," a nationally syndicated article on the town's plight eulogized. "In a sense, Cape Kennedy has the earmarks of a sun-washed Appalachia."[6] Carl McIntire, many residents hoped, would be the man to change all that.

Fundamentalist Christian minister Dr. Carl McIntire was a relatively well-known, albeit controversial public figure whose religious radio show, the *20th Century Reformation Hour*, broadcasted across the country and had been on the air for fifteen years by the time news of his Brevard County acquisitions hit local central Florida media outlets in late December 1970.[7] The 64-year-old preacher had built his reputation during the height of Cold War tensions with the Soviet Union in the 1950s as a virulent anticommunist. His uncompromising message of nationalism combined with biblical literalism drew in like-minded listeners from coast to coast. From his home state of New Jersey he published his own periodical, the *Christian Beacon*; pastored a local church in the town of Collingswood; created his own fundamentalist denomination, the Bible Presbyterian Church; headed his own separatist church organization, the International Council of Christian Churches; founded several Bible colleges and seminaries; and owned a patriotic-themed hotel, the Christian Admiral Hotel, and several other tourist spots in the historic seaside town of Cape May, all in addition to his radio presence. McIntire's identity revolved around his unabashed political and religious conservatism. Stubborn to a fault and a powerfully charismatic preacher, McIntire never backed down from a fight, always prepared to stick to his principles in the face of insurmountable odds, rallying his supporters behind him every step of the way. He was, as media scholar Heather Hendershot describes him, "God's angriest man."[8] To anyone who lived in or visited the Cape Canaveral area during the Apollo boom years, a figure like Carl McIntire would likely have seemed an undesirable misfit at best. Yet with the stroke of a pen and a multimillion-dollar lease agreement, McIntire laid claim to a large swath of prime Brevard property and revealed plans to transform it into something that would nearly rival the rocket launches in grandeur: a new Jerusalem.

"The name of Cape Canaveral was catapulted to international fame a decade ago when man began reaching skyward from its sandy shoreline," read *Florida Today*'s first editorial on the McIntire project. "The name Cape Canaveral will again be heard and seen almost daily across the nation and around the world as a different sort of mission to the heavens is launched from our area."[9] What McIntire promised was a revival, both economic and spiritual, for a town many residents feared would soon be forgotten as the space program waned and potentially monumental tourist attractions like the soon-to-open Walt Disney World in Orlando siphoned vacationers farther away from Brevard County. On December 29, 1970, McIntire announced his grand vision for a "religious—educational—retirement development" in Cape Canaveral. His plan included the Cape Kennedy Hilton and its adjacent conference facilities, two former aerospace industry buildings, a 216-unit apartment complex, and several other plots of land for future projects. The most notable feature was a historically accurate scale replica of biblical Jerusalem.[10] It was going to be a mammoth undertaking, yet the reward of a consistent tourist base quieted any local skeptics who might have otherwise raised doubts about McIntire's ability to follow through on such lofty expectations. The preacher's enthusiasm for his Cape Canaveral vision was palpable. He had a devoted nationwide following and a shrewd knack for raising money. Cape Canaveral residents thus placed their collective faith in Carl McIntire's message, believing his Gateway to the Stars—as the complex would soon be christened—could be a miracle in the making.[11]

The troubled, decades-long presence of Gateway to the Stars in Brevard County is a strange, forgotten moment in the history of Cape Canaveral's symbiotic relationship with the space program. Situated in NASA's backyard, the properties McIntire owned—the Hilton and two former aerospace buildings—were intimately connected to the golden period of American space exploration. His desire to turn those places into sites of religious significance left many Floridians at the time bewildered about the minister's motive. Why Cape Canaveral? Was it merely cheap real estate and a foolhardy gamble that drove him there? Carl McIntire justified all his views through the lens of Christian prophecy and a literal interpretation of Scripture. Scholars consider McIntire to be one of the standard-bearers of Cold War fundamentalism, yet his enthusiasm for space exploration is absent from academic works on him.[12] For McIntire, the space race served a divine purpose: it was a way for Christian America to ascend into the heavenly realm and prevent the atheist Soviet Union from staking claims there. Space, he believed, was where the holy war against global communism would be won.

McIntire's obsession with the spiritual significance of the space program was never more evident than Christmas Eve 1968, when astronauts Frank Borman, William Anders, and James Lovell broadcast their Christmas greetings to the people of Earth from lunar orbit via satellite television, reading from the first chapter of Genesis and ending with a prayer. McIntire heralded the Scripture reading from space as a sign of God's blessing on the United States. Apollo 8 and the American space program became regular topics of discussion on his radio broadcasts and in the *Christian Beacon*. He also undertook two different mobilization campaigns inspired by Apollo 8: one to place the first four words of Genesis, "In the beginning God," on the United States Postal Service Apollo 8 commemorative postage stamps; and a second to send letters to NASA in support of the astronauts' reading the Bible and praying from space.[13] From McIntire's apocalyptic view of the space race, American astronauts like the Apollo 8 crew were "in a true sense ministers and messengers of God."[14]

Carl McIntire envisioned his Gateway to the Stars complex as a spiritual complement to NASA's presence in Cape Canaveral—as a base camp of sorts for Christian tourists to experience and understand what was happening at nearby Kennedy Space Center. It represented a point of unity between a strict fundamentalist understanding of the world and a scientific endeavor that placed American footsteps on the surface of the moon. While McIntire certainly intended this project to infuse some much-needed tourist dollars into the local economy, he saw his Florida ventures primarily as a place for revival, for doing the necessary work of interpreting the Space Age from a fundamentalist Christian world view. In McIntire's mind, a visit to Gateway to the Stars would reveal to guests that there was a divine purpose to US efforts in space exploration.

Cape Canaveral Reimagined

I'd rather see the Cape sink into the sea than see Carl McIntire come in here.

Anonymous local minister quoted in Charles Reid, "Thunder on Right Rolls into Town," *Florida Today*, January 24, 1971, 1A

More than one million visitors made the trip to Cape Canaveral to tour the museum at Kennedy Space Center in 1970.[15] Launch tourism had hardly dropped off, either; local reporters described massive six-hour traffic jams across Brevard County after the Apollo 14 launch in February.[16] But all was not well in Cape Canaveral. It was a town that already felt frozen in time, hopelessly shackled to an earlier, more prosperous period and peering warily ahead at an uncertain future. As space historian Roger Launius points out, even NASA itself failed to fully comprehend plan for the situation that was to come after the thrill of the launches had faded. "Apollo had not been conducted under normal political circumstances," he writes, and the "exceptional circumstances surrounding Apollo would not be repeated."[17] In late 1972, the *New York Times* remarked in an editorial on the town that one of its most distinguishing features was "its own porno district."[18] "You get the feeling very quickly that the big party is breaking up," the *Sunday Mail* had said in 1971. "It is typically American that Cape Kennedy's best hope for the future should come from a defrocked Presbyterian minister and Walt Disney Productions."[19]

Regarding a massive tourism venture like Walt Disney World in Orlando and Carl McIntire's Gateway to the Stars as equally important for Brevard County's economic revival would seem naïve soon enough to Floridians, but in 1971, the year Walt Disney World opened and McIntire took control of his Cape Canaveral properties, the prospect was indeed a possibility, especially to local business leaders. McIntire's acquisition of the Hilton hotel—which he promptly renamed the Freedom Center Hotel—was particularly promising, as it meant visitors to the hotel could use Brevard as a home base for exploring other central Florida attractions. Using double-decker buses emblazoned with advertisements promoting Gateway to the Stars, McIntire provided a free shuttle service to and from Orlando as a way of enticing more guests. "Those people who go to Disney World will say, 'Let's go to Cape Canaveral to see what Jerusalem looked like,'" he predicted.[20] It was all part of his plan, he explained in an early press conference, to make the area "a stairway to the stars instead of the backdoor to space."[21] With promises like that it was difficult for residents to not feel some excitement over what was soon to come for their area, and reports from the time echo the optimism McIntire projected. *Florida Today* commented that "having a large-scale religious center located at Brevard's beaches should help in changing the area's 'sex and booze' image."[22]

That optimism faded quickly, however, as the press began to print stories that questioned McIntire's intentions and cast doubt on the benefit he might bring to the area. One syndicated story included quotes from several Brevard County ministers, who expressed frustration at the thought of Carl McIntire establishing a base in their town. Rev. L. R. Lindsay, a Presbyterian minister, lamented that Cape Canaveral would soon become "the world capital of anticommunism. If the orange trees don't blossom, it will be a communist plot."[23] The article included an interview with an unnamed official at Kennedy Space Center regarding McIntire's recent visit with them on January 22, 1971. Tour guides showed an exuberant McIntire everything he asked to see but had specific instructions to not introduce the preacher to anyone "whose name meant something." The anonymous official added that he supposed NASA would eventually have to welcome him to town one of these days, "but I'm damned if I'm going to do anything about it right now."[24] By late January 1971, only a month after McIntire announced his plans officially, *Florida Today* had printed several pieces that were critical of his motives, his lack of financial transparency, his beliefs, and the impact of his controversial reputation on the town. One article charged that McIntire routinely lied about the number of radio stations that broadcast his program and the size of his donations, saying that he claimed a far larger audience than he actually had and was not open about the amount of money he made from his ministry. "If he says he has 653 radio outlets, that means he has 500," the article said, quoting a source described as an anonymous "former colleague" of McIntire's "who's afraid of him."[25]

Florida Today was only one of several newspapers to publish articles based on inquiries about McIntire's life and work. Some of the earliest investigative reporting outside central Florida came from the *Miami Herald*, where journalist Mike Baxter wrote two consecutive stories about McIntire in February 1971. Baxter wrote that his purported audience of 12 million "may be an exaggeration, but whatever audience he does have he appeals to them repeatedly."[26] He also poked fun at McIntire's verbal mannerisms, accusing him of vanity for his tendency to frequently refer to himself in third person. His critics, Baxter noted with amusement, liked to play a game of counting "Carl McIntire references to Carl McIntire on his broadcast. The claimed record is 33. Another counted 27. A novice scored only a dozen."[27] For the most part, McIntire refrained from responding to his Florida critics. According to his fundamentalist beliefs, the secular world would invariably show resistance to his efforts to spread the Gospel. Mike Baxter's work in the *Miami Herald*, however, prompted a response from the angry minister. Two weeks after Baxter's articles were published, a letter from McIntire was published in the paper. "In my own mind all of this is connected with the over-all worldwide communist assault against the anti-communists and those who believe in God," McIntire wrote. "The press is on trial: not Carl McIntire."[28] Skepticism toward McIntire was not limited to the press and local religious leaders. When a *Florida Today* pollster asked a woman from Merritt Island what she thought about McIntire moving into the area, she answered bluntly: "I hate it, it just makes me ill to think about it."[29]

Despite the hype McIntire created about his grand vision, when he arrived in Brevard County in early 1971, the most noticeable manifestation of his presence locally was the change in ownership of the Cape Kennedy Hilton. The rest of his proposed innovations—the Jerusalem Museum, the retirement center, the seminary, and the condominiums—existed only on paper. But the Hilton's new, far more pious life as the Freedom Center Hotel was a radical enough change for Cape Canaveral residents. As the former Hilton's large convention center began to host anticommunist religious rallies and industrial buildings that once manufactured technologies for America's space program were renovated into classrooms and a library for a fundamentalist Bible college that had recently lost its accreditation in New Jersey, the town's post-Apollo identity slowly began to take shape. "Bye-bye Alan Shepard and Wild Turkey on the rocks," read one *Florida Today* editorial, "The Hilton's bar will be closed. The jam sessions, short skirts and long-hairs will be gone. Even the Gideon Bibles may be replaced by the Gospel of Carl McIntire."[30] Cape Canaveral's religious conversion had commenced.

Gateway to the Stars Christian Community

> When God opens doors, we enter. This Cape Canaveral development has
> given us a gateway to the stars. We are a people who look to the Heavens.
> We believe that the Lord is coming in the clouds with power and glory.
>
> Carl McIntire, "At the Gateway to the Stars," Bible conference promotional bro-
> chure, n.d., Carl McIntire Collection, Princeton Theological Seminary Archives

An unusual billboard twenty feet tall and thirteen feet wide greeted drivers on
Florida's State Road A1A as they drove through the town of Cape Canaveral
on January 28, 1971. The new sign was an unusual choice for an area struggling to construct an identity independent of launch-day tourism at Kennedy
Space Center. This was not a typical billboard designed to entice vacationers
to make an unscheduled detour on their way to the beach. It was a giant replica of the Apollo 8 commemorative stamp, emblazoned with the words "In
the beginning God" in gold lettering above the mission's famous *Earthrise*
photograph. It was the stamp Carl McIntire had dreamed of, the one he had
organized a letter-writing campaign to the postal service about to pressure it
to keep the words from the Book of Genesis in its final design.[31] McIntire's win
with the postal service was a major triumph for Christian America, in McIntire's opinion, and seeing the stamp's image from the side of the busy roadway
would undoubtedly "remind millions that Genesis 1:1 was read from about
the Moon." It was a tribute, McIntire explained during the ceremony dedicat-
ing the billboard, to Apollo 8 astronauts Borman, Anders, and Lovell, who
had "like the trumpet of Gabriel speaking to the Virgin Mary told mankind
that God, who began all things, would consummate all things for His glory."[32]

The giant stamp billboard was a visible manifestation of McIntire's plan to
make Cape Canaveral a site of worldwide religious revival. Cape Canaveral
was soon to become a center of pilgrimage for Christians across the nation
and beyond, a marriage of timeless faith and modern technological wonders
that could occur in only one place. McIntire envisioned an all-encompassing
development for fundamentalist Christians that included hotel accommoda-
tions, worship facilities, seasonal and permanent housing, a college, a retirement center, potential work opportunities, shopping and dining, and un-
obstructed views of rocket launches. At the heart of it all would be a Bible
conference, an ongoing assortment of lectures, church services, motivational
talks, and themed events that would showcase the connection between fundamentalist Christianity, conservative anticommunist politics, and the space
program.

From the moment he secured his first Cape Canaveral properties, McIntire expressed both a sense of pride and a sense of duty that came from his connection to a location so deeply involved with American efforts in space. Owning those properties was both a personal victory and a grave responsibility. "We must be No. 1 in space," he declared in an interview after touring Kennedy Space Center.[33] For McIntire, the scientific work done at Kennedy Space Center was as important as and intimately connected to his own work of preaching the Word of God, as he referred to his ministry. Both efforts would support the same goal. When *Florida Today* asked him if he had any difficulty reconciling space exploration with his literal reading of the Bible, McIntire answered, "None whatever." He added, "We're all going to be in space very soon when the resurrection takes place and the Lord's coming in the clouds of Heaven."[34] Space flight was a deeply spiritual pursuit, he believed, and visitors to his Gateway to the Stars were immersed in McIntire's unique merging of fundamentalism and religious reverence for NASA's accomplishments.

"Located along side of the U.S. Space Center where man went to the Moon, the new conference will offer the visions of the Scriptures and the prophecies of God's Word to the Bible-believing people," McIntire explained.[35] One of the many brochures McIntire produced proclaimed that Gateway to the Stars would be a community where "Christian people can live together in Christian fellowship enjoying the beautiful, clean air of the Florida suncoast."[36] McIntire's promotional materials presented his vision of a self-contained religious enclave in the shadow of NASA as a kind of microcosm of the fundamentalist movement. Here fundamentalist Christians would be separate from the world while still living within it. Here they would hear a painstaking adherence to biblical literalism while simultaneously reveling in the scientific advancements that would ultimately reveal the truth of the Scriptures.

There was no better place than Cape Canaveral, a 1970s tourist book published by McIntire declared, for a Christian conference that "exalts God's Word and deals with the revelation that God has given of the last days."[37] More than any other part of McIntire's Gateway to the Stars master plan, the Bible conference was to be the centerpiece of his ministry in Brevard County. It would enable McIntire and like-minded preachers and political figures to present a conservative Christian, anticommunist, apocalyptic-filter world view. The conference presented NASA's launches as technologically sublime and presented a spiritual overlay to an experience that routinely inspired "awe and wonder, often tinged with an element of terror."[38] The space program elicited such moments of sublime wonder among those who traveled to Cape Canaveral to watch rockets blast into space. Historian of technology

David Nye writes that the communal aspect of launch viewings and the na-
tionalistic goals of the space race made rocket launches both a spiritual and
patriotic event, transforming each individual's sense of awe into a collective
"belief in national greatness."[39] Launch events were, Nye argues, "less a matter
of spectatorship than a pilgrimage to a shrine where a technological miracle is confidently expected."[40] The promotional material McIntire produced
about his Bible conference and the rest of the complex reveals this sense of
participation—almost ownership—of America's space accomplishments and
his belief that watching launches was a primary element of experiencing Cape
Canaveral as a site of Christian pilgrimage. One glossy pamphlet promoting
the hotel boasted that it was the building where astronauts, including Neil
Armstrong, had stayed before their mission launches. A list of their room
numbers appeared alongside a large image of the hotel's post-Hilton lobby,
which included a wall filled with framed photographs of individual astronauts
like icons of saints lining the interior of a church.[41] Christians who visited
would be surrounded by things that reminded them of the connection McIntire made between their faith and the space program.

The first several months of McIntire's vacation packages to Gateway to the
Stars brought in a small but steady stream of Christian visitors from northern
states eager to escape the winter cold in sunny Florida. Around 250 guests a
month traveled together in designated buses from New Jersey in February
and March 1971. But by May, hotel employees who had stayed on when McIntire took control had begun to express frustration about stagnating numbers.
One man who ran a concession in the hotel told *Florida Today* at the beginning of the summer that his business was down over 50 percent. Another
claimed that McIntire had had "only . . . one wing" of the large hotel complex
open since he took over.[42] A month earlier, in April 1971, local newspapers reported that McIntire's troubled Bible seminary, Shelton College, had given up
trying to convince the state of New Jersey to reverse its decision to revoke its
accreditation and would be moving operations to Cape Canaveral soon.[43] In
June, McIntire floated rumors of a high-rise condominium complex in addition to the apartment buildings he was already leasing, but he admitted when
asked that he had not yet filed the proper permits for such a project.[44]

Despite frequent assurances that setbacks were only temporary and that
crowds of fundamentalist vacationers would soon flock to Brevard County,
attendance at the hotel and at conference sessions remained dismally low
throughout 1971. "He hasn't been able to get his people to come here," an
unnamed source told *Florida Today* in September. "After all, it is difficult to
introduce something new to people set in their ways."[45] This statement was
prescient. The connection between fundamentalist Christianity and the space

program was not as obvious in the minds of McIntire's followers as he imagined, despite his frequent pronouncements on the subject. In addition, his listeners were not enough of a tourist base to keep Gateway to the Stars running. Moreover, residents of Brevard County had not warmed up to McIntire's plans, leaving the Bible conference and McIntire's hotel facilities strictly to the tourists. The most interest McIntire garnered locally with his Bible conferences was in relation to the Apollo 15 launch, when he invited residents to hear him preach a sermon at the complex on the theme of how space exploration was supported by the Bible.[46] However, much to the dismay of his critics, donations from McIntire's followers continued to pour in despite the poor showing of the Cape Canaveral project's first year. McIntire never seemed to show discouragement, either, asking his radio audience for continued support and printing appeals in the *Christian Beacon* for his followers to "pray and to give, to rewrite wills" in order to fulfill the vision of Gateway to the Stars.[47] McIntire told one listener in 1971 that "good people like you and others who realize that if the battle is lost in this country, it is lost everywhere" were his only source of funds to pay the mortgages in Brevard County.[48]

By far the most ambitious component of Gateway to the Stars was the Jerusalem Museum, which McIntire described in interviews as early as December 1970 as an interactive full-scale exhibition that would accurately reproduce Solomon's Temple and other sacred sites. This museum was destined to be Cape Canaveral's *other* landmark; it would be "the Space Center and the Moon on one side, and Cape Canaveral, the Temple, and the literal unfolding of the Scriptures on the other side," an arrangement that would enable visitors to "be truly confronted with Christ."[49] It would be the first of its kind—a way for Christians to experience a biblical site of tremendous importance in the shadow of the site where America launched rockets into space. Yet by November 1971, nearly a year after McIntire presented his idea for a Jerusalem Museum, the proposed site, a former aerospace industry building once owned by Chrysler, sat empty. "Maybe it is just a dream," a *Florida Today* editorial mused. "Sometimes when a man really desires something with his soul as well as his heart, then reality vaporizes."[50]

On November 5, 1971, a small group of McIntire's followers came to Florida for a ceremony that dedicated the conference center within the Freedom Center Hotel to the memory of Representative L. Mendel Rivers, the former chair of the House Armed Services Committee and a McIntire supporter.[51] At this event, McIntire revealed a model of Solomon's Temple inside the Chrysler building that was to serve as a placeholder until the full-scale version could be constructed. This replica was already well known, having been the featured exhibit at the Pavilion of Judaism at Expo 67, Montreal's 1967 World's Fair, where it was reportedly the event's second-most-visited feature.[52] The model which sat on a large table in an otherwise empty display room, was hardly comparable to the full-scale, immersive landmark he had promised Cape Canaveral. But as with everything else in McIntire's grand vision for Gateway to the Stars, money and tourism were not the primary objectives. "Our purpose in building the temple will be to get people to search the Scriptures," Stan Rittenhouse, McIntire's public relations spokesperson, explained. "We want to turn America back to God, to focus the people on Godly things."[53]

The Decline and Legacy

> We were certain that God wanted us to have a witness to Him in Cape Canaveral, and we were confident that we could occupy it as Israel occupied her land, by the help of God. And so the miracles came.
>
> Carl McIntire, "Canaveral Saved," *Christian Beacon*, January 2, 1975, Carl McIntire Collection, Princeton Theological Seminary Archives

What began with such enthusiasm in early 1971 had transformed into something unrecognizable by late 1974—a host of unkept promises and poorly implemented plans contrasted starkly with the picture of a bustling tourism haven Carl McIntire painted several years earlier. "The first landmark snaring the camera-ready eye of Space Coast tourists is not an Atlas rocket, but a mammoth postage stamp," said a disparaging article in *Florida Today* that year.[54] Vacant buildings, an unused conference center, and the seemingly deserted former Hilton hotel constituted the harsh reality of Gateway to the Stars, which the Apollo 8 postage stamp billboard still heralded. "The Freedom Center Hotel ages quickly now, a grand old dame whose family no longer comes to visit," lamented *Florida Today* in another piece in October 1974, adding that the hotel was a good spot to visit only if you "want to be alone."[55] But Carl McIntire did not share in the disappointment local residents felt. To his mind, the Cape Canaveral project was a miracle, one that a slow start in attracting guests could hardly dim. All Gateway to the Stars needed was time and patience. Despite the fact that few tourists were traveling to Florida in search of a religious awakening, McIntire still intended to lure them away from the excitement of Walt Disney World and give them precisely that.[56]

"This is a historic hour as we await the second coming of Christ and 'occupy' in places where it means the most," wrote McIntire in his typical bombastic style.[57] The tenets of McIntire's fundamentalist Christian beliefs indeed instructed the faithful to "occupy" the world in anticipation of the End Times and the Lord's imminent return, a theme he invoked repeatedly when he spoke about his Cape Canaveral project. His conviction that the space program was not just blessed by God but was integral to the ultimate triumph of Christian America over Soviet communism added eschatological urgency to his mission and ultimately justified—in his eyes, at least—his unwavering dedication to what most of central Florida by the mid-1970s viewed as a sinking ship. Tourist dollars merely helped pay the bills for another month, McIntire reasoned. The real business of Gateway to the Stars was saving souls, a focus that was evident in the themes of his Bible conferences. The events at his conference center pushed a very specific ideology that consisted largely of worship services and anticommunist seminars taught by McIntire and his close allies. In 1972, for example, the conference hosted a lecture about the dangers of rock music given by a professor from Bob Jones University, a fundamentalist Bible college.[58] Beginning in 1975 and continuing annually for several years, one of the organizations affiliated with McIntire, the 20th Century UFO Bureau, held conferences in which conservative evangelicals discussed alleged UFO sightings and attempted to reconcile them with the Bible.[59] These types of events hardly helped to assure Brevard County residents that McIntire's lofty goals for the complex would ever come to fruition. One look at the infamous Apollo 8 stamp billboard, a local journalist remarked, "may partially explain what has and has *not* happened since 1971. The sign was a six-cent stamp in a 10-cent stamp economy. Faded, rain-stained and solitary—yet, to beach residents a constant reminder of the pervasive McIntire presence on 300 acres of some of the choicest Florida soil money could buy."[60]

This commentary encapsulated the local mood by the mid-1970s, as McIntire's Cape Canaveral holdings lay dormant and negative remarks in central Florida newspapers replaced the steady stream of optimistic "what-ifs" that characterized early stories on Gateway to the Stars. A few articles appeared beginning in late 1971 on the state of Shelton College, McIntire's Bible college that he had recently relocated to Cape Canaveral, that offered colorful, tongue-in-cheek descriptions of the student body and the college's curricula. "Bumper stickers read *Honk if You Love Jesus* instead of *Make Love Not War*," one article wrote.[61] Discussions of the ill-fated Jerusalem Museum revealed a strong sense of collective resentment about the hype surrounding what McIntire pitched to the town and the lackluster reality of what existed three years later. One visitor to the museum in October 1974 asked a museum tour guide when the full-scale Solomon's Temple would finally be built there. The guide answered: "When they get the money."[62]

McIntire's ongoing financial woes were well known among local residents by 1974, especially when news leaked of disputes between the minister and the company who held the lease for the properties, Shuford Mills, a textile firm based in North Carolina. The owner of Shuford Mills was reportedly unhappy with the "considerable opposition" in Brevard County to McIntire's presence.[63] The *New York Times* reported in August 1974 that McIntire saw "sinister forces at work." He claimed that efforts to discredit him were not local but were instead a government plot to remove him from Cape Canaveral so he would not "embarrass anyone when Russian space experts arrive next year" to participate in the Apollo-Soyuz space docking missions.[64] His well-known anticommunist stance, he felt, meant that his presence so close to Kennedy Space Center was a liability to NASA. Cocoa Beach mayor Lee Caron remarked in the same article that it was "very depressing" to see the former Hilton hotel complex, which was not even seven years old and was once the nicest in the area, already "badly mildewed from neglect and misuse."[65] "They're not selling a damn thing out there," said the director of the Cape Kennedy Chamber of Commerce, George T. Cone. "It isn't even run like a professional hotel business. All they're doing is recruiting people for McIntire's dog-and-pony show."[66] In spite of everything, however, *Florida Today* ran a story on October 1, 1974, that revealed that McIntire had reached a deal with Shuford Mills to purchase the 300 acres, including the Freedom Center Hotel and several of the buildings used by Shelton College, for a total of $14 million. This agreement marked the culmination of a month of negotiations between McIntire and Shuford Mills and was a surprising about-face for Shuford, which had heavily criticized their lessee and had rejected a previous cash offer from him. "I think God changed Mr. Shuford's heart," McIntire commented.[67]

Shuford Mills required a down payment of $1 million to by December 27, 1974.[68] For the last three months of 1974, central Florida newspapers closely followed McIntire's quest to the necessary cash, reporting on every new development. McIntire, who had initially asked 40,000 of his dedicated radio listeners to each send a minimum donation of $100 to cover the costs, continued to appeal to his fundamentalist audience by emphasizing how crucial his mission in Cape Canaveral really was. "If Cape Canaveral is lost, it will be one of the great disasters of this century to God's people," he warned.[69] On December 23, 1974, *Florida Today* announced that McIntire was $450,000 short of his goal with only four days left.[70] By December 26, the total had risen to $910,000. "I believe the Lord is going to give it to me," he stated in a telephone interview the day before the deadline, his faith as steadfast as ever.[71] The next morning, December 27, Carl McIntire flew into the local airport to hand a check for the full $1 million to Shuford Mills representatives in person. As he confidently strode into the lobby of what was now *his* Freedom Center Hotel to the applause of around fifty supporters, he looked poised to give one of his trademark sermons. "A miracle of miracles," he declared to the small crowd. "The good Lord, God of heaven and Earth, has given it to us."[72]

After he had secured the $1 million and had purchased the properties he had been leasing in Cape Canaveral, McIntire declared that "we are going to keep it until we go up into the clouds," conjuring an image of the End Times and Christ's Second Coming that evangelicals would instantly identify with. "When that trumpet blows, I'm going higher than the rockets and I won't need any help from Cape Canaveral."[73] In the pages of the *Christian Beacon* and on his radio show, McIntire described the process of raising the down payment as a series of miracles that unfolded one by one as evidence of reveal God's blessing on his Cape Canaveral vision. Financing for Gateway to the Stars, he said, came from the "Bank of Heaven."[74] McIntire continued doing what he always did: appealing to his base for spiritual and financial support and waiting for God to deliver a triumph. His unwavering faith in the importance of Gateway to the Stars to Christian history never faltered.

However, controversies dominated news of the complex for the rest of its existence. In October 1975, a fire broke out in the former Chrysler building that had housed the sparsely stocked Jerusalem Museum, destroying the model of Solomon's Temple that had once been the main attraction.[75] A state fire marshal reported that the fire was unquestionably arson.[76] What was potentially more distressing was the fact that visitors who traveled to Gateway to the Stars as part of the vacation packages McIntire offered to like-minded Christians were less than enthusiastic about their experiences there. *Florida Today* reporters seemed to take pleasure in writing editorials about the increasingly dismal vacations of visitors to the hotel and conference facilities. An article from October 1976 told how four busloads of senior citizens from Pennsylvania had left Cape Canaveral unhappy with the hotel. They said that their "rooms were dirty, their food poor" and that they had been misled about the activities planned for them during their stay.[77] One woman who visited as part of the group labeled the entire experience as "a ripoff—a very carefully organized ripoff."[78]

In September 1975, less than a year after McIntire's "miracle" purchase of Gateway to the Stars, local newspapers reported that McIntire had decided to "quietly abandon" a sizeable portion of his real estate holdings in Cape Canaveral, allowing Shuford Mills to repurchase them. This included the former Chrysler building that housed the fire-damaged remnants of the Jerusalem Museum as well as some apartment buildings and a large swath of vacant land.[79] The Freedom Center Hotel and its adjacent conference center, the parking lot of which sat empty and overgrown with weeds, remained in Mc-

Intire's possession.[80] Both buildings remained in operation under McIntire and various associates for a total of two decades. After the mid-1970s, when Project Apollo reached its end and the local economy had failed to reap any financial rewards from McIntire's presence there, the hotel and conference center largely faded from view, at least in terms of media coverage. What had once held so much promise for a Cape Canaveral revival became a seldom-used shell of its former life as the popular Cape Kennedy Hilton, flanked by weather-beaten signs advertising attractions that were long gone, such as the ill-fated Jerusalem Museum.[81] Rumors of revitalization occasionally surfaced, most notably in 1986 when a former navy engineer proposed plans to work alongside McIntire's organization in order to establish a Space Camp–style facility at Gateway to the Stars similar to the popular children's camp at NASA's facilities in Huntsville, Alabama. But this plan failed to gain traction with investors and it quickly fell through.[82]

In the late 1970s and early 1980s, the hotel and conference center primarily hosted worship services led by ministers affiliated with McIntire, religious lectures, Christian films, and the annual UFO Conference.[83] On May 25, 1991, the *Orlando Sentinel* ran a story announcing that McIntire had finally sold the Freedom Center Hotel and L. Mendel Rivers Conference Center to a new owner, ending his lengthy tenure in Brevard County.[84] Gateway to the Stars was no more, and McIntire's vision of a Christian community that would reveal the biblical precedent for America's missions to the heavens was put to rest with it. "Once upon a time, when the locals were getting drunk on Mercury and Gemini journeys to the stars, Brevard was supposed to be what Disney World became," the *Orlando Sentinel* reminisced in 1986. "Advertising Bible studies and Christian movies, the Gateway was once much more."[85]

Conclusion

> My mission in Cape Canaveral is to point people to Jesus Christ so they might be saved and have eternal life.
>
> Carl McIntire quoted in Keefer, "McIntire Leads 'Cape Crowd'"

In its last years, Gateway to the Stars continued to serve as a physical manifestation, albeit a depressing one, of Carl McIntire's vision for Cape Canaveral as a great center for Christian pilgrimage. McIntire believed to the end that the combination of strong fundamentalist preaching at the Bible conference and the incomparable sensory experience of rocket launches from Kennedy Space Center would move hearts, even as the 1980s drew to a close and the world continued to change around him. Radio ministries such as his had largely been replaced by the celebrity and scandal of mass-marketed televangelism. Carl McIntire's with his continued warnings about the threat of global communism and his lack of attention to flashy presentation made him an anachronism. He was a relic from another era, the *Orlando Sentinel* said in June 1988, who contrasted sharply with the new breed of fundamentalist televangelists. "There are no Rolex watches visible, no precious rocks on his fingers. Today, the reverend looks just like a Sunday school teacher," the article said. "His black shoes are frayed around the edges. He could use a new pair."[86]

From virtually every angle—in terms of popularity, commercial success, or financial gain—the Gateway to the Stars was a resounding failure. Tourists never really materialized, the theming of the properties likely confused or even alienated potential guests, few of McIntire's supporters wished to take a leap of faith on a down payment on a Florida condominium, and a massive conference center that prohibited alcohol and smoking dried up any real opportunities to use the space for anything other than Christian-themed events or worship services. But McIntire's vision of Gateway to the Stars as a tourist attraction and a community that offered a fundamentalist Christian, politically conservative world view set the stage for later developments along similar lines. In 1978, popular televangelists Jim and Tammy Faye Bakker opened Heritage USA, a Christian theme park in South Carolina that eventually included a water park, television studios, a church, a hotel, conference facilities, and residential timeshares. By 1986, five years before McIntire's Gateway to the Stars closed for good, Heritage USA was the third most visited theme park in America, ranking behind only Disneyland in California and Walt Disney World in Florida.[87]

American patriotism and conservative evangelicalism also proved a successful tourist draw in 2001, when the Holy Land Experience theme park—described at the time as "the most ambitious religious tourist attraction in the nation"—opened down the road from Brevard County in Orlando.[88] Similar to what Carl McIntire envisioned for his Jerusalem Museum, Holy Land Experience offered visitors the chance to immerse themselves in full-scale reproductions of biblical locations that were intended to lead to spiritual awakening in a theme park environment. Patriotism also played a large role as Holy Land Experience, where dramatic stage shows that honored the military were featured alongside shows on the life of Christ.[89] Additionally, since 2007, the Creation Museum of the Answers in Genesis ministry in Petersburg, Kentucky, has presented a staunchly biblical and literalist Christian interpretation of a scientific subject—the origins of the universe. Like McIntire's vision of an all-encompassing visitor experience at Gateway to the Stars where tourists could understand the scriptural basis for space exploration, the Creation Museum offers visitors a heavily curated selection of scientific subjects though the filter of biblical prophecy, in the hope that a visit will lead people to the conclusion that young-Earth creationism is truth and that humans and dinosaurs lived alongside each other.[90]

Above all, Gateway to the Stars embodied Carl McIntire's dedication to separatist fundamentalism and its principle of existing in the world but not being of the world. The complex and the concept behind it seemed timeless, ahead of its time, and hopelessly outdated all at once. It offered visitors a way to experience the groundbreaking modern technologies of space exploration in a controlled, biblically minded environment, where McIntire could immediately interpret every rocket launch or tour of Kennedy Space Center through the lens of scripture and apocalyptic revelation. Hotel or conference center guests could opt to live in this setting full-time by purchasing a condominium, they could send their children to a fundamentalist Bible college right across the street, and they could attend daily worship services in the conference center. All of this existed in the shadow of America's "moonport," where humanity first extended its reach into God's heavens.

Despite apathy and disdain from the local community regarding his presence and the steadily dwindling number of guests, McIntire's passion managed to win over some of his harshest critics. When former Cocoa Beach mayor Lee Caron was interviewed in 1986, twelve years after commenting to the *New York Times* that McIntire's project was a disappointment for Brevard County, he told the *Orlando Sentinel* that he had had a change of heart as well as a religious conversion. "I never met a man I loved and respected as much," he said of McIntire, whose sermons at Gateway to the Stars led Caron to become a born-again Christian. "I never knew anyone could preach like that."[91] More than tourist money or economic revitalization, what Carl McIntire wanted in Cape Canaveral was to bring his interpretation of the Gospel to people in order to save their souls. If stories like those of the former mayor of nearby Cocoa Beach are any indication, he likely considered his mission accomplished.

Notes

1 Mary Ann Hill, "'The Hilton' Is Dead; Its 'Wake' Was Merry," *Florida Today*, February 1, 1971, 4A.
2 Carl Hiaasen, "McIntire's Canaveral: The Promised Land?" *Florida Today*, October 13, 1974, 10.
3 Hill, "'The Hilton' Is Dead," 4A.
4 Carl McIntire, "'Gateway to the Stars': The Miracle of Cape Canaveral," promotional pamphlet, n.d., Carl McIntire Collection, Princeton Theological Seminary Archives.

5 Thomas O'Toole, "Blues at the Cape," *Washington Post*, January 31, 1971, B2.

6 O'Toole, "Blues at the Cape."

7 Heather Hendershot, "God's Angriest Man: Carl McIntire, Cold War Fundamental-ism, and Right-Wing Broadcasting," *American Quarterly* 59, no. 2 (2007): 373–396.

8 Hendershot, "God's Angriest Man," 375.

9 "Stairway to the Stars," *Florida Today*, December 30, 1970, 4A.

10 Cecil Foister, "'Jerusalem' to Rise: Hilton to Become Religion Complex," *Florida Today*, December 29, 1970, 1A.

11 "London in Brevard," *Florida Today*, January 25, 1971, 1B. This article marks the earli-est mention of the name Gateway to the Stars for the complex outside of McIntire's promotional materials on the project.

12 Most historiography on the fundamentalist movement during the Cold War tends to dismiss McIntire as, in the words of George Marsden, "marginal to the national scene"; *Fundamentalism and American Culture*, 2nd ed. (Oxford: Oxford University Press, 2006), 232. Because of this, while McIntire warrants some mention in many scholarly works on the subject, he receives little serious attention. Matthew Sutton's *American Apocalypse: A History of Modern Evangelicalism* (Cambridge, MA: Belknap Press of Harvard University Press, 2014) briefly details McIntire's heavily publicized split with the National Association of Evangelicals (p. 288) as well as his anticommunism. One biography of McIntire exists by historian Markku Ruotsila (*Fighting Fundamental-ist: Carl McIntire and the Politicization of American Fundamentalism* [Oxford: Ox-ford University Press, 2016]), but it omits any discussion of McIntire's interest in the American space program, the space race, or his Cape Canaveral properties. Heather Hendershot's *What's Fair On the Air? Cold War Right-Wing Broadcasting and the Public Interest* (Chicago: University of Chicago Press, 2011) deals heavily with McIntire but is focused on his radio show and his activism surrounding it.

13 For a discussion of the Apollo 8 letter-writing campaigns on behalf of prayer in space and McIntire's involvement in them, see Kari Edwards, "Prayers from on High: Reli-gious Expression in Outer Space during the Apollo Era, 1968–76," *European Journal of American Culture* 39, no. 3 (2020): 297–316.

14 Carl McIntire, *The Biblical Significance of the Trip to the Moon*, 1969, brochure, Carl McIntire Collection, Princeton Theological Seminary Archives.

15 "Millions Visited Space Museum during 1970," *Florida Today*, January 16, 1971, 3B.

16 "Record Million Visitors Snarl Brevard Traffic," *Florida Today*, February 1, 1971, 1A.

17 Roger D. Launius, "Abandoned in Place: Interpreting the U.S. Material Culture of the Moon Race," *Public Historian* 31, no. 3 (2009): 13.

18 Boyce Rensberger, "Cape Hoping for the Best in Wake of Apollo's Boom," *New York Times*, December 8, 1972.

19 Ross Mark, "Epitaph for a Moonport," *Sunday Mail* (Glasgow, Scotland), February 21, 1971, 10.

20 Charles Reid, "Jerusalem Museum Outlined," *Florida Today*, January 23, 1971, 1A.

21 "Stairway to the Stars," *Florida Today*, December 30, 1970, 4A.

22 "Stairway to the Stars."

23 Thomas O'Toole, "Blues at the Cape," *Washington Post*, January 31, 1971, B2.

24 O'Toole, "Blues at the Cape."

25 Charles Reid, "Thunder on Right Rumbles into Town," *Florida Today*, January 24, 1971, 1A.

26 Mike Baxter, "Stormy Petrel of the Far Right," *Miami Herald*, February 7, 1971, 229.

27 Baxter, "Stormy Petrel of the Far Right."

28 "McIntire: 'Sacrilegious' Article Brings $10,000," *Miami Herald*, February 20, 1971, 13.

29 "Inquiring Photographer: How Do You Feel about fundamentalist minister, Carl McIntire, Coming to Brevard?" *Florida Today*, January 31, 1971, 3B.

30 Reid, "Thunder on Right Rumbles into Town."

31 In the wake of the Apollo 8 mission, the United States Postal Service announced a commemorative postage stamp honoring the historic flight. According to an article in the *Philadelphia Inquirer* from January 26, 1969, the stamp originally included the phrase "In the beginning God . . ." printed over the Earthrise image, but the post office removed the quote out of fear of protest. McIntire saw the article and urged listeners to his radio program to write President Nixon imploring him to have the Genesis quote restored. On February 27, 1969, after correspondence between McIntire and Post Office Director of Philately Virginia Brizendine, the post office relented and included the quote on the stamp's design. The stamp was released on May 5, 1969. Letters between McIntire and the Post Office as well as copies of letters sent by listeners to the president can be found in the Carl McIntire collection at Princeton Theological Seminary.

32 Suzi Wilson, "McIntire Unveils Giant Billboard: Apollo 8 Postage Stamp Replica," *Florida Today*, January 29, 1971, 1B.

33 "Cape 'Uplifts' McIntire," *Florida Today*, January 23, 1971, 6A.

34 "Bible Supports Space Exploration—McIntire," *Florida Today*, September 20, 1971, 1B.

35 Carl McIntire, "Gateway to the Stars," *Christian Beacon*, January 7, 1971, 1, Carl McIntire Collection, Princeton Theological Seminary Archives. McIntire imagined his Cape Canaveral property as a place where a city within a city would emerge that was full of neighborhoods and had all the amenities residents would need.

36 Carl McIntire, "Gateway to the Stars Christian Community," promotional brochure, n.d., Carl McIntire Collection, Princeton Theological Seminary Archives.

37 "Gateway to the Stars: Cape Canaveral Bible Conference," promotional pamphlet, 1978, Carl McIntire Collection, Princeton Theological Seminary Archives. This pamphlet includes a welcome message from the director of the complex at the time of printing, Rev. James Blizzard. The text of the pamphlet may have written by Blizzard, not McIntire.

38 David E. Nye, *American Technological Sublime* (Boston: MIT Press, 1996), xvi.

39 Nye, *American Technological Sublime*, 43.

40 Nye, *American Technological Sublime*, 239.

41 "Gateway to the Stars: Cape Canaveral Bible Conference."

42 Al Marsh, "Ante, Stakes Raised in McIntire Gamble," *Florida Today*, July 25, 1971, 1A.

43 Marsh, "Ante, Stakes Raised in McIntire Gamble."

44 Jerry Lipman, "McIntire to Build High-Rise," *Florida Today*, June 21, 1971, 1B.

45 Charles Reid, "The Rev. Carl McIntire: New Man in Our Town," *Florida Today*, April 28, 1971, 15E.

46 Marsh, "Ante, Stakes Raised in McIntire Gamble."

47 McIntire, "Gateway to the Stars," 1, Carl McIntire Collection, Princeton Theological Seminary Archives.

48 Carl McIntire to Dorothy Plack, September 28, 1971, Carl McIntire Collection, Princeton Theological Seminary Archives.

49 Carl McIntire, "Onward, Christian Soldiers," *Christian Beacon*, 39January 2, 1975, 2.

50 John Winter, "McIntire's Temple of Jerusalem," *Florida Today*, December 19, 1971, 14.

51 "Dedication to Honor Rivers," *Florida Today*, November 6, 1971, 10C.

52 Edwin E. Riley Jr., "McIntire Outlines 'New Jerusalem' Plans," *Florida Today*, November 6, 1971, 1B.

53 Winter, "McIntire's Temple of Jerusalem."

54 Carl McIntire, "Canaveral Saved," 1.

55 Hiaasen, "McIntire's Canaveral."

56 "Bible Supports Space Exploration—McIntire," *Florida Today*, September 20, 1971, 1B.

57 Carl McIntire, "The Miracle of Cape Canaveral," promotional pamphlet, 1974, Carl McIntire Collection, Princeton Theological Seminary Archives.

58 "'Dangers of Rock' Lecture Set," *Florida Today*, November 16, 1972, 6D.

59 The earliest mention of the UFO conferences and the 20th Century UFO Bureau in Cape Canaveral appeared in the fall of 1972 in an announcement of the first session of what UFO Bureau director Robert Barry hoped would be a yearly conference on the subject of UFOs, extraterrestrial life, and Christian prophecy at Gateway to the Stars. See "UFO Seminar Planned," *Florida Today*, November 12, 1972, 5B. The final mention of the annual conference in *Florida Today* came in 1981; Billy Cox, "Blips, Discs and Weird Lights Discussed at UFO Symposium," *Florida Today*, May 8, 1981, 9.

60 Hiaasen, "McIntire's Canaveral."

61 Hiaasen, "McIntire's Canaveral."

62 Hiaasen, "McIntire's Canaveral."

63 Donald Janson, "Right-Wing Cleric May Face Eviction," *New York Times*, August 23, 1974, 36.

64 Janson, "Right-Wing Cleric May Face Eviction."

65 Janson, "Right-Wing Cleric May Face Eviction."

66 Hiaasen, "McIntire's Canaveral."

67 Carl Hiaasen, "McIntire Group To Buy Property," *Florida Today*, October 1, 1974, 1A.

68 Hiaasen, "McIntire Group To Buy Property."

69 Carl Hiaasen, "Freedom Center Sale Deadline Nears," *Florida Today*, December 23, 1974, 1B.

70 Hiaasen, "Freedom Center Sale Deadline Nears."

71 Charles Keefer, "McIntire $90,000 Off for Gateway Control," *Florida Today*, December 27, 1974, 1B.

72 Charles Keefer, "McIntire Buys Gateway," *Florida Today*, December 28, 1974, 2B.

73 Charles Keefer, "McIntire Leads 'Cape Crowd,'" *Florida Today*, January 5, 1975, 1B.

74 Carl McIntire, "The Bank of Heaven," *Christian Beacon*, January 2, 1975, 1.

75 Glenn Singer, "State to Report on McIntire Fire," *Florida Today*, October 28, 1975, 3B.

76 "Museum Fire Set, Officials Tell McIntire," *Orlando Sentinel*, November 19, 1975, 20.

77 Glenn Singer, "Tourists Criticize Stay at McIntire Motel," *Florida Today*, October 22, 1976, 1B.

78 Singer, "Tourists Criticize Stay at McIntire Motel."

79 Glenn Singer, "McIntire To Yield Part of Complex," *Florida Today*, December 9, 1975, 1A.

80 Bill Belleville, "Work Begins on Vietnamese Village," *Florida Today*, September 4, 1975, 1B.

81 Billy Cox, "All Is Not Classy in Fair Brevard," *Orlando Sentinel*, August 17, 1986, 67.

82 Michael Lafferty, "Supporter Launches Plan for Youth Space Facility," *Florida Today*, July 18, 1986, 2.

83 Gateway to the Stars regularly ran advertisements in local newspapers inviting the public to events held there. They included public showings of Christian films and regular worship services run by Rev. James Blizzard, the director of the complex for many years, by Blizzard's family and associates, and by McIntire himself. Lectures in the style of the Bible conference events McIntire envisioned in 1971 were also held infrequently. The UFO Conference ran annually until at least 1981.

84 Tony Boylan, "Island Vacation in Cape Canaveral," *Orlando Sentinel*, May 25, 1991, 34.

85 Cox, "All Is Not Classy in Fair Brevard."

86 Billy Cox, "The World According to McIntire," *Orlando Sentinel*, June 5, 1988, 39.

87 Megan Rosenfeld, "Heritage USA & the Heavenly Vacation," *Washington Post*, June 15, 1986, H1.

88 Mark Pinsky, "New Theme Park Has That Old-Time Religion," *Washington Post*, January 20, 2001, B9.

89 Helena Chmielewska-Szlajfer, "'Authentic Experience' and Manufactured Entertainment: Holy Land Experience Religious Theme Park," *Polish Sociological Review* 4, no. 200 (2017): 549.

90 Mark Pinsky, "Inherit the Spin," *Orlando Sentinel*, June 10, 2007, F3.

91 Cox, "The World According to McIntire."

11

"Put Us on Display"

The Making of the Alabama Space and Rocket Center, 1958–1970

EMILY A. MARGOLIS

Amid celebrations of the fiftieth anniversary of the Apollo moon landings, another semicentennial quietly passed in Huntsville—the half-century since the opening of the U.S. Space & Rocket Center. Since March 1970, the center (known until the 1980s as the Alabama Space and Rocket Center) has showcased the missile and rocketry research of the nearby Redstone Arsenal and Marshall Space Flight Center to an audience of 17 million visitors and counting.[1] Consisting of a museum, an outdoor rocket garden, and camp facilities for aspiring astronauts and aviators, the U.S. Space & Rocket Center is Alabama's most visited ticketed tourist attraction.[2] To the residents of "Rocket City," the center is much more than a museum. It is a source of pride and generator of profit, a workplace, a destination for school field trips, and even a wedding venue. To date, little has been written about the history of the center, an institution as integral to Huntsville's economy and identity as Redstone and Marshall.

This chapter traces the origins of the U.S. Space & Rocket Center to one enterprising citizen, Walter V. Linde, who recognized in Huntsville's emerging reputation as Rocket City an opportunity to grow the local economy through tourism. His twelve-year quest to build a space museum reached every corner of the state, from the University of Alabama's Denny Stadium to the Governor's Mansion. The ultimate form and function of the museum reflected a decade of debate and dreams from diverse stakeholders at the local and state levels. The making of the center also reveals how Alabamians instrumentalized their relationship to national space activities for ends that often had little to do with spaceflight.

Becoming Rocket City, 1952–1958

At the turn of the twentieth century, the northern Alabama town of Huntsville was known, if at all, as the "Watercress Capital of the World."[3] By mid-century, Huntsville had embraced an emerging identity as space capital of the universe. National attention on the missile and rocketry research at the US Army's Ordnance Rocket Center and later the Army Ballistic Missile Agency at Redstone Arsenal catalyzed the creation and adoption of this new identity. The activities of Wernher von Braun, the former Nazi who had developed the V-2 rocket for Germany and who now directed technical operations for US army missile research, focused the national spotlight on Huntsville as an epicenter of research and development for US space technologies.[4]

From 1952 to 1958, von Braun used his expertise, office, and charisma to elevate the status of spaceflight from fictional to feasible in the public consciousness. Working alongside a cohort of "space boosters" that included scientists, engineers, science fiction authors, and space artists, von Braun created educational and inspirational content for print media and television.[5] One of the boosters' most visually stunning and impactful activities was a collaboration with *Collier's*, a magazine with 12 million weekly readers, on a series of special issues dedicated to the topic of spaceflight. From March 1952 to April 1954, von Braun contributed seven articles to this series under the byline "Dr. Wernher von Braun, Technical Director of the Army Ordnance Guided Missiles Development Group in Huntsville, Alabama."[6] Inspired by these popular writings, Walt Disney invited von Braun to consult on and present three space-themed episodes for the *Walt Disney's Disneyland* television series from 1955 to 1957.[7] More than 42 million people tuned in for each episode, nearly one-quarter of the population of the United States.[8] Through his work to popularize spaceflight, von Braun became a household name and put Huntsville on the map, as evidenced by the thousands of letters from children and adults alike, that he received at his office in Huntsville.

The spotlight on von Braun and Huntsville intensified on January 31, 1958, when the United States launched its first successful satellite atop a Jupiter-C rocket that had been designed and manufactured under his supervision. The following day, newspapers carried a photo of von Braun and collaborators grinning widely at a 2 a.m. press conference at the National Academy of Sciences in Washington, DC, triumphantly holding a model of the Explorer satellite overhead. *Time*, one of the most respected and widely read news magazines of the day, profiled him in its February 17, 1958, issue. His portrait appeared on the cover above the caption "Missileman von Braun."[9] The issue also featured a brief introduction to Huntsville in a one-page article titled "Rocket City, USA."[10] Huntsville's renown, cultivated on screen and in the pages of *Collier's* and *Time*, quickly manifested on the streets of Madison County.

Proposing the Space Museum, 1958–1960

One enterprising resident took notice of and inspiration from Huntsville's newfound attention. In his role as the proprietor of a chain of Phillips 66 service stations, Walter V. Linde had his finger on the pulse of Huntsville's arterial roadways. Shortly after the successful launch of the Explorer satellite, he and his staff began receiving inquiries from vacationing motorists who wanted to "see missiles and other signs of the space age" as they passed through Huntsville.[11] In these requests Linde recognized an opportunity to make Rocket City into a destination rather than a detour for American tourists. He conceived of a space museum to attract visitors from near and far and spent the next twelve years enlisting the most prominent Alabamians and Alabama institutions in his cause. Linde's idea would change Huntsville and the nation's perception of it for generations to come.

Linde's line of work uniquely positioned him to witness the growth of the American tourism industry, which centered on the family road trip. The postwar years marked the golden age of the family vacation, as millions of newly affluent and mobile middle-class Americans sought relaxation and recreation along the nation's expanding interstate highway system.[12] By the mid-1950s, vacationers were spending almost $25 billion annually. Linde began chasing tourist dollars while living and working in Tuscaloosa, Alabama. In 1956, he chaired the Publicity and Tourism Committee of the Tuscaloosa Chamber of Commerce.[13]

Later that year, Linde relocated his young family to Huntsville, where he took the helm of a chain of Phillips 66 service stations. Business promised to be brisk in Huntsville. As the army continued to grow its missile facilities at the Redstone Arsenal in 1950–1960, the city's population increased by 340 percent.[14] According to his daughter Debbie Linde, he was also drawn to Huntsville because of his fascination with spaceflight and his sense of affinity with von Braun (Linde believed they shared a German Lutheran heritage).[15] In Huntsville, Linde's personal interest in spaceflight intersected with his pursuit of tourism.

In late 1958, Linde conceived of a museum of spaceflight and rocketry for Huntsville. The queries at his service stations and the popularity of a display of US Army missiles and rockets at the Alabama State Fair in the summer of 1958 convinced Linde of the demand for such an attraction.[16] He approached the Huntsville-Madison County Chamber of Commerce in 1958, noting that at present, "There is no place in the United States for youngsters to go to in order to keep up with the missile development."[17] He believed that Huntsville, proud home of the Ordnance Rocket Center at the Redstone Arsenal, should provide a place for America's youth to learn about national efforts in space.[18] Linde explained that Rocket City had an obligation and opportunity to educate the next generation of scientists and engineers on whose shoulders Cold War lawmakers placed the burden of national security. As science education became codified as a national imperative with the National Defense Education Act of 1958, museums increasingly asserted their importance as informal science classrooms.[19] While meeting this national need, the space museum would also bring financial rewards. Linde estimated that "a space museum would be the equivalent of [adding] a new industry" to the Huntsville economy.[20]

The chamber of commerce, desperate to diversify an economy that had been precariously dependent on defense spending since the opening of the Redstone and Huntsville Arsenals in 1941, was receptive to Linde's idea. Tourism was an underdeveloped industry that had interested locals for years. As early as 1951, the chamber attempted to promote Huntsville's hospitality sector through a "Highway Holiday" program. Its purpose was to convert passing motorists into tourists. Law enforcement officers would pull the motorists off the highway and then "arrest" and "jail" them overnight in one of Huntsville's finest inns. The unsuspecting tourists received a complementary dinner and gifts from fifty merchants.[21] This unusual program was most certainly intended for white travelers, as traffic stops could be sites of harassment and violence against Black motorists.[22] However, the "Highway Holiday" program failed to establish Huntsville as a desirable destination for vacationers.

The chamber welcomed Linde's suggestion and in late 1958, President James Walker established the Space Museum Committee and installed Linde as chair. The committee's first major action was to commission a feasibility study with the financial support of the City Council.[23] In spring 1960, the chamber let a contract with the Southern California Laboratories of Stanford Research Institute (SRI), a nonprofit research arm of Stanford University that specialized in economic analyses in the arts and entertainment sector.[24] Tasked with estimating the market potential, identifying a size that would be profitable, and forecasting financial returns for the proposed space museum, SRI dispatched a team of economists and "techno-economists" to Huntsville. The Stanford experts also conducted site visits at twenty-two comparable attractions across the southeastern United States and contacted twelve museums, planetariums, and other facilities throughout the country, including Disneyland.[25]

SRI had previously consulted for the Walt Disney Company and looked to its successful Anaheim, California, theme park as inspiration for the proposed space museum in Huntsville. Whereas Linde imagined a static display of missiles modeled after the state fair exhibition, SRI encouraged the Space Museum Committee members to consider the new vogue in exhibit design. In his report from a visit to the SRI offices in Pasadena, California, Linde wrote, "Stanford Research was most anxious to see us change our concept to an educational participation type and steer away from the static type." During their visit to California, SRI researchers facilitated a meeting between the Huntsvillians and Dick Irvine, chief designer for Disneyland, and a visit to Disneyland to experience interactive exhibits in situ.[26]

In June 1960, the SRI experts published their findings in a 190-page report titled "Economic Feasibility of a Proposed Space Science Exhibit in Huntsville, Alabama," which validated and quantified Linde's assertions that a space museum would be both popular and profitable. SRI researchers estimated that an attraction featuring educational exhibits as well as amusements would draw more than half a million tourists to Huntsville annually. They projected that the visitors would spend approximately $1 million in the city each year. The report projected how tourist dollars would flow throughout the service, retail, and entertainment sectors of the economy. The researchers estimated that a family would spend only 4 percent of its vacation budget on admission to the space museum; they would spend the remainder at Huntsville's restaurants, shops, theaters, hotels and motels, service stations, and laundromats.[27]

In all, the proposed space museum promised "the same financial benefit to the community that would come from a small factory."[28] In the recent past, Huntsville had been a town of factories; from the 1880s through the Depression, textile mills that processed locally grown cotton had dominated the city's economy.[29] Since the opening of the Redstone and Huntsville Arsenals in 1941, the livelihoods of most Huntsvillians had been supported, directly or indirectly, by defense work.[30] Although a space museum would showcase aerospace research and development in Huntsville, the tourism industry it supported would be "largely independent" of that industry.[31] In that way the museum was positioned to diversify the economy while celebrating its most important pillar.

Huntsville's elite received SRI's findings with enthusiasm. Mayor R. B. Searcy exclaimed of the proposed space museum, "This is the greatest thing that can happen to this community." Senator John J. Sparkman was "intrigued by it—greatly so," and Representative Robert E. Jones Jr. found it "most exciting." Endorsements also rang out from the Alabama Section of the American Rocket Society and the leadership of the Redstone Arsenal.[32] "Missileman" von Braun, the man whose renown paved the way for a space museum, even briefed members of Alabama's congressional delegation in Washington about the importance of this project.[33]

Selling the Space Museum, 1960–1965

Despite early excitement, stakeholders from the chamber and from city government deliberated for nearly five years about the best course of action following from the economic feasibility study. During that time, Linde worked to maintain momentum and build broad interest in the project. He was eager to engage the entire community in the effort, understanding that Huntsville's residents were both prospective financiers of the museum and part of its intended audience. To illustrate the principles of the SRI report, Linde's alter ego, "Mr. Tourist," paid surprise visits to civic events and social gatherings throughout Madison County. Wearing a suit covered in 500 $1 bills, "Mr. Tourist" dramatically tossed "his hard-earned money to and fro for food, lodging and leisure during his annual vacation" (fig. 11.1).[34] Linde's schtick amused his audiences and was widely reported on in the *Huntsville Times* and *Huntsville News*.

Another Huntsville entrepreneur, Hubert R. Mitchell, was eager to capitalize on SRI's findings, as Drew Adan details in this volume. In 1959, Mitchell, a keen observer of leisure culture who operated businesses in the entertainment sector, conceived of an amusement park in Huntsville to serve vacationing families across the Southeast.[35] Mitchell selected space as his amusement's central theme because "every space launching brings complete coverage throughout America in newspapers, radio and television" of the space activities in Huntsville.[36] Like Linde, Mitchell intended to capitalize on this free and abundant publicity for his attraction, Space City USA.

As Mitchell imagined it, Space City USA would be as ambitious an undertaking as the US space program itself. It would include an amusement park to rival the most popular in the nation and an expansive complex of lodgings, convention facilities, camping and fishing grounds, retail stores, and a permanent exhibition of aerospace hardware.[37] Yet even though this enticing vision attracted investors from across the United States, it failed to launch.[38] Significantly, the chamber of commerce, whose support helped local businesses grow and thrive, was not involved in the effort. Its attention was elsewhere.

As Mitchell courted investors, the chamber debated how to finance and organize a space museum. A solution manifested in Mobile. In 1962 a group

Figure 11.1. Walter Linde dressed as "Mr. Tourist." Source: *Decatur Daily*, October 16, 1963. Courtesy of *Decatur Daily*.

of citizens, led by the Mobile Area Chamber of Commerce, rallied together to build a permanent home in Mobile Bay for the USS *Alabama*, a retired World War II–era battleship. Their efforts grew into the "Mighty A" campaign, which began with a statewide solicitation for donations and culminated in the establishment of a nonprofit educational institution overseen by a state-run commission. The success of the USS Alabama Battleship Memorial Park, which opened on January 9, 1965, before a crowd of 2,000, convinced the Space Museum Committee that a state-run commission was the way forward for Huntsville, too.[39]

To set up such a commission, the committee needed to pursue an amendment to the state constitution. Through a quirk of Alabama politics, legislation through amendment, the governor could form a state-run commission only by writing it into the state constitution.[40] Huntsville legislators championed the space museum in Montgomery. In summer 1965, state representative Harry Pennington introduced a bill to the house that proposed that the state form an eighteen-member commission to oversee the space museum and issue $1.9 million in bonds to finance the project.[41] The bill passed the legislature on August 26, 1965, with the senate consenting unanimously.[42] Now the fate of Huntsville's space museum rested in the hands of Alabama voters.

In preparation for the general election on November 30, Linde formed a steering committee to lead the campaign for Amendment 3, as the space museum proposal was known. To this role he brought organization, creativity, and a flair for public relations. Linde emphasized personal interactions and recruited 166 volunteers from across the state to promote Amendment 3 in their communities.[43] The volunteers arranged speaking engagements for experts from Redstone Arsenal and Marshall Space Flight Center with "practically every club, organization and association existing in Alabama."[44]

Linde also engaged "all news media, newspapers, magazines, radio, television, billboards, etc." to urge Alabamians to "Vote Yes."[45] He arranged for WAFG-TV, the local ABC affiliate in northern Alabama, to broadcast an informational panel about the proposed space museum.[46] One of Huntsville's two daily papers, the Huntsville Times, often published favorable articles on the space museum and later endorsed the amendment.[47] Directing voters to "Put Us on Display," editors of the Huntsville Times wrote that the space museum "deserves the support of the whole community" because it "would be North Alabama's most popular tourist attraction."[48]

Dave L. Christensen, the vice-president of the failed Space City amusement park and a member of the Space Museum Steering Committee, developed a promotional pamphlet for statewide circulation in advance of the November 1965 election (fig. 11.2). The trifold brochure included endorsements from prominent Alabamians, including von Braun; Governor George C. Wallace; Major General John G. Zierdt, commanding general of the Army Missile Command; and Paul "Bear" Bryant, head coach of the University of Alabama football team. Bryant was the official chair of the committee for Amendment 3, although Linde did most of the legwork.[49] Bryant, who had led the Crimson Tide to national titles in 1964 and 1965, was beloved throughout the state (except, perhaps, in Auburn).

The brochure outlined two advantages that Huntsville's space museum would bestow on Alabama. First, the exhibitions would foster "pride in Alabama's growth and national responsibilities." Bryant heralded the attractions as "a public show place to tell the world of Alabama's great accomplishments in perfecting space and missile technology." Writing as "a citizen of Alabama" rather than as an expert rocket engineer, von Braun celebrated that the pro-

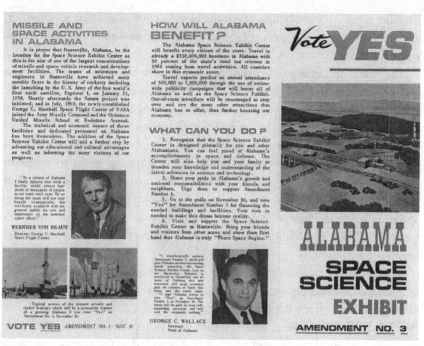

Figure 11.2. "Vote YES" brochure in support of Amendment 3, 1965. Courtesy of U.S. Space & Rocket Center Archives.

posed museum would "firmly establish with the general public [Alabama's] role and importance in the national space effort."[50] These endorsements reveal a significant amplification in the museum's audience and message. Whereas Linde aimed to educate America's youth about the science of spaceflight, the Steering Committee suggested that the museum would educate citizens of America and the world about the achievements of space-age Alabama.

Second, the brochure highlighted the financial benefits of the proposed space museum. Governor Wallace assured voters that the museum would bring "economic gain for citizens of South Alabama and the entire state."[51] Readers learned that Rocket City would become a gateway for tourists to explore Alabama's other historic sites and points of interest and would thus increase spending and tax revenue statewide.

Coded within the brochure was an acknowledgment of Alabama's dim national and global reputation. At this time, Alabama had the ignoble distinction as the home of more incidents of racially charged violence than any other state. When civil rights activists peacefully organized against segregation and other forms of oppression in Alabama, their efforts were frequently met with violence from the state and from white supremacists, including the terrorist

Ku Klux Klan. To outside observers, the bloody encounters that appeared on television and in newspapers suggested that Alabama was a hopelessly backward place.[52]

The space museum would challenge these impressions by "informing the many visitors of our progress"—in space, if not on the ground.[53] Huntsville was the ideal host city for such an attraction, having integrated more quickly than the rest of the state. As historian Brian Odom explains, Huntsville's elected leaders and business community "conceded to the demands of the [civil rights] movement when it became necessary to do so" in order to maintain federal patronage. Yet they did so "in ways that failed to address deep-seated structural economic and social inequality."[54] In fact, the growth of the aerospace industry in Huntsville "produced new disparities" between the city's white and Black residents due to direct and indirect discrimination in hiring.[55]

Governor Wallace understood the potential economic benefits of rehabilitating Alabama's image. His infamous commitment in his 1963 inaugural address to "segregation now, segregation tomorrow, segregation forever!" had "placed added pressure on industrial recruitment."[56] Civil rights activists organized demonstrations and protests that objected to the role of industrial expansion in perpetuating racial inequality and oppression. Students from Alabama A&M College, a historically Black college in Huntsville, brought this message to the epicenter of the American financial system, the New York Stock Exchange, with fliers that said, "To bring in new plants and businesses to Huntsville aids segregation and subjects additional employees to racism."

Other activists demonstrated at the factory gates. In 1965, Martin Luther King Jr., who had recently been released from jail following his arrest during Wallace's violent crackdown of the Selma to Montgomery march for voting rights, joined students to protest a new Hammermill Paper Company facility in Selma. After that protest, "many companies declined to risk the kinds of criticism, boycotts, or demonstrations directed against Hammermill because of its willingness to locate in an area where racial prejudice was so blatant."[57] A manager from Northrop Corporation explained his difficulty in attracting talent to the aerospace company's Alabama outpost in a letter to Governor Wallace. He described how "two engineers with graduate degrees who agreed to move to Huntsville changed their minds" after observing the "racial mess" in the American South.[58] Despite subsidies and tax incentives, industrialists hesitated to invest in a state where social conditions threatened their corporate image and their efforts to recruit and retain employees.

Wallace may have understood the proposed space museum as an opportunity to reconcile his problematic racial politics with his economic agenda.

The museum would affirm that despite the ongoing struggle for civil rights, Alabama could produce and attract sufficient talent to conquer the most significant technical challenge in the nation's history. By extension, if Alabama could support and sustain the high-tech aerospace industry, any industry could thrive there.

Making the Alabama Space and Rocket Center, 1966–1970

The majority of Alabama voters found the "Vote Yes" appeal compelling. Over 75 percent of the more than 120,000 people who voted favored the ballot measure.[59] After the election, Wallace convened the Alabama Space Science Exhibit Commission to oversee the financing, construction, and operation of Huntsville's space museum. He appointed eighteen members, including Lieutenant Governor James B. Allen, state representative Harry L. Pennington, Huntsville mayor Glenn Hearn, and Wernher von Braun.[60] As one of their first acts, the commissioners asked Economics Research Associates, a Los Angeles–based firm, to advise it on implementing the space museum concept.[61] Linde, an honorary commissioner, continued with much of the heavy lifting.[62] He successfully negotiated the donation of 400 acres of Redstone Arsenal real estate for the site of the space museum.[63]

The Economics Research Associates report that landed on commissioners' desks in August 1966 provided a road map for realizing Linde's vision. Echoing the sentiments of the "Vote Yes" brochure, the researchers acknowledged that the "principal purpose" of the space museum, now officially named the Alabama Space and Rocket Center, was "to commemorate and project to the people of Alabama, the United States and the world, the contributions of the U.S. Army Missile Command at Redstone and NASA's Marshall Space Flight Center." Recognizing the inherent tension between military and civilian applications in space, Economics Research Associates suggested organizing the museum around the unifying theme of freedom. "However unrelated these goals may seem," the report's authors explained, "they come together in the pursuit of freedom—the Army in the military defense of freedom and NASA in search of free horizons."[64] However, Economics Research Associates failed to recognize the problems with centering the new attraction on the theme of freedom while many Alabamians continued to struggle for freedom from racial oppression.

The task of realizing the recommendations of Economics Research Associates fell to Edward O. Buckbee, public affairs officer at Marshall Space Flight Center. At von Braun's suggestion, the commission named Buckbee founding director of the Alabama Space and Rocket Center in December 1967.[65] With expertise in communicating and interpreting NASA's activities to general audiences and strong connections to Marshall and Redstone, Buckbee was well suited for the job. He spent the next three years acquiring hardware from his colleagues at NASA and Redstone Arsenal, contracting exhibition designers, hiring staff, and promoting the museum throughout the Southeast. At the recommendation of Economics Research Associates, Buckbee sourced part of the center's collection from objects displayed at Marshall's modest in-house exhibition, known as the Space Orientation Center, which had served official visitors and employee families since 1963.[66] A particular challenge, and one that Buckbee was very persistent about, was cobbling together components of the Saturn V moon rocket from NASA centers and contractors across the country. When Buckbee succeeded, it was the only example of this rocket on display in the United States at the time and a point of pride for the center.[67]

Advertised as the "biggest space museum in the free world," the Alabama Space and Rocket Center opened to the public on March 17, 1970 (fig. 11.3). The center included a museum building modeled after a blockhouse (a partially underground structure that housed launch operations) and an outdoor "rocket garden" with displays of rockets and missiles. The visitor experience commenced with an eleven-minute film, *Freedom to Explore*, that highlighted Huntsville's essential role in securing and advancing the American way of life. This film established a framework for understanding military and civilian space activities as two sides of the same coin.[68]

The first two galleries offered primers in astronomy and space science. The Space Dimensions gallery took visitors on a walking tour of the solar system in miniature. The next room, Military and Space Sciences, featured exhibits and hands-on activities that explained the fundamentals of rocketry, such as the principle of propulsion. That gallery opened onto the museum's central atrium, where visitors encountered space hardware, including the Apollo 6 command module and Saturn V instrument unit. The atrium offered points of access to the Military and Space Science Applications gallery, a shrine to Project Apollo and the work of Redstone Arsenal, and a mezzanine-level exhibit that celebrated Wernher von Braun.[69] The rocket garden featured NASA rockets and army missiles displayed in vertical launch positions to stunning effect. The behemoth Saturn V moon rocket, which was displayed horizontally, was the central attraction, along with a mockup of the Apollo 11 landing site, Tranquility Base, complete with a lunar module and, at designated times, demonstrations of a lunar rover, or "moon buggy."[70]

In the first ten months of operation, the museum welcomed over 200,000 visitors from across the United States.[71] Since then it has remained one of Alabama's best-attended and highest-earning attractions, thanks, in part,

Figure 11.3. Aerial view of the Alabama Space and Rocket Center, ca. 1970. Courtesy of Alabama Department of Archives and History.

to the opening of Space Camp at the museum in 1982 (and with the added attraction, Alabama Space and Rocket Center rebranded itself as the U.S. Space & Rocket Center) and to the refurbishment and expansion of the exhibits. On January 31, 2008, in celebration of the fiftieth anniversary of the launch of the Explorer 1 satellite, the U.S. Space & Rocket Center opened the 70,000-square-foot Davidson Center for Space Exploration, which proudly recounts Huntsville's role in the history of Project Apollo.[72]

Conclusion

At the time Linde conceived of the space museum in 1958, America's presence in space consisted of a few small satellites. By the time the Alabama Space and Rocket Center opened to visitors in 1970, Huntsville's engineers had placed four astronauts on the moon. During the intervening twelve years, as the Huntsville-Madison County Chamber of Commerce engaged stakeholders across Alabama, pride and profit from tourism and industrial development joined education as essential objectives for this attraction. Today, residents, businesses, and civic leaders continue to embrace and promote Huntsville's identity as Rocket City, an identity now informed as much by the U.S. Space

& Rocket Center as by Marshall Space Flight Center and Redstone Arsenal. Yet the origins of this important and impactful attraction remain obscured by myth. As the center recounts on its website, at the same time that von Braun was perfecting the Saturn V moon rocket, "he was also preparing to launch another important project: a permanent exhibit to showcase the hardware of the space program."[73] Although von Braun's celebrity created demand for a space attraction in Huntsville, the role he played in the campaign for Amendment 3 was that of a supporter. Credit for the concept and the momentum behind the project belongs to Linde alone.

Misattributing the museum's origins not only erases the contributions of Linde and his collaborators, it also removes the U.S. Space & Rocket Center from the social, economic, and political context out of which it formed. The history of the center shows how a community instrumentalized its connection to the space agency to advance aims wholly unrelated to spaceflight. Viewing the attraction through the eyes of Rocket City residents rather than through those of famed rocket engineers reveals new dimensions in spaceflight's relationship to tourism, the civil rights struggle, industrial expansion, and science education in Cold War Alabama.

Notes

1 "About Us: History and Overview," U.S. Space & Rocket Center, https://www.rocketcenter.com/aboutus#ourstory.

2 Ken Roberts, "Alabama Ranks Top Tourist Attractions," *Tuscaloosa News*, January 28, 2019.

3 Christopher Lang, "The Huntsville Depot and Dennis Watercress," *Huntsville Historical Review* 31, no. 1 (2006): 35.

4 Michael J. Neufeld, *Von Braun: Dreamer of Space, Engineer of War* (New York: Vintage, 2008) is the definitive biography of von Braun.

5 De Witt Douglas Kilgore, "Engineers' Dreams: Wernher von Braun, Willy Ley, and Astrofuturism in the 1950s," *Canadian Review of American Studies* 27, no. 2 (1997): 104. For more information on space booster Willy Ley, see Jared S. Buss, *Willy Ley: Prophet of the Space Age* (Gainesville: University Press of Florida, 2017).

6 Von Braun penned or contributed to articles for *Collier's* on March 22, October 18, and October 25, 1952; February 28, March 14, and June 27, 1953; and April 30, 1954. Catherine L. Newell, "The Strange Case of Dr. von Braun and Mr. Disney: Frontierland, Tomorrowland, and America's Final Frontier," *Journal of Religion and Popular Culture* 25, no. 3 (2013): 419.

7 For more information on Walt Disney's approach to science programming, see J. P. Telotte, "Disney in Science Fiction Land," *Journal of Popular Film and Television* 33, no. 1 (2005): 12–20.

8 Newell, "The Strange Case of Dr. von Braun and Mr. Disney," 421.

9 "Reach for the Stars," *Time*, February 17, 1958, 23.

10 "Rocket City, U.S.A.," *Time*, February 17, 1958, 25.

11 "Space Museum Study Started by Committee," *Huntsville Times*, newspaper clipping, n.d., collection of Debbie Linde, Huntsville, Alabama. News coverage of human space-flight launches similarly drew curious vacationers to the gates of NASA's launch complex. See Emily A. Margolis, "'See Your Spaceport': Project Apollo and the Origins of Kennedy Space Center Tourism, 1963–1967," *European Journal of American Culture* 39, no. 3 (2020): 249–274.

12 Susan Sessions Rugh, *Are We There Yet? The Golden Age of Family Vacations* (Lawrence: University Press of Kansas, 2008), 17.

13 "Chamber Cites Walt Linde," *Tuscaloosa News*, October 4, 1957, collection of Debbie Linde; Leon Jaworski, "Let's Meet the Challenges of the Surging 'Sixties . . . ," *Houston*, April 1960, 11–14, Houston Metropolitan Research Center, Houston, Texas.

14 Monique Laney, *German Rocketeers in the Heart of Dixie: Making Sense of the Nazi Past during the Civil Rights Era* (New Haven, CT: Yale University Press, 2015), 47.

15 Debbie Linde, conversation with author, February 15, 2017.

16 "Space Display and Museum Urged in City," *Huntsville Times*, December 12, 1958, collection of Debbie Linde.

17 "Space Display and Museum Urged in City."

18 For a detailed history of Redstone Arsenal, see *The 75th Anniversary of Redstone Arsenal, 1941–2016* (Redstone Arsenal, AL: US Army Materiel Command, 2016).

19 See Karen A. Rader and Victoria E. M. Cain, *Life on Display: Revolutionizing U.S. Museums of Science and Natural History in the Twentieth Century* (Chicago: University of Chicago Press, 2014), 175–207.

20 "Space Display and Museum Urged in City."

21 James Record, *A Dream Come True: The Story of Madison County and Incidentally of Alabama and the United States*, vol. 2 (Huntsville, AL: John Hicklin Printing Company, 1978), 294.

22 Gretchen Sorin, *Driving while Black: African American Travel and the Road to Civil Rights* (New York: Liveright, 2020), 284–285.

23 "Study Asked for Space Museum Here," *Huntsville Times*, November 13, 1959, collection of Debbie Linde.

24 For a history of the Stanford Research Institute, see James D. Skee, "By the Numbers: Confidence, Consultants, and the Construction of Mass Leisure, 1953–1975" (PhD diss., University of California, Berkeley, 2016); Weldon B. Gibson, *Stanford Research Institute: A Story of Scientific Service to Business, Industry and Government* (New York: The Newcomen Society in North America, 1968); and *SRI: The Founding Years: A Significant Step at the Golden Time* (Los Altos, CA: Publishing Services Center, 1980). Donald L. Nielson offers an overview of SRI's most significant projects in *A Heritage of Innovation: SRI's First Half Century* (Menlo Park, CA: SRI International, 2006).

25 Stanford Research Institute, "Economic Feasibility of a Proposed Space Science Exhibit in Huntsville, Alabama," June 1960, 1, David L. Christensen Collection, box 2, folder 2013.009.052, U.S. Space & Rocket Center Archives, Huntsville, Alabama.

26 Walter V. Linde, "Report of Walt Linde on Space Science Exhibit," typed speech, n.d. [ca. spring 1960], collection of Debbie Linde.

27 Stanford Research Institute, "Economic Feasibility of a Proposed Space Science Exhibit in Huntsville, Alabama," 6, 25.

28 Stanford Research Institute, "Economic Feasibility of a Proposed Space Science Exhibit in Huntsville, Alabama," 25–28.

29 Huntsville City Planning Commission, "Huntsville, Alabama: Population and Economy, Background Trends, Report 1," July 1963, 2–3, David L. Christensen Collection, box 2, folder 2013.009.048, U.S. Space & Rocket Center Archives.

30 By 1963, the Huntsville City Planning Commission found that "either directly or indirectly, the jobs of nearly 90 per cent of Huntsville's citizens are dependent on aerospace research and development activities." Huntsville City Planning Commission, "Huntsville, Alabama: Population and Economy, Background Trends, Report 1," iii.

31 Stanford Research Institute, "Economic Feasibility of a Proposed Space Science Exhibit in Huntsville, Alabama," 25.

32 Konrad Dannenberg to Walter V. Linde, July 27, 1960, collection of Debbie Linde.

33 "Museum Report Is Given Solons," *Huntsville Times*, February 19, 1960, collection of Debbie Linde.

34 Linde picked up this gimmick during his time with the Tuscaloosa Chamber of Commerce. See "Typical Tourist," *Decatur Daily*, October 16, 1963; and "Tourist Dollars Count," *Tuscaloosa News*, January 29, 1956, collection of Debbie Linde.

35 Space City USA, "Space City USA Prospectus for the Securities and Exchange Commission," February 14, 1964, 7; Tec-Productions, Inc., "Space City USA Theme Park, Huntsville-Madison, Alabama," Fact Sheet, November 11, 1963, 1. Both in collection of J. P. Ballenger, Huntsville, Alabama.

36 Advertisement for "Space City USA" stock, n.d., collection of J. P. Ballenger.

37 Tec-Productions, Inc., "Space City USA Theme Park," 1.

38 See chapter 12, this volume.

39 "Park Complete History," *USS Alabama Battleship Memorial Park*, 2017, http://www.ussalabama.com/park-complete-history/.

40 Howard P. Walthall Sr., "Options for State Constitutional Reform in Alabama," in *A Century of Controversy: Constitutional Reform in Alabama*, ed. Bailey Thomson (Tuscaloosa: University of Alabama Press, 2002), 160–161.

41 Mike Marshall, "Harry Pennington Remembered as the 'Quintessential Southern Gentleman' Who Was 'Larger than Life,'" *Huntsville Times*, June 7, 2012; "Harry L. Pennington, Sr., September 3, 1919–June 2, 2012," *Huntsville Times*, June 5, 2012. Both in collection of Debbie Linde. See also "Proclamation 3," *Huntsville News*, November 1965, 10, Dave L. Christensen Collection, box 1, folder Space Science Exhibit, U.S. Space & Rocket Center Archives.

42 Jerry Hornsby, "Space Museum," *Huntsville Times*, August 27, 1965, Dave L. Christensen Collection, box 1, folder 2013.009.028-.044, U.S. Space & Rocket Center Archives.

43 Guy L. Jackson, "Exhibit Bond Vote: Drive Started for Passage," *Huntsville News*, n.d., Dave L. Christensen Collection, box 1, folder 2013.009.028-.044, U.S. Space & Rocket Center Archives; "Amendment 3 Committee Announced," newspaper clipping, November 4, 1965, 22, Dave L. Christensen Collection, box 1, folder Space Science Exhibit, U.S. Space & Rocket Center Archives.

44 Jackson, "Exhibit Bond Vote."

45 Jackson, "Exhibit Bond Vote."

46 "Jaycees Present," newspaper clipping, n.d., collection of Debbie Linde.

47 "Amendment No. 3—Yes," Huntsville Times, November 21, 1965, Huntsville-Madison County Public Library.

48 "Put Us on Display," Huntsville Times, October 10, 1963, Dave L. Christensen Collection, box 1, folder 2013.009.047 (Huntsville Space Science Exhibit), U.S. Space & Rocket Center Archives.

49 Draft of press release from governor's office, n.d., Dave L. Christensen Collection, box 1, folder Space Science Exhibit, U.S. Space & Rocket Center Archives.

50 "Vote YES Alabama Space Science Exhibit, Amendment No. 3" brochure, Dave L. Christensen Collection, box 1, folder 2013.009.047 (Huntsville Space Science Exhibit), U.S. Space & Rocket Center Archives.

51 "Vote YES Alabama Space Science Exhibit."

52 For information about the relationship between Wallace's racial politics and economic stagnation, see William Warren Rogers, Robert David Ward, Leah Rawls Atkins, and Wayne Flynt, Alabama: The History of a Deep South State (Tuscaloosa: University of Alabama Press, 1994), 549–569.

53 "Vote YES Alabama Space Science Exhibit."

54 Brian C. Odom, "Civil Rights and Economic Development in Space Age Huntsville, Alabama," in NASA and the Long Civil Rights Movement, ed. Brian C. Odom and Stephen P. Waring (Gainesville: University Press of Florida, 2019), 115.

55 Brenda Plummer, "The Newest South: Race and Space on the Dixie Frontier," in NASA and the Long Civil Rights Movement, ed. Brian C. Odom and Stephen P. Waring (Gainesville: University Press of Florida, 2019), 66.

56 Rogers et al., Alabama: The History of a Deep South State, 581.

57 James C. Cobb, The Selling of the South: The Southern Crusade for Industrial Development, 1936–1980 (Baton Rouge: Louisiana State University Press, 1982), 139, 145.

58 Richard Paul and Steven Moss, We Could Not Fail: The First African Americans in the Space Program (Austin: University of Texas Press, 2015), 134, 147.

59 Mike Marshall, "Space Center Took Flight on Coattails of Mickey Mouse," Huntsville Times, November 28, 1999, collection of Debbie Linde.

60 "Alabama Space and Rocket Center Story," booklet, n.d., Dave L. Christensen Collection, box 2, folder 2013.009.050, U.S. Space & Rocket Center Archives.

61 Nielson, A Heritage of Innovation, 14–20.

62 George C. Wallace to Walter V. Linde, December 13, 1965; "Linde Named on Center Commission," newspaper clipping, n.d. Both in collection of Debbie Linde.

63 Marshall, "Space Center Took Flight on Coattails of Mickey Mouse"; "Congressman Announces Passage of Measure to Award Land for Exhibit," Redstone Rocket, March 20, 1968, Army Materiel Command History Office, Huntsville, Alabama; Record, A Dream Come True, 775.

64 Economics Research Associates, "Implementation Analysis of a Space Science Center," August 1966, Section II, 203, collection of Debbie Linde.

65 Ernst Stuhlinger and Frederick Ordway III, Wernher von Braun: Crusader for Space (Malabar, FL: Krieger Publishing Co., 1996), 284.

66 Brian C. Odom, "Evelyn Falkowski," March 20, 2019, NASA website, https://www.nasa
    .gov/centers/marshall/history/evelyn-falkowski.html.
67 Edward O. Buckbee, conversation with the author, February 14, 2017.
68 Alabama Space and Rocket Center, "World's Largest Missile and Space Exhibit,"
    news release, n.d. [ca. 1970], Alabama Department of Archives and History, Mont-
    gomery, Alabama; Alabama Space and Rocket Center, "Plan for the Alabama Space
    Science Center," n.d. [ca. 1969], 6–7, Dave L. Christensen Collection, box 2, folder
    2013.009.050, U.S. Space & Rocket Center Archives.
69 Alabama Space and Rocket Center, "Plan for the Alabama Space Science Center," 10–
    26.
70 Alabama Space and Rocket Center, "Plan for the Alabama Space Science Center," 22.
71 Unsigned form letter to members of the Alabama Space and Rocket Center Science
    Advisory Committee, n.d., Alabama Space Science Exhibit Commission Collection,
    Marshall Space Flight Center History Office, Huntsville, Alabama.
72 "Davidson Center for Space Exploration," State of Alabama Engineering Hall of Fame,
    https://engrhof.org/members/davidson-center-for-space-exploration/.
73 "About Us: History and Overview."

12

Space City USA

A Theme Park of the South that Failed to Launch

Drew Adan

Today little remains of Space City USA, an attraction once touted as "the fabulous theme park of the South," save for a few patches of concrete, rebar, and earthworks. The land surrounding Lady Ann Lake outside Huntsville, Alabama is now the site of the Edgewater housing development, and only a handful of inconspicuous vestiges of the park can be seen around the owners' association complex. The concrete foundation of a caveman-themed roller coaster track winds through the forest like a serpentine sidewalk. Small concrete slabs that served as the foundation of an enormous decorative volcano are visible, dotted along the lakefront. An earthen bridge intended to support nineteenth-century steam locomotive cars full of eager park visitors bisects the now-serene lake. In the years immediately following the project's declaration of bankruptcy and the subsequent public auction, park buildings were left to "rust into ruin" and the only visitors to the site were the "occasional curiosity seekers" who left behind a "supply of cans and rubbish . . . the result of more than a few beer parties."[1]

However, in recent years, artifacts from this once-promising venture have begun to resurface. In June 2017, several large promotional signs and other items turned up in a deteriorating barn in Chickamauga, Georgia, and were brought back to Huntsville by a private collector. In the fall of 2019, a local researcher found investment ledgers and other documentation and shared them with Special Collections and Archives at The University of Alabama in Huntsville for scanning. *Huntsville Times* staff writer John Ehinger called these remnants "a silently sarcastic testimony to the strange winds of fortune that affect American business." The story of Space City USA, which unfolded in an era of postwar prosperity and significant federal investment in the region, is one of failure to commodify Huntsville's growing national renown as Rocket City, USA.

Mid-twentieth-century Huntsville experienced an unparalleled level of growth that transformed a sleepy cotton town once known as the watercress capital of the world into a hub of the aerospace industry and aerospace culture. This transformation began with an injection of over $47 million for the construction of a chemical munitions plant at Huntsville Arsenal in 1941 that promised to swell "the channels of trade of this city and section."[2] By the end of 1941, 10,000 civilians were working at the arsenal, earning wages that surpassed other labor positions in the region's traditional industries of farming and textiles.[3] Postwar demobilization briefly threatened Huntsville's economic development, but the city again found itself the recipient of significant federal investment in 1950, with the arrival of roughly 500 military personnel, 65 civilian employees, and 100 German rocket scientists to staff the newly formed Ordnance Guided Missile Center at Redstone Arsenal. The massive influx of federal funding was evidence of continuing federal investment in the Tennessee Valley region that began with Tennessee Valley Authority dam projects earlier in the century. The new federally financed industries precipitated a profound alteration of the physical and economic landscape. By 1951, nearly $700,000 in contracts had been awarded to North Alabama businesses. Eighty-one percent of Redstone's contracts went to 1,500 small manufacturing businesses across the region.[4] From 1950 to 1960, the population of Huntsville grew 340.3 percent, from 16,437 to 72,365,[5] and the population continued to swell throughout the 1960s after the formation of NASA and Marshall Space Flight Center.

Huntsville's new reputation as Rocket City, USA, attracted more than just rocket scientists as the prospect of human space flight captured the public imagination and permeated all aspects of American culture. Early celebrities and advocates of rocket science Willy Ley and Wernher von Braun did much to cultivate this budding popular interest in space flight. Both men had seen how popular entertainment could generate enthusiasm for space exploration as members of the influential Verein für Raumschiffahrt (Society for Space Travel) that helped propel the Berlin "rocketry fad" in the 1920s via films such as Fritz Lang's *Frau im Mond* (Woman in the Moon; 1929) and books like Hermann Oberth's *Die Rakete zu den Planetenräumen* (The Rocket into Interplanetary Space; 1923). Ley fled Nazi Germany in 1935 and pursued a career in speculative and science fiction writing in the United States. Von Braun went on to develop rockets for the Nazi regime at Peenemünde and eventually emigrated to the United States under the Operation Paperclip intelligence program. Ley and von Braun resumed their efforts to engage the public with the exciting prospect of space travel by writing for widely read speculative science fiction periodicals such as *Galaxy* and *Collier's*. Former Marshall Space Flight Center historian Mike Wright observed that von Braun "believed that America's devotion to space fiction in the early 1950's could be channeled into space fact."[6] Of particular success was their "Man Will Conquer Space Soon!" series published in *Collier's*, which served as the basis for a popular and influential television show. Von Braun, Ley, and other early rocket pioneers were featured in the 1950s "science factual" series *Walt Disney's Disneyland* on episodes entitled "Man in Space," "Man and the Moon," and "Mars and Beyond." These episodes combined technical explanations of manned space flight with Disney's flair for animation, creating programming designed to both inform and entertain audiences of all ages. They provided viewers with a history of rocketry, explanations of the challenges and opportunities of human spaceflight, a discussion of the possibility of life on other planets, and dramatizations of what an expedition to the moon or Mars might look like. The episodes had a broad impact on American society, as President Eisenhower allegedly ordered a copy of the film for viewing by top Pentagon officials. Disney animator Ward Kimball described these screenings as an "educational space primer."[7] An estimated 42 million Americans viewed the first episode of the series.[8] In addition to television appearances, von Braun, Ley, Heinz Haber, and Ernst Stuhlinger served as technical consultants for the Tomorrowland section of the Disneyland theme park in Anaheim, California, thus establishing the first connection between the burgeoning space industry and theme park tourism.[9]

The story of Space City USA is an expression of this growing popular interest in space travel and an ill-fated attempt to monetize it. Park planners sought to cash in on space fever by building a traditional theme park with space-themed attractions. They also envisioned a park that would serve to both educate and entertain, like Disneyland's Tomorrowland. Space City USA sought to provide an "amusement, recreation, entertainment, and educational center for the Southeast" while simultaneously serving "as a permanent showcase for our space activities" that would seek to "gain the support of the American people for our space programs."[10] The Space City USA project was ambitious in its scope; the founders intended to build a park "of comparable caliber to such parks as Disneyland, Legend City, and Six Flags" in "the space capital of the free world."[11] Although developers envisioned a theme park attraction of national and international prominence, it was also to be a regional showpiece (see figs. 12.1 and 12.2). Promotional literature spoke of a Southeast "that had entered the space age"[12] and an advertisement in a Huntsville phone book called Space City USA the "fabulous theme park of the South."[13]

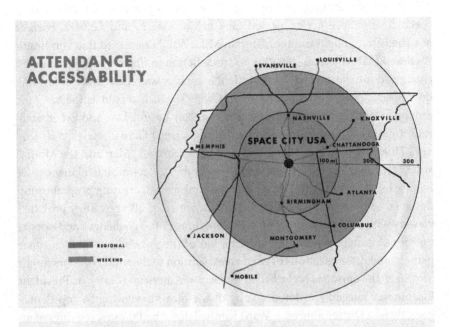

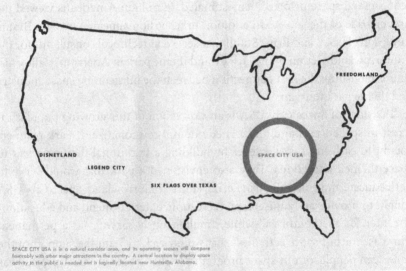

Figures 12.1 and 12.2. Maps showing the favorable regional and national location of Space City, USA. Images courtesy of UAH Archives and Special Collections, Space City Collection.

Promoters emphasized the novel appeal of such a park in a region of the country that did not yet have a similar entertainment complex. Planners described Space City as "completely different from a children's park or typical amusement park" and noted that "nothing of this nature exists in the Southeast."[14] Park founders assumed that the novelty of the project would enrich more than just developers and investors. Space City USA president Hubert Mitchell claimed that the park would have "a great economic impact" on an area that was already experiencing massive growth.[15] The planning for Space City was done at the height of Huntsville's rise to economic and cultural prominence as Rocket City, USA, which coincided with the postwar boom in middle-class vacation, leisure, and travel culture. By all metrics, Space City USA seemed like a promising venture.

Although the initial planning and fund-raising for Space City USA began in late 1959, the enterprise was not incorporated as a business until September 3, 1963, and construction did not begin until January 1964.[16] In the preliminary planning stages of Space City USA, founders consulted various feasibility studies to ascertain the market and the infrastructure for such an amusement park in North Alabama. In fact, park founders first reviewed a study conducted by Stanford Research Institute of California for a different venture that would later become the largest paid tourist destination in the state of Alabama.[17] In 1960, the Huntsville-Madison County Chamber of Commerce commissioned a study to explore the possibility of "establishing a space museum to house some of the original Redstone Arsenal 'hardware' used in various historic missile launches."[18] This initiative would eventually lead to the establishment of the US Space & Rocket Center in 1965, a potentially competing attraction that may have played a role in the demise of Space City USA.

Space City USA founders also reviewed a feasibility study conducted by Tec-Productions, Inc., owned by David Christensen, vice-president of Space City's board. The study opened with an introduction that listed the themes and features of the park, then divided the geographic market for Space City into three parts. The first was the regional market, which consisted of the areas and populations within a 150-minute drive of Huntsville. This sector included the Birmingham, Chattanooga, and Nashville metropolitan areas, which had an estimated population of 2.8 million and an estimated $3.5 billion in disposable income. The second was the weekend market, which included potential visitors within a five-hour drive; that radius included the urban areas of Atlanta, Montgomery, Memphis, and Knoxville and added an additional 4 million inhabitants to the regional market. Finally, the tourist market included the roughly 2.5 million people who passed through the Ten-

nessee Valley each year. The study estimated that 35 percent of tourists passing through the area would stop for at least three hours in Space City and
spend an average of $3.80. Annual attendance was estimated at roughly 1.3
million visitors per year, which would generate a gross income of $4.8 million
a year.[19] The optimistic estimates of this feasibility study may have been biased
by Dave Christensen's leadership role in the project.

Park planners looked to other recently constructed amusement parks such
as Six Flags Over Texas and New York City's Freedomland U.S.A. as blueprints
for financially sustainable and profitable theme park attractions. While Space
City projections were more modest than Six Flag's 1.6 million visitors and
$6.1 million in gross revenue, stockholders still hoped that Space City could
follow Six Flag's commercial success and fully recover the original investment
in as little as twenty-six months after the park's opening day.[20] Space City
board members and designers also looked to the recently opened Freedomland U.S.A. in the Bronx for thematic and stylistic inspiration. Freedomland,
which operated for only four years, from 1960 through 1964, was a US-
history-themed $65 million project billed as the "The World's Largest Enter-
tainment Center." Early documentation and feasibility studies from 1959 refer
to the proposed park in Huntsville as "Time World," but the name "Space City
USA" appears on all subsequent promotional materials the following year. It
is possible that this name change sought to capitalize on the contemporary
buzz surrounding Freedomland U.S.A.[21] Space City and Freedomland had a
similar park aesthetic and overall layout based on historically themed areas.
Freedomland even included a "Satellite City," which had a space-travel motif,
complete with a reproduction of an authentic NASA launch control room
in which visitors could view a simulated launch. This area also included a
section for exhibits that showcased developments in aerospace science and
technology.[22]

While the original stockholders and board members of Space City USA
conceptualized a park in line with national trends in theme park tourism,
the park's leadership consisted of prominent local businessmen from across
the North Alabama Tennessee Valley region. The president, primary investor,
and creator of Space City USA was Hubert R. Mitchell, an industrialist from
Hartselle, Alabama, who was involved in a wide variety of commercial enter-
prises that sought to capitalize on federal development in the Tennessee Valley region. Descriptions of Mitchell in modern accounts run the gamut from
"shrewd businessman" to "wheeler-dealer."[23] His previous business ventures
included a North Alabama hamburger chain called Branded Burgers, an ice
cream wagon, minstrel shows, bowling alleys, a theatrical stage equipment
company in Hartselle, a few drive-in movie theaters, a small theme park on

leased Cherokee lands in Western North Carolina (possibly Cherokeeland or Frontier Land), and a furniture factory in Huntsville.[24] It is rumored that in the 1930s he claimed to have found the outlaw Jesse James imprisoned in Linden, Tennessee, and booked him on a promotional tour across the South.[25] With such a varied résumé, it is clear that Hubert Mitchell had a tenacious entrepreneurial spirit and a knack for showmanship. Mitchell was also a major investor in the failed Keller Motor Corporation that produced eighteen wooden automobiles from 1947 to 1950 in surplus chemical-manufacturing buildings on what is now Redstone Arsenal. Mitchell's interest in the car company stemmed from his defense contract work producing wooden seats for light military aircraft during World War II. When the war ended, he was left with an enormous surplus of seats. He approached car manufacturer George Keller as a potential buyer and ended up investing heavily in his fledgling business. The Keller Automobile Corporation died along with its namesake, George Keller, who suffered a heart attack at the age of 56 on October 5, 1949. Other Space City USA executives included board chair Nelson Weaver, a Birmingham real estate developer, and vice-president David L. Christensen, a Huntsville aerospace executive.[26]

Although these initial investors and large stakeholders provided most of the capital used to purchase the 250 acres, which had been appraised at $1.5 million, a public stock issue of $5 million financed both the acquisition of a 99-year lease on an additional 600 acres of land and construction costs.[27] According to the SEC News Digest, Space City USA, Inc. registered two million shares of common stock offered to the public at the price of $2.50 per share. The plan was to use $500,000 from sales of stock to build the park's motel, which was estimated to cost $1,500,000, and to use the remainder for the construction costs of the actual park, which were estimated at $4,850,000.[28] One million shares were registered with the Alabama Securities Commission for sale in state from an office in downtown Huntsville at 2400 Clinton Street. The remaining one million shares were allocated for out-of-state sales.

According to ledger sheets recently digitized by Special Collections and Archives at The University of Alabama in Huntsville, investors in Space City USA came from varied socioeconomic groups and disparate geographic areas across the country. While most investment came from residents of Alabama and inhabitants of the Tennessee Valley region, stockholders from as far away as California and New York made sizeable investments. Typical investments ranged from $100 to $250. This was a sizable sum in 1964; accounting for inflation, it was the equivalent of roughly $1,000–$2,500 in 2022 dollars. The upper threshold of investments topped out around $5,000–$6,000, or $50,000–$60,000 in 2022 dollars.[29] Among these investors were prominent national figures in the emerging aerospace field. Frederick Ordway III, a notable space scientist, historian, author, and biographer, invested an undisclosed sum in the project, as did German aeroballistic engineer Oscar Holderer.[30]

To advertise the sale of these stocks and generate public interest in the park, promoters used various traditional marketing tactics and paid for advertisements across media outlets available at the time. To kick things off, Space City USA, Inc. threw press parties at the Huntsville Elks Club and the Officer's Club on Redstone Arsenal in January 1964. Local dignitaries attended the Elks Club event such as actor Dean Jones, a native of Decatur; James Record, chair of Madison County Commissioners; and Homer Witt, president of the Huntsville City Council.[31] The following day, local papers ran stories that stoked public excitement for Space City with lines such as "if you would like to hurtle through space in a flying saucer, stop off at Mars and have lunch on the moon, you may not have to wait much longer." In addition to lavish parties and a media blitz, an open house and a skydiving exposition were scheduled for 1964.[32] Board members also encouraged stockholders to act as advertising agents. In an undated letter Hubert Mitchell offered Space City stockholders the option of paying $1 for a "five-color decalomania to be applied to your automobile windshield or [to] use in anyway [sic] you see fit."[33] Such lavish marketing campaigns speak to Mitchell's flair for showmanship. According to a stockholders' report, in 1964, Space City USA spent over $70,000 on salaries, travel expenses, legal fees, promotional materials, and other expenses.[34] While an aggressive advertising campaign was necessary to secure investors, this emphasis used funds that were needed to complete the park's construction.

Although local contractors were responsible for the construction of Space City USA, Academy Award–winning visual effects artist Thomas Glen Robinson of Skylim Corporation (a name derived from the phrase "the sky's the limit") was chosen to implement the overall design of the park. Robinson was involved with other successful and renowned amusement parks such as Disneyland, Legend City, and Freedomland U.S.A. and had also worked on special effects for some of the biggest Hollywood films of the era, such as *The Brothers Grimm*, *Ben-Hur*, *Forbidden Planet*, *Mutiny on the Bounty*, and *How the West Was Won*. The Space City leadership seems to have prioritized attracting investors via connections to successful franchises instead of focusing on the more practical issues of park construction.

The land purchased for Space City USA lay just west of Redstone Arsenal around the 175-acre spring-fed Lady Ann Lake. Park planners envisioned visitors accessing the park along a 2,600-foot strip beside Alabama Highway 20, which is now the frontage road for Interstate 565, or Madison Boulevard.

Initial plans for the park entrance area included the Space City Motor Hotel, designed by Atlanta architectural firm Wise, Simpson, Aiken and Associates, which was to have ten stories and 200 "luxury-type units and suites." Other plans for the entrance area included shopping centers, restaurants, a convention center, information kiosks, and parking facilities to accommodate 1,875 cars.[35] A proposed spaceship-inspired monorail system was to connect the various areas of the park. The train would transport guests from the parking, retail, and lodging areas located along the highway to the attractions situated around Lady Ann Lake.

The proposed park was divided into four thematic sections that progressed chronologically through the history of scientific progress and innovations that were seen as milestones in human aeronautics (fig. 12.3). Developers envisioned a time-travel motif, stating that "a time machine approach will be used to transport the visitor into other ages and events from our own."[36] Thematic sections of the park included the Land of Oz, which would feature a fantasy theme inspired by *The Wizard of Oz* and classic fairy tales. The Lost World was planned to have a prehistoric aesthetic complete with an enormous volcano made from "the largest order of dental plaster ever placed in the world" as well as dinosaur- and caveman-themed rides. The Old South area promised the feel of the antebellum South complete with rides on a fully restored 1897 H. K. Porter wood-burning narrow-gauge steam locomotive named the *Melodia B.* The train originally hauled sugar in Louisiana from the Laurel Valley Plantation but would now haul five observation coaches and run a roughly two-mile track around the park and cross Lady Ann Lake on an old-fashioned trestle.[37] The contract to restore this vintage locomotive fell to Crown Metal Products of Pittsburgh, Pennsylvania, at a cost of $60,000.[38] However, the flagship section of Space City USA would be Moon City, or the World of the Future. This area would showcase concepts familiar to most fans of science fiction, including jet-car and flying-saucer rides. Across the four sections, twenty-three rides were originally scheduled to be operational when the park opened in 1965. These would be able to accommodate 14,430 riders per hour.[39] Other proposed features and attractions of the park included a pirate-themed island called Dead Man's Island that was to have been constructed in the center of the lake by lowering water levels by approximately four feet, a picnic and camping area, an amphitheater with a floating stage, a grass airstrip and hanger with antique aircraft and a hot-air balloon, a skyway cable-car ride with views of the park, and a restaurant in the middle of Moon City that would place diners in a simulated moonscape.[40]

The themes represented in these park sections and attractions mirrored motifs of successful amusement parks of the time. Obvious parallels can be

Figure 12.3. Artist's rendering of an aerial view of Space City, USA. Image courtesy of UAH Archives and Special Collections, Space City Collection.

drawn between the ray-gun gothic aesthetics of Space City's World of the Future and Disneyland's Tomorrowland. The historically based Old South and Lost World sections were likely inspired by the American history narrative of New York's Freedomland U.S.A. and Arizona's Old West–themed Legend City. The Land of Oz concept was eventually fully realized in 1970 when the Land of Oz theme park opened in Beech Mountain, North Carolina. Although the idea of space travel was central to the thematic vision of Space City given its Huntsville location, the proposed attractions were actually an awkward amalgamation of amusement park clichés. This lack of a unifying space-centered vision would probably have perplexed both visitors and investors.

Unfortunately, none of these grand plans materialized, as the project eventually "fell flat on its face and took some $2 million in capital with it."[41] While the venture showed promise in the beginning, it was plagued with a multitude of financial, interpersonal, and meteorological problems. According to a letter Hubert Mitchell wrote to stockholders in 1964, sales of Space City stock got off to a good start; the first public offering of company common stock was quickly sold on February 13–17, 1964. In the following months, Space City's attorney, Mr. Mayer U. Newfield, filed a registration statement with the US Securities and Exchange Commission of the park's intention to offer additional shares for public sale.[42]

However, these additional stock offerings did not generate the same enthusiastic support as the first round and Space City USA investments topped out at roughly $2 million. This was well short of the estimated $5 million in stock sales needed to finance the park. Surviving ledgers provide separate quarterly lists for paid-up stockholders and those on a "deferred payment plan" for the second through fourth quarters of 1964. The earliest investment records date from the second quarter of that year. Stock sales got off to a strong start; 107 investors contributed amounts that ranged from $250 to $1,500. Third-quarter investment in Space City greatly accelerated, with 506 individual investors and 88 deferred payment backers. The third quarter also saw several sizable donations of up to $10,000. However, it appears that enthusiasm for the project began to wane quickly; only 132 people bought stock in the fourth quarter.[43]

The first public sign that the venture was losing momentum was the September 1965 resignation of Hubert R. Mitchell, the project's founder and former president. In his letter of resignation, Mitchell stated that he had "had this in mind for a considerable length of time" and cited the reasons he had reached the point where he felt he could not continue to serve on the board of directors. Of primary concern was a lack of communication between the new president of the corporation and the board of directors. According to Mitchell, the board of directors was not "kept up to date and properly informed of all the overall business operations of the corporation and consulted before any action is taken." He said that even though "Space City, U.S.A. was my idea, and it was me that conceived the idea and developed the same and promoted this idea for many, many months alone" the "corporation has been run completely by the president, without the consent or concurrence of the board of directors." He went to say that many other factors contributed to his decision to resign from the board but that he could not elaborate "because of the length of time that it would take to set each one out."[44] Whatever the exact reasons were for Mr. Mitchell's resignation, it is clear that interpersonal issues and potential mismanagement were present from the very inception of the project.

Hubert Mitchell and his descendants blamed political interference from Alabama's attorney general, Richmond Flowers Sr., for the ultimate demise of Space City USA. Citing Flowers's 1969 conviction for extorting life insurance companies for licenses to conduct business in the state, these critics suggest that Flowers approached Space City leaders for payment under the table for permission to proceed with park construction.[45] To date, no evidence has surfaced that corroborates the accusation or connects Richmond Flowers Sr. to the Space City project. Flowers maintained that the charges against him were politically motivated by his opposition to Governor Wallace's stance on segregation and noted that he was officially pardoned by Jimmy Carter for extorting life insurance companies in 1978.[46]

This is not the only accusation of criminality associated with the project. Many in the Huntsville area remember Space City not as a failed business venture but rather as a calculated scam intended to defraud investors. This sentiment appears to be pervasive among both investors and community witnesses. A Madison resident who rode the train at Space City as a child remembered that the "general consensus among most people here was that it was a scam." He remembers hearing as a child that "they weren't really going to do that. . . . They put up enough for a show and brought people in to get their money."[47] Like the allegations against Attorney General Flowers, the idea that Space City USA was an intentional fraud has not been confirmed by any known sources. However, this reaction speaks to how promising the venture appeared to a community experiencing such explosive growth and prosperity.

By 1966, Space City was in dire financial straits. The *Huntsville Times* ran a piece in the business section exposing the project's "money woes" and its numerous pending lawsuits. Two firms were suing Space City USA for $3,300 in unpaid bills, and landowners were threatening to terminate their lease on rented property. Board chair Nelson Weaver admitted to "some rough spots" and "some problems in raising the necessary funds" but assured the press and investors that "the cases will be taken care of" and said that "we have no thoughts whatsoever of it being abandoned."[48]

The following year, on October 17, 1967, Space City USA, Inc. declared bankruptcy and auctioned off any remaining assets.[49] The auction began at 10 a.m. in the corporate office built on the proposed park site, and a full list of items to be auctioned appeared in that day's edition of the *Huntsville Times*. The assets up for bid included the steam engine and five observation cars and the rail and track they ran on. Proceeds generated by the sale of these items would "cover incurred court costs and 'go towards the payment of priority liens as determined by the court and then to the payment of creditors.'"[50]

Over the years many have theorized about why such a seemingly promising enterprise had failed, but the exact reasons for the demise of Space City have remained a mystery. Even at the time, the reasons for the failure were unclear. However, in a September 1967 piece for the *Huntsville Times*, reporter John Ehinger wrote that "it appears as though Space City was an ill-fated venture from its conception."[51]

Over the summer of 2012, a business professor at The University of Alabama in Huntsville proposed a class project to examine the causes that led to Space City's bankruptcy. Dr. J. P. Ballenger and his students spent the summer gathering data and documents to support their theories but could not point to one specific cause for the failure of Space City.

Ballenger's group proposed that climate and weather were a major concern to park developers and may have been partially to blame for the park's failure. While Alabama is known to have generally mild winters, temperatures regularly drop below freezing in the winter, and periodic weather anomalies have been known to drop temperatures well below zero. Park developers consulted U.S. Weather Bureau data and settled on an operating season that would have run from April through November.[52] Early construction of the park was plagued by numerous weather-related delays and setbacks. Just as construction began on the park in January 1964, Huntsville received a record-breaking snowfall of 17.1 inches that remained on the ground for a week. That year, a record 23.3 inches of rain fell in March and April. This weather anomaly could explain the first postponement of the park opening. Originally slated to open in April 1965, park owners first postponed the opening date to the spring of 1966 and then delayed it indefinitely.[53] During the final winter of construction in 1967, temperatures reached -11 degrees and 7 inches of snow fell.[54]

The actual terrain and hydrogeology of the site also posed serious challenges for construction. The spring that feeds Lady Ann Lake is the second biggest spring in Madison County and the immediate area is home to numerous other smaller springs. This created an excessively swampy terrain with large trees that required expensive removal and stump grinding. The springs further complicated the task of lowering the water level four feet to create "Dead Man's Island." In addition, area contractors claimed that the area was infested with snakes.[55]

Conspicuously absent from Space City's promotional material, marketing campaign, and business model was any direct connection to NASA or Dr. Wernher von Braun, the figurehead of the popularization of space travel. Von Braun thought that the attraction that highlighted Huntsville's contribution to the exploration of space should be a museum. Planning began on what would become the US Space & Rocket Center in 1960. Von Braun pitched the idea to the Alabama legislature as a collaborative project with the US Army Missile Command and NASA. Aware of the local culture and the power of public opinion, he persuaded college football coaches Bear Bryant of Alabama and Shug Jordan of Auburn to appear in a television commercial supporting the project.[56] Although Space City would not have been a direct competitor with the US Space & Rocket Center, there was significant overlap in scope as the mission of Space City included an educational component and a "permanent aerospace industries exhibit."[57] With the full backing of the state of Alabama,

NASA, the US army, and to some extent Walt Disney and ABC, von Braun clearly had an advantage over a team of investors attempting to swim in his wake.

Space City is a lesson in mismanagement, the impracticality of a pay-as-you-go financing model, and the failure of an early attempt to monetize Huntsville's cultural capital as an aerospace hub. Of the $2 million raised via public stock sales, only an estimated $500,000 was spent on construction, which raises the question of how that other $1.5 million was squandered. Consensus seems to be that it was frittered away on advertising, promotional materials, planning, business trips, and site preparation. Even if Space City had succeeded in securing funding to finish the project, it seems unlikely that the park would have been able to take full advantage of Huntsville's reputation without a developed brand that was linked to the space program, as Disney had, or the authenticity of being a NASA-affiliated attraction, like the U.S. Space & Rocket Center. The failure of Space City USA demonstrates that success in the emerging Huntsville tourism market required more than just a thin veneer of space theming.

Notes

1 John Ehinger, "Space City USA in Court Battle—Costly Park Lies in Decay," *Huntsville Times*, September 1967.

2 "Huntsville Gets Chemical War Plant," *Huntsville Times*, July 3, 1941.

3 "10,000 Persons Now Work on War Plant," *Huntsville Times*, December 17, 1941.

4 "Chance for Small Business," *Huntsville Times*, July 8, 1951.

5 US Bureau of the Census, *Census of Population: 1960*, vol. 1, *Characteristics of the Population*, part II, *Alabama* (Washington, DC: Government Printing Office, 1963), Table 8, Population of All Incorporated Places and of Unincorporated Places of 1,000 or More: 1940 to 1960, page 2-15, https://www2.census.gov/library/publications/decennial/1960/population-volume-1/vol-01-02-c.pdf.

6 Mike Wright, "The Disney-Von Braun Collaboration and Its Influence on Space Exploration," Marshall History, n.d., last modified August 3, 2017, https://www.nasa.gov/centers/marshall/history/vonbraun/disney_article.html.

7 David R. Smith, "They're Following Our Script: Walt Disney's Trip to Tomorrowland," *Future*, May 1978, 57.

8 Willy Ley, *Rockets, Missiles, and Space Travel* (New York: Viking Press, 1968), 331.

9 Frederick I. Ordway, Randy Liebermann, and Ben Bova, *Blueprint for Space: Science Fiction to Science Fact* (Washington: Smithsonian Institution Press, 1992), 135.

10 Press release, n.d., MC-95, box 1, folder 1, Space City Collection, University of Alabama in Huntsville Archives and Special Collections, Huntsville, Alabama (hereafter Space City Collection).

11 Press release.

12 Briefing slides, n.d., MC-95, box 1, folder 1, Space City Collection.

13 Advertisement for Space City in the 1964 *Huntsville Telephone Book*.

14 Fact Sheet: Space City USA Theme Park, November 11, 1963, MC-95, box 1, folder 1, Space City Collection.

15 "'Space City' Park Planned," *Huntsville Times*.

16 Fact Sheet.

17 Alabama Department of Tourism, "U.S. Space & Rocket Center Was Alabama's Most Visited Tourist Attraction in 2018," press release, News4, January 23, 2019, https://www.wtvy.com/content/news/US-Space-Rocket-Center-was-Alabamas-most-visited-tourism-attraction-in-2018-504753721.html.

18 Preliminary prospectus, February 14, 1964, MC-95, box 1, folder 1, Space City Collection.

19 Feasibility study, n.d., MC-95, box 1, folder 1, Space City Collection.

20 Fact Sheet.

21 Feasibility study.

22 Paul D. Naish, "Fantasia Bronxiana: Freedomland and Co-op City," *New York History* 82, no. 3 (2001): 259–s85.

23 "The Keller Car," n.d., Huntsville Rewound, http://www.huntsvillerewound.com/HSVkellercar.htm.

24 "Built to Last," *Birmingham News*, December 25, 1998.

25 "Five Things to Know about Keller Cars: State's Oldest Auto, Hercules Engines, Connection to Jesse James (Odd Travels)," AL.com, May 15, 2013, https://www.al.com/breaking/2013/05/5_things_to_know_about_keller.html.

26 Fact Sheet.

27 Fact Sheet.

28 "Space City Proposes Offering," *SEC News Digest*, February 17, 1964, 3.

29 CPI Inflation Calculator, US Bureau of Labor Statistics, https://www.bls.gov/data/inflation_calculator.htm.

30 Investment ledgers, 1963–1964, MC-95, box 1, folder 1, Space City Collection.

31 "Space City Launched," *Huntsville Times*, January 18, 1964.

32 Deborah Storey, "Space City, Amusement Park Proposed in 1960s, Was Never Completed," AL.com, March 18, 2012, https://www.al.com/breaking/2012/03/space_city_amusement_park_prop.html.

33 Mitchell to stockholders, January 3, 1964, MC-95, box 1, folder 1, Space City Collection.

34 Space City Stockholder Annual Report 1964, MC-95, box 1, folder 4, Space City Collection.

35 Press release, n.d., MC-95, box 1, folder 1, Space City Collection.

36 Feasibility study.

37 "Space City Land-Grading Work Starts," *Huntsville Times*.

38 "Space City Gets 19th Century Train," *Huntsville Times*.

39 Press release, n.d., MC-95, box 1, folder 1, Space City Collection.

40 Press release, n.d., MC-95, box 1, folder 1, Space City Collection.

41 "Space City Assets to Be Sold Oct. 17," *Huntsville Times*.

42 Mitchell to stockholders, January 3, 1964, MC-95, box 1, folder 1, Space City Collection.

43 Investment Ledgers, Space City Collection.

44 Mitchell resignation letter, September 8, 1965, MC-95, box 1, folder 1, Space City Collection.

45 Hubert Mitchell biography, Space City Collection.

46 "Richmond Flowers Is Dead at 88; Challenged Segregation and Klan," *New York Times*, August 11, 2007.

47 Oral History Interviews, Space City Collection.

48 "Money Woes Befall Space City Inc.," *Huntsville Times*. January 23, 1966.

49 Bankruptcy sale, October 17, 1967, MC-95, box 1, folder 1, Space City Collection.

50 "Space City Assets to Be Sold Oct. 17," *Huntsville Times*, n.d. [ca. October 1967].

51 Storey, "Space City, Amusement Park Proposed in 1960s, Was Never Completed."

52 Press release, n.d., MC-95, box 1, folder 1, Space City Collection.

53 "Space City Construction," *Huntsville Times*, n.d. [ca. 1964].

54 Storey, "Space City, Amusement Park Proposed in 1960s, Was Never Completed."

55 Oral History Interviews, Space City Collection.

56 Shaila Dewan, "When the Germans, and Rockets, Came to Town," *New York Times*, December 31, 2007.

57 Fact Sheet.

Part 4

Reflections in Regional Art and Architecture

13

Architectural Modernity in NASA's Remote South

JEFFREY S. NESBIT

An Architect Unknown

In 1999, the Historic American Buildings Survey listed the John F. Kennedy Space Center Headquarters Building in Merritt Island, Florida, on the National Register of Historic Places. In the report, the Kennedy Space Center Headquarters Building is described as having architectural significance for the "Federal Government's use of the International Style" through the building's "integrity of design, materials, workmanship, feeling, setting, location, and association as an administrative center of the Apollo Program."[1] And yet, the architect is listed as "unknown." The designers of these early NASA facilities were not anonymous. Through a deeper reading of its architectural history, one learns that NASA wanted its architecture to *appear* bureaucratic, economical, and ubiquitous. Architecture, building aesthetics, and questions of preservation did play a role in the early development of NASA's Kennedy Space Center and Johnson Space Center. Both sites were constructed on the American landscape of the remote South—the wetlands of central Florida and the prairies of the Texas Gulf Coast. No precedent existed for how to build such a complex, house its personnel, and plan for future growth. A modern aesthetic needed to be established and *designed* as background to the technologically celebrated objects. Architects were hired, aesthetics were designed, and the models of American urbanization were made. This chapter suggests NASA's rapid design planning deliberately produced a public image of efficiency and formed its political prowess supported by an architecture presence.

Figure 13.1. The Kennedy Space Center Headquarters building, 1993. Courtesy of NASA.

NASA had an architect for the Kennedy Space Center Headquarters. His name was Charles Luckman—an architect most widely known for inadvertently spurring the rise of the architectural preservation movement in the United States.[2] Coincidentally, in 1999, the same year as Luckman's death, the Kennedy Space Center Headquarters Building was registered as a nationally recognized historic place. Even more coincidental, the Kennedy Space Center Headquarters, which had supported the administrative functions for a successful Apollo mission, was scheduled for demolition in 2019, overlapping with the fiftieth anniversary year of the Apollo 11 moon landing. Kennedy Space Center Headquarters played a key role in the projection of politics—and of architectural modernity generally—but due to its seemingly standard office building presence, preservationists failed to identify or maintain its significance.

After graduating with an architecture degree in 1931, charismatic Charles Luckman scrambled to find a job in the middle of the Depression. Launching his career in advertising design with Colgate-Palmolive-Peet Company, Luckman discovered marketing and management, experience necessary for successfully running a business empire. As early as 1946, 32-year-old Luckman became president of Lever Brothers, leading the company into a world of advanced advertising in consumer goods. During his leadership at Lever Brothers, Luckman cultivated an opportunity to align his business ethos for efficiency and economics with the modernization of architecture's International Style. In the same year, President Harry S. Truman appointed Luckman to serve on the President's Committee on Civil Rights. His ability to integrate into politics at an early age would prove to be of great prominence.

Luckman increasingly advocated for the celebration of new materials and modern construction techniques. In fact, Luckman appointed architecture firm Skidmore Owings and Merrill (SOM), notorious for modern aesthetics, to design Lever House in New York City. Working in Chicago as a draftsman for two years before attending architecture school proved influential. In a time when Chicago was growing with projects by notable Modernists such as Mies van der Rohe, Ludwig Hilberseimer, and SOM, it comes as no surprise that his experience in Chicago shaped an architectural perspective, or possibly a career-lasting philosophy.

Under Luckman's tutelage as president of Lever Brothers, the headquarters building known as Lever House in New York City incorporated the finest of modern architectural techniques: constructed stainless-steel mullions, spandrel glass, and an applied, green-tinted curtain wall. Lever House has since become a landmark of Modernism in the American metropolis. According to a *New York Times* article in 1999, Lever House is a "symbol of corporate America" that "began the transformation of much of Park Avenue from a posh residential address to a corridor of corporate prestige."[3] Luckman insisted the building's ground floor remain accessible to the public and resisted using the full block footprint offered by New York City's building codes—another indication of urban planning principles borrowed from the Chicago-based Modernists.

In his autobiography titled *Twice in a Lifetime: From Soaps to Skyscrapers*, Luckman self-proclaims his success to be associated with "ingenuity" and "tenacity." Aware of his influence on two professions, Luckman exemplifies an American "entrepreneurial spirit" and supposedly inspired individual potentiality. The story of Lever House represents something more, "a building that would look as machine-tooled as American industry, one that would reflect American productive genius."[4] Just as the building became an image of material technology and capital progress, it failed in duplicating its isolated success. Instead, it was Luckman's facilities commissioned by NASA that endure the most anonymity and dismissal. As argued by architect Rem Koolhaas, "Modernism" failed to urbanize by "abstraction and repetition," because the multiplication of population has rendered the city practically illegible.[5] And since we have seen globalization considerably attempt to duplicate, paradoxically Modernism was unable to continue progress. If "the city no longer exists" and Koolhaas aligns the irony of city growth with the decline of urbanism as a field, massive failures in responding successfully to urbanization challenges persist, such as airports, infrastructure, and urban development. Kennedy Space Center Headquarters is precisely that—a modern office architecture "without urbanism." So, how can we consider the role of architecture and image in parallel with national technological and infrastructural progress?

In the winter of 1950, Charles Luckman left Lever Brothers due to a policy dispute, shifting his focus to an architectural practice. Luckman quickly partnered with California-based William Pereira. Specializing in office buildings, airports, and the planning of Air Force facilities, the two successfully landed significant corporate and government contracts. Luckman's experience in governmental policy at an early age under the Truman administration paid off, and quickly Pereira and Luckman received federal commissions for the design of government facilities during the Eisenhower era.[6] By the late 1950s, with a list of military projects in their portfolio, Pereira and Luckman continued to receive government contracts following the transition from military complex to NASA's administrative facilities in Southern California, Houston, and Florida.

Science and technology historian Stuart Leslie described aerospace design and production within Southern California as "the industry in the sky," fundamentally changing the culture of the region—equal to the industry of oil or of Hollywood during their respective peaks.[7] Pereira and Luckman designs exemplified a change in technology not through their exteriors—which were often seen as a military-industrial aesthetic—but through their interior, secured spaces, shaping new archetypes for scientists, engineers, and military officers. In doing so, the interior spaces claimed to directly correspond with a cultural context, including a connection between Southern California, new technologies, and the designs themselves, to encourage prospective employees, in particular, scientists, to work in the region.[8]

While Pereira focused on military projects in California, a new architectural opportunity came to Luckman in New York City. Taking up two Manhattan blocks, Pennsylvania Station, originally designed by McKim, Mead, and White, was completed in 1910. Monumentally sited with quasi-Classical forms of symmetry, Penn Station's real significance was its connectivity to distant cities, including Chicago, St. Louis, and Miami. Enclosed by a massive volume used for passenger waiting at its center, the shape, size, and appearance of Penn Station operated more like an "elaborate shed"—a gateway into the great metropolitan core.

New York City modernized, Penn Station went broke, and airplanes and automobiles overshadowed railway activities, leaving the Neoclassical Roman shed slated for demolition. But the decision came with resistance. In an article titled "Farewell to Penn Station," the *New York Times* reported on October 30, 1963, that "until the first blow fell no one was convinced that Penn Station really would be demolished or that New York would permit this monumental act of vandalism against one of the largest and finest landmarks of its age of Roman elegance." Emotionally charged, the article continues by crying out, "we will probably be judged not by the monuments we build but by those we have destroyed."[9]

With Penn Station undergoing demolition, plans for Madison Square Garden Center and Two Pennsylvania Plaza were well under way—and former Lever president Charles Luckman won the architectural commission. Suppressing Penn Station below street level, Luckman's newly modern alternative included a hotel, 30-story office building, and two arenas. Still today, the heterogenous elements of such planning and architectural production allow continued connectivity in and out of the city, simultaneously driven by new forms of economic stability and by business incubation. The president of Madison Square Garden at the time, Irving Felt, assured that "the gain from the new buildings and sports center would more than offset any aesthetic loss," undoubtedly responding to criticism of tearing down the classical giant.[10] It is ultimately Charles Luckman who is heavily criticized for lack of historical sensitivity to the Penn Station debate.

Charles Luckman was criticized heavily for his neglect of preservation when designing Madison Square Garden, a building considered to accrue economic gain but with an aesthetic loss. Its precursor, Penn Station, constructed in 1910, had been considered the "finest landmark" in bustling New York City.[11] By 1963, the historic railroad station had been demolished, making way for Luckman's Madison Square Garden multipurpose complex.[12] Forty years later, in an obituary, Luckman was "remembered less for the buildings he designed than for inadvertently spurring architectural preservation to become a major national movement."[13] Still today Luckman is largely known not for his impressive portfolio, but for provoking an architectural preservation movement at a national scale.

In 1999, the same year as his death, a strange reversal occurred: the Kennedy Space Center Headquarters Building was added to the National Register of Historic Places, marking its architectural and historical period of significance from 1965 to 1975. The Headquarters Building was representative of the federal government's use of the International Style. Although the Kennedy Space Center Headquarters Building underwent interior renovations with the advent of new technology, it maintained an integrity of design, materials, workmanship, feeling, setting, location, and association as an administrative center of Project Apollo.[14]

Curiously, as mentioned, the Historic American Buildings Survey lists the architect/builder as "unknown," with no mention of Charles Luckman—and the Kennedy Space Center Headquarters is now demolished. It's difficult to know whether the survey report fails to list Luckman due to a clerical error or to a political oversight. Regardless, rather than being recognized as a histori-cally significant landmark capable of adaptation in the face of technological change, the building has been folded into NASA's ongoing history. As systems and facilities become obsolete, immediately after new advancements in rocket technology and changes in federal budgets, consolidation is necessary.

In 1999, Luckman died, and the Kennedy Space Center Headquarters was registered as a historical building. In that same year architect and author Mark Wigley pointed out that "replicas of permanent and unique objects have be-come more important than the originals."[15] President Kennedy's 1961 charge for a lunar landing by the end of the decade required the design of massive, centralized facilities. NASA's administrative buildings and facilities planning, including flexible and adaptable interiors, are indications of the rising image of architectural modernity. The argument for design of the Space Center's fa-cilities was not about its technology, symbol, or even permanence, but more concerned with political positioning for a nation in transition.

Becoming Bureaucratic

President John F. Kennedy's "decision for an accelerated Moon landing was ultimately a political decision made in terms of cold war strategy" and more importantly, he knew the Soviet Union could not "compete with Western economies."[16] In a tense political climate in the United States, it was reported in 1966 that NASA received an astonishing 2.66 percent of the US federal budget (compared with a mere 0.5 percent in 2014). This considerable *eco-nomic* ambition enabled the new space complex to be conceived and con-structed at a massive scale. And thus, "to Congress, the moon program meant money," but for NASA, "perhaps the biggest challenge was organization" of not only building construction, but also systems at an unprecedented scale.[17] In alignment with the necessity for such rapid pacing, increased requirements for administrative facilities tended to also reject assumptions of historical aes-thetic. During the Presidential Address, on May 25, 1961, Kennedy encour-aged development in space, calling it an "urgent national need," attempting to evoke emotional enthusiasm in a national audience.[18] Political pressure, systems innovation, and coordination of large-scale labor, including substantial financing, led to novel techniques for organizing the space complex. To successfully carry out its planning, NASA needed architects who understood management, logistics, and fiscal constraints all the way from business marketing down to the technological elegance in the construction of space.

Immediately following their respective presidencies, US government facilities are renamed, and public visitor's centers and presidential libraries are constructed in their honor. For the spaceport, presidents are consistently associated with its technology and political imagination. On November 28, 1963, six days after Kennedy's assassination, President Lyndon B. Johnson renamed the space complex Cape Kennedy. Although he made only short visits on both occasions, it is these brief moments in history that tend to record presidential power and an image directly linking that power to past and future successes in the space program. At first, office and operation facilities at the Kennedy Space Center didn't seem to reflect this overt appearance in politics—based largely on the fact that the historical image of NASA had been associated with science and technology. On one hand, monuments are highly orchestrated architectural results of legislative powers for commemorating presidential influences, while on the other, the government is committed to adopting seemingly anonymous facilities to honor presidential names. The Kennedy Space Center Headquarters Building—an office building (not Hangar S, the historically rich assembly facility)—soon became background for a political image of power. The blurring of military missiles with the burgeoning civilian space administration had been a strategy since Kennedy's predecessor.

Eisenhower was not interested in space exploration, but neither was Kennedy. In fact, President Kennedy used the space program for political maneuvering to distort international attention on national security.[19] Even though it was Eisenhower who established NASA, in the end, it is Kennedy who is praised for early US space programs. President Johnson's dedication of "Cape Kennedy" helped secure Kennedy's political legacy, and his image, tied to NASA's political pursuits. Just as with President Kennedy's association with national progress and aspirations for reaching the moon, NASA's modern architectural forms in the remote wilderness were just another attempt to manufacture a kind of frontispiece—similar to cultural historian Martin Jay's "linking the human desire to mold the environment for aesthetic purposes with the domination of nature."[20]

Insistence on speed, efficiency, and modernization in launching of rockets in space led to expansion of earlier versions of facilities designed on Cape Canaveral Air Force Station: on one hand, emotionally charged nationalism wrapped into politics and management of funds, and on the other, a rise of modernization, technology, and science. This opportunity led to businessman and architect Charles Luckman's interest and eventual involvement in design and planning for the new space complex. In its technologically infused patriotism, Kennedy Space Center epitomized America's image of modernity. For Luckman, "the project that evoked my deepest emotional involvement as an architect and concerned citizen was the race to the moon, launched in May 1961, when President John F. Kennedy stirred the nation."[21] The culture of architectural Modernism spreading out from American cities such as New York and Chicago mirrored the production of a space complex. For economist John Kenneth Galbraith, this transition of the bureaucracy's administrative apparatus required a "technostructure"—the administrative body that contains the power to control its economic goals—enabling the technical landscape to become fully operational.[22] This corresponded with Luckman's philosophy of architecture, and of business in general.

One of the first buildings constructed in this new industrial area was the Kennedy Space Center Headquarters.[23] As part of the larger launch complex, the Headquarters and its associated assembly of centralized facilities were not "anonymous" shells. Located five miles from the famous Apollo launchpads 39A and 39B, the Kennedy Space Center Industrial Area was the embodiment of the Cold War fear of nuclear conflict.[24] The possibility of atomic bombing would have seemingly led to the decentralization and dispersal of military facilities,[25] but here the architecture of NASA's offices placed them in a central facility in the once remote South of the highly militarized landscape.

The John F. Kennedy Space Center Headquarters Building was completed in 1965. Known then as "M6-399," the building housed two primary administrations: government offices (NASA) and aerospace defense (United States military). The enormous facility enclosed 319,000 square feet to accommodate 2,031 employees. By 1968, an additional 120,000 square feet had been added for 1,100 more people. The building was held up by reinforced concrete frames and enclosed by masonry walls. Vertical shallow fins on the facade mimicked the glass and steelwork of Mies van der Rohe in Chicago, an architect Luckman admired greatly. The design of the building anticipated future growth and planned for modular expansion, forming U- and L-shaped masses. Construction was incremental (not phased) so the administrative staff could keep up with rocket development next door. The first U-shaped structure was completed in 1964, followed by the second phase in 1965 with two L-shaped wings on either side. By 1968, the third phase had added two more L-shaped wings. While the design anticipated change, the shell of the total 439,000 square feet remained the same. Demolished in 2019, the Kennedy Space Center Headquarters found itself unable to keep up with technological progress *and* unable to sustain its historical significance.

* * *

Following the cultural and political "measure of certainty," in the book *Rescuing Prometheus* by Thomas Hughes, "construction began on launching facilities at Cape Canaveral, Florida, before the engineers had a final design for the missile."[26] The rapid pace of coordination and large-scale systems management, illustrated by Hughes, offers a kind of technological embrace—a positive reliance on technological elegance. In 1962 at Rice University, President Kennedy delivered a speech, uttering the famous lines, "We choose to go to the moon in this decade and do the other things, not because they are easy, but because they are hard."[27] It has been suggested that "Kennedy approved plans to go to the moon because he—and perhaps peculiarly and particularly he—knew that the single image, however arduously achieved, could be magnified and extended globally, and, in an instant, change the world."[28]

Examining the less-celebrated architecture of NASA in the 1960s reveals a certain technological aesthetic of modernization that seems opposed to preservation. The memorialized infrastructures at Cape Canaveral and the Kennedy Space Center are forgotten histories, iconic forms of economic prosperity during the height of Cold War conflict. After various delays in the process, the old four-story building was demolished in 2019—a strange tension with its historical registration, office bureaucracy, and Luckman's modern design anticipating adaptability. Kennedy Space Center Headquarters left us with a deep history in architectural modernity yet was seen as a ubiquitous (and expendable) administrative building typology in the shadows of NASA's political and technological success.

Space Center as Modern Ubiquity

The culture of architectural modernization from American cities, such as Houston, became implicitly connected to the production of the space complex—a ubiquity of urbanization processes in American culture—through form, technology, and rational order. The influence of defense-related science originated from World War I and stretched into the next decade, drastically shaping the science disciplines and their associations with city form.[29] For example, the Astrodome constructed in 1965 followed both the image of NASA's concurrent space program and its innovation in engineering technology—erected as the largest dome in the world at the time of its construction and in-

Figure 13.2. West court of Kennedy Space Center Headquarters, 2019. Photo by Roland Miller.

Figure 13.3. Front elevation of Kennedy Space Center Headquarters, 2019. Photo by Roland Miller.

Figure 13.4. Corridor at
Kennedy Space Center
Headquarters, 2019.
Photo by Roland Miller.

stalling the first artificial grass, known as "AstroTurf." Projects like the Astrodome greatly influence how we might consider the relationship between the
construction of American cities tangential to the expansion, and "enclosure,"
of NASA's administrative bureaucracy.

Historically, infrastructure is often associated with supporting urban development and correlated civil engineering disciplines to aid in the management of population growth in dense urban regions.[30] Yet, in the NASA Space
Center, infrastructure is acknowledged as an extra-urban, or perhaps even
an extraterrestrial, military support system launching Cold War scientific in-

novation, and therefore becoming a surrogate for territory, power, and false reliance on technological innovation. As revealed in the formulation of territory in the Cold War, including its transformation and abandonment of rapid construction of infrastructure supporting early space programs, the military redefined the relationship to urban boundaries.

The ubiquity of modern architecture aligned with the speed of political pressure in the space complex. NASA Manned Spacecraft Center in Clear Lake, on the outskirts of Houston (now known as Lyndon B. Johnson Space Center), was also designed by Charles Luckman. *Time* praised Luckman for supposedly planning and designing 49 buildings in 48 days. The planning and design for such an inconceivable task would require the use of repeatable forms and distributions of ubiquitous design criteria.

On January 3, 1962, Charles Luckman, with Houston local architects Brown & Root, submitted design guidelines titled, "Master Plan and Architectural Concept: Manned Spacecraft Center." The Space Center would be located on 1,620 acres in Clear Lake, Texas. The report offers direct recommendations for the US army's Corps of Engineers to incorporate into its construction. The document includes a master plan scheme and an architectural concept that proposes "modules" to form a NASA-based architectural vocabulary. Luckman's team had established criteria for distributing such modules, and the report offers specific suggestions "with respect to building areas and functional relationship."[31] Planning for NASA's new home in Houston included four primary land use elements: antenna test range, lunar landing area, barge berthing facility, and office buildings. Unsurprisingly, the Manned Spacecraft Center is occupied predominately by infrastructural elements, with little indication of an architectural aesthetic, even though this initial planning comprises 1,045,500 square feet of enclosed architectural space. The large footprint of climate-controlled space is a combination of programs, including laboratories, testing facilities, spacecraft research, and project management offices to house more than 3,300 personnel in total (whereas the original Kennedy Space Center Headquarters accommodated more than 3,000 people in a single 439,000-square-foot building).

The site, situated adjacent to the coast for ease of shipping materials by barge, would seem to be challenged by obstruction of the Gulf Coast Plains natural bayous; however, for the architects, two primary objectives were identified as core to the planning. When describing land development, "the successful scheme must provide a maximum of simple-shaped developable land for future construction" and "should be arranged so that certain important facilities, such as Project Management, are easily accessible."[32] The report indicates the site's relationship with the barge-loading stages but also describes its relationship to Ellington Air Force Base and the Houston International Airport (now known as William P. Hobby Airport). Although much of the planning required logistical coordination between the various testing facilities and material transmission, the architects prioritize "simple-shaped" for repeatable order connected to other forms of air travel. Anxious to meet hard-pressed deadlines set by NASA's leadership in Washington, the architectural decisions tend to follow Luckman's philosophy of architectural modernity at the intersection of business and economic responsiveness. The ubiquity of form follows an increase of bureaucratic funding—unlike the launch-complex facilities constructed in the prior decade at Cape Canaveral with a leaner and more assertive space administration.

The spatial recommendations for the arrangement of facilities on site include building orientations according to sun angles and establishing a module from the interior dimension. Based on typical office interior measurements, "office planning modules" become the primary architectural feature for maximizing planning flexibility. The interior sets the exterior enclosure, thereby extending the interior dimension across the entire master planning grid—an "architecture without urbanism." General modules are set to 4 feet 8 inches as the "desirable minimum" office dimension—supposedly allowing for the greatest flexibility.[33] Precast panels are then used in this dimension and deployed across all facilities with varying degrees of orientation for buildings and design systems. The administration facilities are generally one or two stories, long horizontal facades covered by overhanging roofs and supported by thin *pilotis*. Otherwise, the various testing and laboratory facilities allow for large volumes minimizing obstruction to overhead cranes.

Contrary to the typical claims of decentralization in the era of American Modernism, President Kennedy's 1961 congressional charge in Washington for lunar landing by the end of the decade required significant design and construction of massive, centralized facilities. These facilities are largely left out of both architecture and aerospace histories. At first, office and operation facilities at Johnson Space Center don't seem to play a role in the projected image of space exploration—based largely on the fact that the "image" of NASA is more directly associated with science and technology. On the surface, facilities in Houston seem like anonymous and decentralized industrial complexes. Yet, successes of American space programs relied heavily on strategies of logistical coordination, including centralization of administrative management—depending on the scale of perspective and the boundary of ubiquity. Can the planning of NASA's administrative facilities be seen uniquely derivative of the military complex, or do operational facilities signify other forms of mid-century urbanization regardless of technological "progress"?

With interests in redefining office building interiors in the 1960s, "office landscaping . . . embodies systems analysis, human factors and an intimate understanding of environmental control."[34] The adoption of the "buroschaft" as an office made from "interdependence of elements" produces "a layout which is natural to the work to be accomplished" involving a kind of modification of planning, for ultimate flexibility. Landscape "should be seen not as representing actual countryside but rather as a vision of an ideal scene"; then the "burolanschaft," in the case of optimizing office workspace, illuminates a peculiar overlap of flexibility and control.[35] Such rationale, or ideology embedded in logics of spatial organization, became tools for blurring the hidden possessions of control. The "burolanschaft" illustrates a new machine-like systematic scene projecting images of perception tied to capital, and ultimately the state itself. However, the spatial implications of office design in this model do not reduce user separation. On the contrary, the user is deeply embedded within labor powers of NASA's bureaucracy, including the link to political maneuvering, and mechanizing the office landscape, based on optimal performance for Cold War defense and geopolitical power. Therefore, considerations of functional aspects of office operations and work-flow processes do not participate in the means by which landscape is *actually* played out, but further exacerbate its projected power and participation in the expanded urban landscape.

During Project Apollo, the space complex "supported the inspection, check-out, and integration of the spacecraft modules," including a long list of storage, transmission, and testing facilities. Just as vital to the space program as the highly celebrated Vehicle Assembly Building and launch pad, the Space Center achieved "an optimum of order in the office" at an unprecedented level of "office park" coordination.[36] Interestingly, this tends to correspond with Charles Luckman's philosophy on architecture, and on business in general. Luckman's design of the Manned Spacecraft Center facilities is a tool—maybe not anonymous, but it could be understood to attain a kind of *technological elegance*—situated between economics and bureaucratic ubiquity for a national agency. The Manned Spacecraft Center is ultimately an expression of architectural modernity—a representation of values and requirements for centralizing operations, networking logistics, and optimizing the mechanization of NASA's image.

Planning decisions for the Manned Spacecraft Center were made on the basis of function first, cost second, and aesthetics much further down the line as buildings were constructed in a hurry. Similar to the Air Force at Cape Canaveral, NASA was financially aggressive in the first few years. But processes quickly became occupied by a large bureaucracy years before ever reaching

Figure 13.5. Corner of Building 1 at Johnson Space Center, 2019. Photo by Jeffrey S. Nesbit.

the moon in 1969. Today, Johnson Space Center is challenged by a radically different economics in US space history. Contrary to Luckman's planning recommendations, Johnson Space Center today appears disorganized, with adjacent building programs unassociated, and departments decentralized. Several facilities are still standing but not in use. Following the mothball attitude set by the military, a surplus of square footage disaggregates operations inefficiently. In response, Johnson Space Center's facility management assembled a list of bids for demolishing the facilities of greatest maintenance cost—attempting to purge its footprint for greater economic and energy efficiency. Using current replacement value as a criterion, the facility management team identified fifteen facilities and considered the relatively small number of people who would need to be relocated. All fifteen were office buildings, none related to industrial or laboratory purposes. In the end, senior management in Washington resisted, and all these facilities still exist today.

Centralization of administrative processes presents an age of *technological elegance* in the face of modernization. And yet, it is the administrative apparatus of NASA that provides evidence for the entrance of national politics and economics into modernization of the American city. Kennedy Space Center and Johnson Space Center broaden the role of infrastructural urbanization attempting to spatialize efficiencies, reject preservation, and automate

Figure 13.6. Facade of Building 1 at Johnson Space Center, 2019. Photo by Jeffrey S. Nesbit.

a synthetic system of coordination, procurement, and management of federal funds.

Today, Luckman's Kennedy Space Center Headquarters building is replaced. Completed in 2019 directly adjacent to the original building, the new Headquarters is less than half the size of the original building, accommodating 500 employees in 200,000 square feet of office space. Designed by Orlando-based HuntonBrady Architects, the new facility is said to have "an accent color scheme representing a planet in the solar system" and will "embody the optimistic spirit of NASA and [Kennedy Space Center]," without any indication or notion of infrastructural efficiency, or speed.[37] We are now

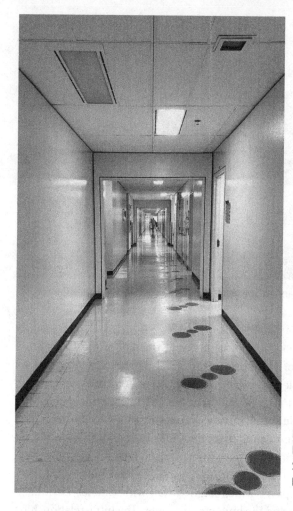

Figure 13.7. Corridor in Building 1 of Johnson Space Center, in 2019. Photo by Jeffrey S. Nesbit.

left with an object of symbolism, signaling a caricature of future optimism—a failure of modernity in the face of global contemporary capital. Charles Luckman, the business savvy architect, with sensitivity to budgets, coordination, and management, greatly shaped the spaceport complex and NASA's industrial area. Kennedy Space Center Headquarters and Johnson Space Center planning leave us a deep history in architectural modernity yet are seen as a ubiquitous (and expendable) administrative building typology in the shadows of NASA's political and technological narratives proclaiming national success.

The memorialized architectural objects in the spaceport complex—including the commemoration of two presidents' roles in space pursuits—are rendered as anonymous office buildings, failed architectural forms during a time of economic prosperity. But perhaps the administrative facilities sup-

Figure 13.8. Master plan of Manned Spacecraft Center (now Johnson Space Center), 1962. Courtesy of NASA.

porting the highly celebrated Apollo launches at the end of the 1960s are a more accurate portrayal of an unbiased historical perspective, rendered by the ubiquity of the urban landscape. NASA's Space Centers, commemorating Presidents Kennedy and Johnson, reinforce NASA's interest in establishing an aesthetic of power while simultaneously rendering the spaceport complex economical. Therefore, the misunderstood architecture of NASA should be considered vital to its impact in aerospace history and in aesthetics of American urbanization, more generally. And the shape of modernity meant leaving behind the historical past—which aligned well with Charles Luckman's philosophy of architecture and the built environment. The intra-relationships of the administrative apparatus, including manufacturing and assembly facilities, inevitably become dismissed, replaced by new economics while purposely preserving their political memory and leaving behind their modern architectural archaeology. The need to invent new organizational structures, architectural design, and planning for Kennedy Space Center and Johnson Space Center led to strategically resisting conflict of communication between project and mission groups through innovative technological progress at a large scale, *and* subdued coordination, including management of infrastructural systems in NASA's remote American South.

Notes

1 Historic American Buildings Survey, "Cape Canaveral Air Force Station, Launch Complex 39, Headquarters Building," in HABS No. FL-581-A (Atlanta, GA: National Park Service, 1999).

2 Ada Louise Huxtable, "Farewell to Penn Station," *New York Times*, October 30, 1963, 38. Huxtable called it a "monumental act of vandalism," adding that "we will probably be judged not by the monuments we build but by those we have destroyed."

3 Herbert Muschamp, "Charles Luckman, Architect Who Designed Penn Station's Replacement, Dies at 89," *New York Times*, January 28, 1999.

4 Charles Luckman, *Twice in a Lifetime: From Soaps to Skyscrapers* (New York: W. W. Norton, 1988).

5 Rem Koolhaas, "Whatever Happened to Urbanism," in *S, M, L, XL: Office for Metropolitan Architecture* (New York: Monacelli Press, 1994).

6 Luckman, *Twice in a Lifetime.*

7 Stuart Leslie, "Spaces for the Space Age: William Pereira's Aerospace Modernism," in *Blue Sky Metropolis: The Aerospace Century in Southern California*, ed. Peter Westwick (Berkeley: University of California Press, 2012).

8 Leslie, "Spaces for the Space Age." According to Leslie, "aerospace modernism appealed as much to the military as to its defense contractors" and it was the "Air Force officers, like their corporate counterparts" who actively embraced the Modernist forward thinking.

9 Huxtable, "Farewell to Penn Station," 38.

10 Susanne Broyles, "Penn Station and the Rise of Historic Preservation," Museum of the City of New York, New York Stories, May 8, 2012, https://blog.mcny.org/2012/05/08/penn-station-and-the-rise-of-historic-preservation.

11 Broyles, "Rise of Historic Preservation."

12 "New Madison Sq. Garden to Rise atop Penn Station," *New York Times*, July 25, 1961, 1.

13 Muschamp, "Charles Luckman, Dies at 89."

14 Historic American Buildings Survey, "Cape Canaveral Air Force Station."

15 Wigley, "Architectural Cult," 39. Wigley continues by explaining, "The discourse about the preservation of monuments is completely entangled with the discourse about reproduction. Reproduction is both a force that erodes the unique condition of the original and a force of preservation."

16 Michael Beschloss, "Kennedy and the Decision to Go to the Moon," in *Spaceflight and the Myth of Presidential Leadership*, ed. R. Launius and H. E. McCurdy (Chicago: University of Illinois, 1997), 51–67.

17 Charles Benson and William Faherty, *Moonport: A History of Apollo Launch Facilities and Operations*, The NASA History Series (Washington, DC: NASA, 1978), 87–107.

18 John F. Kennedy, "Urgent National Needs—Address of the President of the United States (H. Doc. No. 174)," May 25, 1961, in *Congressional Record: Proceedings and Debates of the 87th Congress, First Session, Volume 107—Part 7, May 18, 1961, to June 7, 1961* (Washington, DC: GPO, 1961), 8877–8882.

19 For more information, see Mark Shanahan, *Eisenhower at the Dawn of the Space Age: Sputnik, Rockets, and Helping Hands* (Lanham, MD: Lexington Books, 2016); and John M. Logsdon, *John F. Kennedy and the Race to the Moon* (London: Palgrave Macmillan, 2010).

20 Martin Jay, "No State of Grace: Violence in the Garden," in Jay, *Essays from the Edge: Parerga and Paralipomena* (Charlottesville, VA: University of Virginia Press, 2011), 64–76. For Jay, the garden should be an extension of the countryside, not a formalized extension of an architectural order. Therefore, landscaping has been naturalized and "gardens are precisely the sites in which force and human violence are in tense constellation with collective power."

21 Luckman, *Twice in a Lifetime*, 391.

22 John Kenneth Galbraith, *The New Industrial State* (London: Hamish Hamilton, 1967). Galbraith argues that "it is reasonably certain that a man can be landed on the moon within the next five years," based on modernization's "measure of certainty."

23 Luckman, *Twice in a Lifetime*.

24 Benson and Faherty, *Moonport*, 87–107. When describing in detail the facility construction and operations of Kennedy Space Center for the Apollo mission, *Moonport*, a NASA-published document, focuses primarily on facts of construction, timelines, and procedural organization, rather than identifying derivatives of the decision-making process. It is this "procedural" nature that tends to emerge as a more holistic architectural device for deploying the administrative and industrial complexes found in Kennedy Space Center and the Johnson Space Center.

25 Peter Galison, "War against the Center," in *Architecture and the Sciences: Exchanging Metaphors*, ed. A. Picon and A. Ponte (Princeton, NJ: Princeton Architectural Press, 2003). Galison describes how, in the postwar era, framed by rapid technological progress, cybernetics, and society's anxiety over rising mass consumption, military manufacturing and production became decentralized in the United States as a core product of enclosing processes and distributing military networks across vast geography. Industrial dispersal occurred due to the possible implication of atomic bomb target areas in urban cores; as described by Galison, "city by city, country by country, the bomb helped drive dispersion." Targets became less focused on factories and more on the resources that would greatly destabilize the systems and processes of manufacturing.

26 Thomas Hughes, *Rescuing Prometheus: Four Monumental Projects that Changed the Modern World* (New York: Random House, 1998).

27 John F. Kennedy, "Address at Rice University on the Nation's Space Effort," in Houston, Texas, September 12, 1962. John F. Kennedy Presidential Library and Museum, https://www.jfklibrary.org/learn/about-jfk/historic-speeches/address-at-rice-university-on-the-nations-space-effort.

28 Nicholas de Monchaux, *Spacesuit: Fashioning the Apollo* (Cambridge, MA: MIT Press, 2011).

29 David Kaiser, "The postwar suburbanization of American physics," *American Quarterly* 56 (2004): 851–888. Even within history of science discourse, the relationship between urban pattern and scientific innovations is characterized by physicist David Kaiser as the "Cold War bubble" of American physics during the late 1940s and into

the 1950s; see also, David Kaiser, "The physics of spin: Sputnik politics and American physics in the 1950s," *Social Research* 73 (2006): 1225–1252.

30 See Jennifer Light, *From Warfare to Welfare: Defense Intellectuals and Urban Problems in Cold War America* (Baltimore: Johns Hopkins University Press, 2003). Using a history of atomic bombs to describe the modes of decentralizing city form through a diffusion of military-industrial complex, Jennifer Light effectively illustrates how the United States made attempts to apply military-related activities to solve urban challenges in urban planning and urban life but ultimately failed.

31 Charles Luckman and Browne and Root, *Master Plan and Architectural Concept: Manned Spacecraft Center, National Aeronautics and Space Administration* (Houston: Brown & Root, Inc., 1962), 4.

32 Luckman, Browne and Root, *Architectural Concept: Manned Spacecraft Center*, 9.

33 Luckman, Browne and Root, *Architectural Concept: Manned Spacecraft Center*, 54.

34 Carroll Cihlar, "Landscaping: An Environmental System," in *Office Landscaping: An Open Plan Concept of Office Design*, ed. Carroll Cihlar (Elmhurst, IL: The Business Press, 1969), 13–15.

35 Kenneth Olwig, "Recovering the Substantive Nature of Landscape," *Annals of the Association of American Geographers* 84/4 (1996): 630–653. The epistemology of "landscape" seems to describe not only the vista or perspective of natural scenic views but also the human interaction dependent upon such environmental conditions. As Olwig argues, "the substantive meaning of landscape as a place of human habitation and environmental interaction" enables a more focused definition not only confined to territory. Suggesting a connected relationality between "community and place," the township and country epistemological derivations imply a function of power and representation.

36 Cihlar, "Landscaping," 13–15; See also, Hans J. Lorenzen, "The Economic Reasons behind Landscaping," in *Office Landscaping: An Open Plan Concept of Office Design*, ed. Carroll Cihlar (Elmhurst, IL: The Business Press, 1969), 38–41. The adoption of the *buroschaft* as an office made from the "interdependence of elements" produces "a layout which is natural to the work to be accomplished" and implicates a kind of modification of the plan—ultimate flexibility. However, it is a bit ironic this flexibility is made not for reasons of human independence, but rather for "setting up efficient communication systems" and ultimately "achieving an optimum of order in the office."

37 HuntonBrady Architects, "How Architecture Will Launch NASA into Next Century," February 21, 2017, https://www.huntonbrady.com/post/how-architecture-will-launch -nasa-into-next-century

14

The Race for Space

Creating Modern Landmarks in Postwar Huntsville, Alabama

CAROLINE T. SWOPE

> Those of us who trod the steps of the Courthouse did not even know that
> we were living in a setting already doomed . . . gasping its last breaths. It's
> dead and gone now. The shiny new Courthouse complex rising high in the
> sky will someday be remembered with nostalgia by today's young and yet
> unborn.
>
> "The Old Era Passes," *Huntsville Times*, August 15, 1965, A7

These prophetic words were written in 1965, when Madison County's new modernist courthouse was under construction. The "courthouse that space built," as the press described it at the time, was a modernist departure from the antebellum and Reconstruction-era architecture that had dominated Huntsville and Madison County during the previous century.

Huntsville, Alabama, now known as a space and defense center, had a modest start in the early 1800s. The city was incorporated in 1811, before Alabama was officially a state. Planters from Virginia, Georgia, and the Carolinas moved into the region, enticed by vast swaths of land. Huntsville grew as cotton merchants, lawyers, and bankers congregated around what was known as Cotton Row downtown.

After the Civil War, large textile mills and cotton-seed mills moved into the city, enticed by the ability to pay lower wages than comparable factories in New England. The downtown developed with Richardsonian Romanesque and late Victorian-era commercial architecture, while high-style Queen Anne Victorian residences were built near the commercial core. Building stock in the mill company areas of town could have easily been mistaken for New England mill towns; company houses were arranged in neat rows. The architectural environment, while impressive, was not unique in northern Alabama.

In the twentieth century before World War II, a pedimented Greek Revival courthouse that had been built in 1840 dominated the downtown. Another prominent building was a Greek Revival bank that was built around 1935. The remaining street front was dominated by two- and three-part commercial buildings. Several fine Greek Revival mansions were located in residential neighborhoods. The city's economic drivers—and its architecture—would change drastically after World War II.

Huntsville grew exponentially beginning in 1941, when the United States Army selected almost 40,000 acres of land to the south of the city as a site for two large arsenals—the Huntsville Arsenal (a chemical warfare plant) and the Redstone Ordnance Plant (which manufactured artillery shells).[1] The same factors that made Huntsville attractive for development in the nineteenth century—its proximity to rail and river transport—helped secure its selection for the new arsenal. The availability of large tracts of land and electric power from the Tennessee Valley Authority were also reasons why the army selected Huntsville.[2] The next year, 1942, construction of mustard and chlorine gas plants began. Production facilities for bombs were built as well. In 1949, the army again selected Huntsville, this time as a site of missile production. The Huntsville Arsenal and the Redstone plant merged, and the following year a team of more than 100 German scientists under the leadership of Wernher von Braun moved to Huntsville to support the army's missile program. Members of von Braun's team were responsible for many of Huntsville's architectural and planning designs in the mid-twentieth century. Von Braun's team was ultimately responsible for the Jupiter-C rocket and for the Saturn V rocket that helped astronauts reach the moon.

In 1960, a presidential executive order established the George C. Marshall Space Flight Center in Huntsville. By 1967, the population of what had been a quaint southern town had exploded to an estimated 144,000.[3] Huntsville and its neighboring areas struggled to keep up with the intense demand for housing. The combination of the influx of so many highly trained professionals, many who weren't from the South—or even from the United States—and the presence of the space-age technological sector pushed traditional southern architectural conventions in unexpected ways. This chapter examines the Madison County Courthouse, Huntsville's First Baptist Church, and Edmonton Heights, a Black neighborhood built north of the city, to see how their creation and the contemporary architectural choices that were made when they were built reflected Huntsville's growth as a modern space-age municipality.

Huntsville, which was surrounded by cotton fields interspersed with small farming hamlets in the early 1950s, would have contrasted starkly with the cosmopolitan architectural world that many of the German scientists had lived in before emigrating to the United States. During the 1930s, famed German modernist Ludwig Mies van der Rohe was the director of the Bauhaus, a new school of architecture, in Berlin. Von Braun lived in Berlin during this time and would have seen the new buildings. The school, which the Nazis eventually shut down, had a significant impact on art and architectural trends in the western world. Simple designs, unadulterated natural materials, streamlined aesthetics, and a modern style that was conducive to mass production were some of the architectural hallmarks of the movement. Heinz Hilten (a native of Berlin) and Hannes Luehrsen (who studied in Berlin and Aachen and later served as a city planner in Berlin) were two of the German planners and architects who came to Huntsville to work at Redstone and Marshall.[4] In addition to their government work, they each completed several other architectural projects. Both worked on the master plan for the University of Alabama in Huntsville. Hilten designed or assisted in the design of a number of elementary schools in Huntsville in the 1960s. He also worked on St. Mark's Lutheran church, part of the Von Braun Center entertainment complex, and homes for several of the German scientists. Hilten said of his work, "I was not a space scientist that designed the rockets, but I designed the spaces those scientists worked in."[5] Luehrsen initiated the planning for Memorial Parkway in 1950 and shortly thereafter submitted a master plan for the Heart of Huntsville project, a massive federally funded program that allocated almost $10 million for the acquisition and redevelopment of downtown land.[6]

Huntsville and Madison County's growth challenges were a frequent subject of newspaper articles during the 1950s. Luehrsen's master plan recommended that the city build a civic center surrounded by other public buildings that included a city hall, the courthouse, and an art museum. The centralization of civic amenities was unusual in an era where communities were increasingly being decentralized. Board of Commissioners chair Roy L. Stone described the design for the city center as a "relief"; he was pleased to see that it would preserve the downtown area.[7] In current usage, one would assume that the word "preserve" referred to an element of historic preservation, but that is not how Stone used it. For him, "preserve" clearly meant that functions would be preserved and added to, but the majority—if not all—of the buildings supporting the programs would be new and modern. Luehrsen submitted his plan to the county commissioners and to the Huntsville City Council in a joint meeting with the Huntsville Housing Authority. Federal funding would be needed to accomplish such an extensive building program, and some local housing and commercial centers would need to be condemned. In1958, the proposed plan was given the name Heart of Huntsville. The project, which centered on Big Spring Park, was advertised as a slum clearance and redevelopment program. However, the earliest discussions of the plan did not discuss slum clearance, at least not in sources now available. That portion of the plan was either downplayed at the start or possibly gained additional focus once it became clear that federal funds were available for that use. The plan went through several adjustments, but in its final form, the U.S. Department of Housing and Urban Development razed most of the west side of Courthouse Square in the first "improvement" stage. This was a row of commercial buildings that was anchored by the antebellum-era First National Bank.

A major component of the plan involved clearing Black residential and commercial neighborhoods out of the city center under the guise of slum clearance. Many owners who were displaced by the Heart of Huntsville project were given assistance by the Federal Housing Administration. Of this group, many used federal assistance to move into new single-family homes in Edmonton Heights, a housing project backed by the Federal Housing Administration. The neighborhood was designed by Folmar & Flinn, Montgomery-based developer that specialized in speculative housing. The firm had worked on at least two other Black neighborhoods in the state and was one of the largest speculative housing builders in the Southeast at the time. They developed at least two other speculative housing subdivisions in Huntsville, Meadow Hills and Lakewood Manor, both located on the northern edge of the city west of Highway 231. Edmonton Heights is also located north of the city, but it is on the east side of highway 231, directly south of Alabama A&M, a historically Black university.

The *Huntsville Times* noted that "the Negro man or woman who have their life savings tied up in a home" could be negatively impacted by the project.[8] In November and December 1960, advertisements in the *Huntsville Mirror* and *Birmingham Mirror* (both Black-owned newspapers) compared the modernity of Edmonton Heights houses to the run-down "Shot Gun style" houses that Black families lived in downtown, on land slated for redevelopment by the Heart of Huntsville project.[9] Edmonton Heights is populated with a mixture of Minimal Traditional, Contemporary, and Ranch houses. Newspaper advertisements for Edmonton Heights specifically contrasted the neighborhood's modern-style architecture with the traditional shot-gun–style homes the African American community lived in downtown.

Edmonton Heights was clearly planned to provide housing to replace dwellings demolished in Black neighborhoods of the city center. However, housing wasn't all that was destroyed for the Black community. Some Black business owners lost businesses in prime urban locations. One resident re-called that the city took his dad's BBQ stand on Church Street as part of the Church Street Development Plan. His father took the building apart board by board, to keep what little he could have of the establishment. He retired after he lost his business.[10] Other residents recalled that although Edmonton Heights was a nice place to live and was one of the few neighborhoods in the city where Black Huntsville residents could live, the houses were overpriced. One resident remembered his father, who was a realtor, commenting on the inflated prices for the neighborhood. His father knew the prices were too high, but you had to "bite the bullet if you didn't want to live in the projects."[11] He said that his father struggled to pay the inflated mortgage on their home in the 1970s. Although early residents of Edmonton Heights worked at a variety of jobs, the significance of Redstone and NASA in the local job market is clear. Sixteen percent of the residents worked at Redstone or NASA, while another 9 percent served in a branch of the armed forces.[12]

As Huntsville began to reimagine its downtown, what to do about the Madison County Courthouse became a controversial element of the Heart of Huntsville project. Luehrsen's master plan kept the city's governmental services downtown, but the courthouse (which had been built in 1914) was deemed too small to serve the growing community's needs. Although news-paper reports indicate there was some nostalgia for keeping the 1914 building intact, the majority of those who attended public meetings wanted to raze the old courthouse and build something contemporary and more representative of Huntsville's status as a space-age city. The Huntsville Historical Society was one of the major opponents of a new courthouse, but its members eventually became reconciled to the plan when it became clear that they would not pre-vail. Instead, they focused on softening the appearance of a new courthouse by requesting a "timeless" rather than a "modern" look.[13] Although the *Hunts-ville Times*, which reported on the meeting where Historical Society members expressed this view, didn't explain the difference between the two styles, it is likely that the modern look focused more on large glass curtain walls. The final building design may have been considered more "timeless": its stream-lined columns support the portico, adding a Neo-Formalist detail to the In-ternational Style building. Some of the major arguments for retaining the site centered on keeping the location's historic use for a courthouse. This would help develop the historic portion of the city by continuing to support civic and governmental centers at Courthouse Square. There was also discussion of cost. The consensus was that foundation work, while costly, would still be less expensive than acquiring additional land for a new building.[14] The court-house plan was considered the pinnacle of the Heart of Huntsville project; it

Figure 14.1. Madison County Courthouse. Photo courtesy of Historic Huntsville Foundation.

anchored downtown Huntsville with a contemporary building worthy of the new space-age capital.[15]

Planning of the courthouse formally began in April 1961, when the County Board of Commissioners surveyed the county's needs. Final plans were presented to the public in November 1964.[16] Two architectural firms collaborated on the plans: Jones, Crow, Mann & Associates of Huntsville and Northington, Smith, Kranert & Associates of Florence, Alabama.[17] The firms chose the International Style for the courthouse (fig. 14.1). One of the main tenets of that style was a naïve belief that regional variations in architectural style to accommodate different climates would not be needed. Technology could ensure that all locals could have the same idealized, utopian buildings. For Huntsville, a city that was rapidly becoming a technological center in the southeast, the pull toward this style was inevitable.[18] The courthouse design won a corporate honor award from the Gulf States Regional Conference of the American Institute of Architects in 1967.[19] The design was specifically praised for how it adjusted to unusual site conditions, most notably a cave under part of the courthouse.

Northern Alabama is the site of thousands of caves, the majority of which are in Madison and Jackson Counties. The National Speleological Society is headquartered in Huntsville because of its proximity to so many caves. The city's municipal water supply originated in Big Spring Cave, which was first recorded in 1916. Although the previous courthouse was built over the cave site, the size of the new courthouse and the additional weight from the proposed tower raised concerns. The state's geologist, Dr. Walter B. Jones, referred to Big Spring Cave as one of the most dangerous of the more than 1,500 caves he had entered in his career.[20] When soil borings were taken to determine the stability of the site, a new cave was discovered under the northwest corner of the intersection of Jefferson Street and North Side Square at this time.[21] Unbridled faith in modern technology was tempered with caution in this instance, and adjustments were made to the design. The courthouse is not centered on the lot. In addition, to ensure that the courthouse was properly supported, twelve concrete caissons were installed inside the cave, under the southern portion of the courthouse. Caissons support the steel girders that bridge the cave, forming the foundation for part of the courthouse. The deepest caisson was expected to sink fifty feet underground.[22] The southern part of the courthouse complex is shorter to reduce the weight on the cave's ceiling. The eleven-story tower portion of the building is constructed on solid rock.

The price tag for the finished building was more than $5 million. The local press lauded it as a progressive space-age design that was worthy of Rocket City and as one of the most modern buildings in the South.[23] It was opened at the very end of 1966. The interior is open with double staircases that appear to float between the first three floors (fig. 14.2). A massive three-story mosaic mural showcases the county's history. Of the dozens of figures depicted on the mural, only two were contemporary, John Sparkman, a US senator, and Werner von Braun, the director of the Marshall Space Flight Center at the time. The official dedication was held on March 5, 1967. The usual dignitaries were present, including the commander of Redstone. The festivities focused on both tradition and progress. The president of the Huntsville Historical Society presided at the opening ceremonies. Senator Sparkman said that he hated to see Cotton Row and the old courthouse torn down but that the "changes were signs of progress."[24] Newspaper advertisements proudly described the "courthouse that space built."[25]

As downtown Huntsville was replacing its early twentieth-century Courthouse Square with an International Style modern marvel, one of Huntsville's largest churches had just finished a massive building program that rivaled that of the courthouse in 1966. First Baptist Church dominates the landscape south of downtown on a sprawling twelve-acre site. It is one of Huntsville's most distinctive buildings, in part due to its massive sanctuary that features colorful celestial mosaics. But the rocket-shaped bell tower is perhaps the most tangible reminder of how deeply Huntsville's status as a space-age city

Figure 14.2. Interior of Madison County Courthouse. The mural depicts historic figures of Alabama's past. Photo courtesy of Historic Huntsville Foundation.

permeated almost all aspects of the city's identity in the post–World War II era. The unusual building gained considerable attention, and not all of it was positive. One architect called it "the second ugliest building" in Huntsville.[26] A letter from a member of the public claimed that the mural was garishly colored in "Crayola colors." The tower has been called a "phallic symbol" and a "beer can in the sky." Construction costs for the church were so high that one man picketed outside the church with a sign calling for it to close its doors and spend money on social service programs instead.[27] The massive fund-raising efforts required for the construction project are evidence of the congregation's tenacity and vision. The modern design of the church, especially its unusual space-themed mosaic, stained-glass windows, and rocket-inspired bell tower, is a purely Huntsville institution and clearly a reference to the NASA-permeated environment of the Space Age.

The First Baptist congregation organized in 1809; it is the oldest in the Alabama Baptist Convention. The congregation moved to several different sites over the centuries. By the early 1950s, it was located at the corner of Clinton and Gallatin Streets downtown.[28] It was clear that space was insufficient for the projected needs of the congregation, and a planning committee was formed in 1954. The congregation wanted to purchase additional land at their downtown site, but they were unable to do so. The long-range planning committee felt that it was important that the church stay as close to downtown as possible; they believed that an urban location was part of the congregation's identity.[29] In the spring of 1954, the church decided to purchase about ten acres on Governor's Drive (then known as Fifth Avenue) for $123,144. A building committee was formed in 1957 and selected the Birmingham firm of Lawrence Whitten & Son for the design.[30] The estimated cost of the new sanctuary and support spaces was a jaw-dropping $1.6 million. Financing the project proved to be difficult; the congregation was unable to secure a loan for more than $500,000. The church decided to proceed with an education building and a chapel, postponing construction of the main sanctuary. Bids for the project were taken at the end of 1960. Church legend claims that a turnip was unearthed during the groundbreaking ceremony and locals quipped that the church was built on a turnip patch. The education building encompassed more than 100,000 square feet of space that accommodated twelve nursery departments and twenty-three classrooms for older children. The fellowship hall was designed to seat 600 and the chapel in the new structure could accommodate 350.[31] A prayer room, a conference room, a library, a recreation room, an adult study, a kitchen, and offices were part of the plan as well.

The church was clearly planned for many young families, and the expanding Redstone Arsenal attracted new residents to the Huntsville area. But not everyone felt that the growth was stable. The church contacted more than 300 mortgage companies when it was searching for a financial partner to underwrite $500,000 in loans. None would fund the project. There was considerable concern that a changing political situation could decrease federal support for the arsenal and if that happened, the church would lose a large portion of its members as people moved out of the area in search of other work. Finally, in May 1961, First Federal Savings and Loan Association, located just a few blocks away from the church, agreed to make the loan. The education wing was operational by early 1963. Construction on the sanctuary started before the education wing had been completed. An additional $500,000 loan was secured for the sanctuary and tower. When construction bids came in at more than $1.5 million, significantly beyond the projected budget, the decision was made to defer construction on the bell tower and move forward with the sanctuary. The combined costs for the sanctuary and the education building was more than $2.5 million. The sanctuary, which accommodated 2,200 people, was completed in 1966 (fig. 14.3).[32] In 1953, the church had 1,384 members. By 1963, three years before the sanctuary was finished, the congregation numbered 2,626. Much like the Madison County Courthouse, the church complex, which was spaciously planned, was too small to meet actual needs almost immediately.[33]

Although the education wing was substantially remodeled in the late 1990s and no longer retains its original design, the sanctuary's Neo-Formalist design and seven evenly spaced arches, reminiscent of the barrel vaults used at the Baths of Caracalla in Rome, is mostly intact. The Neo-Formalist style, which emerged in the mid-1950s and peaked in the 1960s, was commonly used for public buildings. Religious institutions, which often selected this modern substyle, were drawn to the abstract symbols and curving forms that seemed more spiritual and traditional than some of the other modernist styles of the era. The technological requirements of the building are impressive. More than 200 tons of reinforcing steel and 65 tons of structural steel were used to frame out the sanctuary, while support elements were sunk almost five stories into the ground. They were engineered to support the longest spanning concrete roof east of the Mississippi at the time.[34] The church's iconic mosaic, which was completed in 1974, fills the entire southwestern façade. The work measures 47 by 154 feet and is composed from more than 2.5 million pieces of glass tile (fig. 14.4).[35] At the time of construction it was one of the largest mosaics in the United States. It weighed more than six tons and covered 6,000 square feet. The design of the mosaic is particularly striking. It was approved in 1966 after the planning committee rejected non-objective drawings. The pictorial program showcases creation and redemption. It has been suggested

Figure 14.3. Sanctuary and bell tower of First Baptist Church, 2022. Photo courtesy of Caroline T. Swope.

that the galaxies are orbiting Christ, representing an atomic structure. The inclusion of the Alpha and Omega signs and the seven stars held in Christ's hand refer to the Book of Revelations.[36] The stained-glass windows continue the cosmic theme, showing order evolving from disorder in the cosmos (figs. 14.5, 14.6).[37]

One of the most striking elements of the building is the bell tower, which wasn't completed until 1987, at a cost of more than $1 million. Original plans called for a masonry and concrete structure, and the footings were poured when the education building was completed in 1962. However, the height of the tower interfered with the flight path for the Huntsville Airport. The budget was also a concern. When the airport relocated southwest of town, the church returned to its tower plans. The building committee in 1980 ap-

Figure 14.4. Detail of the entrance to the sanctuary of First Baptist Church, 2019. Photo courtesy of Patrick Hood.

Figure 14.5. Sanctuary of the First Baptist Church, 2022. Stained-glass windows on each side of the building depict a space scene with orbiting planets and stars. Photo courtesy of Caroline T. Swope.

Figure 14.6. Detail of stained-glass window in the sanctuary of First Baptist Church depicting planets in orbit, 2022. Photo courtesy of Caroline T. Swope.

proved a different tower design that was 225 feet tall. That is the tower that was built. The first ninety feet of the structure is steel, while the remaining portion is aluminum. The entire exterior was coated with zinc that would age to a dull gray patina.[38] At the time, the *Guinness Book of World Records* listed the tower as the world's tallest prefabricated church steeple. What is most noticeable about the steeple, which towers over Huntsville, is its distinct rocket shape. Even the most casual observer can see the clear connection between the space-themed church and the dominance of the local aerospace industry.

The modernity and rapid growth of Huntsville in the postwar era are directly tied to Huntsville's emergence as a center of the aerospace industry. But modernity is an elusive subject because what is considered modern is constantly changing. Trying to codify—much less preserve—futuristic design is challenging. Huntsville continues to stay modern; many of its key industries now focus on technology. As the definition of "modern" changes, some people criticize the city's historic space-age buildings for their dated appearance. The prediction in 1965 that the new International Style courthouse would someday be remembered with nostalgia has not come true. Madison County recently made the decision to raze the courthouse. Plans for a new courthouse, located at another site downtown, are under development. Preliminary plans indicate that the current site, which served for three courthouse buildings before the current building was constructed, will become a green space. First Baptist remodeled its fellowship area decades ago, but for now, the iconic mosaic and bell tower are secure. In 2016, the congregation launched an ambitious program to restore the deteriorating mosaic, paying more than $1.5 million to have the artwork completely rebuilt.[39] The most striking thing about many of the city's newest and most ambitious architectural projects is how they have rejected any attempt to use contemporary design. The city currently seems enamored with poorly proportioned copies of neo-Classical architecture.[40] These new buildings are the antithesis of the sleek, iconic designs of the 1960s that celebrated Huntsville's ascent to dominance as a city shaped by rockets, space, and technology. But once, not so long ago, the city's architecture proudly showcased the modernity of the space race, and echoes of the Jetson-like buildings are still tucked into corners of the still-expanding city.

Notes

1 Before this time, Edgewood Arsenal in Maryland was the only chemical manufacturing site for the Chemical Warfare Service, but that facility had little room to expand.

2 David G. Buchanan and John P. Johnson, "Historic Properties Report: Redstone Arsenal, Alabama, with the George C. Marshall Space Flight Center," July 1964, 14, typescript, https://apps.dtic.mil/dtic/tr/fulltext/u2/a175812.pdf.

3 "Courthouse History," Madison County Alabama, https://www.madisoncountyal.gov/government/about-your-county/history#ad-image-11.

4 "Let Others Express Selves," *Huntsville Times*, June 17, 1957, 1.

5 Historic Huntsville Foundation, "A Life's Work: The Heinz Hilten Collection," *Foundation Forum*, Winter-Spring 2016, 8, http://www.historichuntsville.org/wphhsvf/wp-content/uploads/2016/10/4Newsletterwinter-spring-2016WEB.pdf.

6 The project also included the Heart of Huntsville Mall, located at the intersection of Clinton Avenue and Memorial Parkway, which opened in 2001. This area was razed in 2007 for the Constellation Huntsville project, which includes hotels and is still under development.

7 "County Presented Civic Area Plans," *Huntsville Times*, July 2, 1961, 1.

8 "County Presented Civic Area Plans," 6.

9 Advertisement for Edmonton Heights, *Huntsville Mirror*, November 26, 1960, 5.

10 Caroline T. Swope, "Edmonton Heights," Section 8, National Register of Historic Places Nomination Form, US Department of the Interior, National Park Service, Washington, DC, 2020.

11 Swope, "Edmonton Heights."

12 Swope, "Edmonton Heights."

13 "Facts on City, County, during and after Civil War," *Huntsville Times*, March 4, 1967, 13.

14 Jerry Hornsby, "The Ground Might Cave In—But So What, Those Courthouse Cav-

erns," *Huntsville Times*, n.d., First Baptist Church Vertical Files, Madison County Heritage Room, Huntsville Public Library.

15 Hornsby, "The Ground Might Cave In."

16 "New Madison County Courthouse," First Baptist Church Vertical Files, Madison County Heritage Room, Huntsville Public Library.

17 The Jones in Jones, Crowe, Mann & Associates was Harvie P. Jones, who later became known for his restoration work on historic buildings throughout Alabama. Northington, Smith, Kranert & Associates was responsible for the Von Braun Center, the clubhouses at the Huntsville City Golf Course, and the Louis Crews Stadium at Alabama A&M University in Normal.

18 Lucy Barry, "Meet the 90-Year-Old Architect behind Some of Huntsville's Most Iconic Buildings," AL.com, December 15, 2015, https://www.al.com/business/2015/12/post _277.html.

19 "Courthouse Design Wins Top Honor for Architect," *Huntsville Times*, April 30, 1967.

20 Hornsby, "The Ground Might Cave In."

21 Sarah Cure, "Cave City: Huntsville Is Rich in Caves, with at Least 240 Cataloged and Explored," AL.com, February 13, 2011, https://www.al.com/entertainment-times/ 2011/02/huntsville_is_rich_in_caves_an.html.

22 "Drilling Work Start Nears for Courthouse Foundation," *Huntsville Times*, n.d., Local History Vertical Files, First Baptist and Madison County Courthouse, Huntsville Public Library, Madison County Heritage Room.

23 "Dedication, Open House to Be Held a Month from Now," unlabeled newspaper clipping, n.d., Local History Vertical Files, First Baptist and Madison County Courthouse, Huntsville Public Library, Madison County Heritage Room.

24 Beth Russler, "4th Courthouse Dedicated: Sparkman Addresses Ceremony," *Birmingham Post-Herald*, March 6, 1967.

25 Advertisement for First Federal Savings and Loan, 1967, *Huntsville Times*, newspaper clipping, n.d., Vertical Files, Special Collections, Huntsville-Madison County Public Library. Huntsville continued to grow exponentially. Within five years of the dedication of the new courthouse, newspapers were reporting a significant shortage of space in the new courthouse. See Jerry Berg, "Newest Court Has No Space," *Huntsville Times*, May 28, 1967.

26 Dale James, "First Baptist to Dedicate Its Bell Tower," *Huntsville Times*, September 12, 1987, 6A.

27 James, "First Baptist to Dedicate Its Bell Tower."

28 First Baptist Church, *Sanctuary and Educational Facilities Dedication Month*, May 1966, pamphlet, n.p.

29 First Baptist Church, Long Range Planning Committee, Long Range Planning, to First Baptist Congregation, Huntsville, AL, May 7, 1956, First Baptist Church Vertical Files, Madison County Heritage Room, Huntsville Public Library.

30 James Jones, "Big Huntsville Baptist Sanctuary Is Going Up," *Birmingham Post Herald*, May 18, 1964.

31 First Baptist Church, *Sanctuary and Educational Facilities Dedication Month*, pamphlet, May 1966, n.p., First Baptist Church Vertical Files, Madison County Heritage Room, Huntsville Public Library.

32 First Baptist Church, *Sanctuary and Educational Facilities Dedication Month*.

33 Caroline T. Swope, "First Baptist Church," [Huntsville, Alabama,] SAH Archipedia, n.d., http://sah-archipedia.org/buildings/AL-01-089-0062.

34 First Baptist Church, *Sanctuary and Educational Facilities Dedication Month*.

35 "First Baptist Church Mosaic Work Underway," *Huntsville Times*, October 6, 1972.

36 Bill Frazier, "Inch by Inch, Giant Mosaic Spreads across First Baptist," *Huntsville Times*, October 15, 1972; Alvin H. Hopson, "The Mosaic," manuscript notes, n.d., First Baptist Church Vertical Files, Madison County Heritage Room, Huntsville Public Library. Hopson was the pastor of First Baptist when he wrote these notes.

37 Hopson, "The Mosaic."

38 Susan Still, "Baptists' Tower Fulfills Dream of 25 Years," *Huntsville Times*, October 5, 1985.

39 Anna Claire Vollers, "Deteriorating 'Eggbeater Jesus' Mosaic Can't Be Saved, but Church Has Big Plans," AL.com, May 24, 2016, https://www.al.com/living/2016/05/deteriorating_eggbeater_jesus.html.

40 This is particularly evidenced in the visitors' center at the botanical garden and the faux Roman amphitheater currently under construction. The proposed design for a new city hall is topped with a Greek Revival temple and looks like a late entry to the Chicago Tribune Tower Competition of 1922.

15

Institutional Orbits and the Inflatable Form in the Space Age

On July 20, 1969, Neil Armstrong and Buzz Aldrin planted an American flag on the lunar surface, forever altering our relationship to planet Earth. They were dressed in perfectly sealed personal bubbles that provided them with oxygen, water, and communication and that protected them from the alien landscape. That same morning back on Earth, a half-naked, long-haired engineer named Charley Tilford also raised an American flag. He was perched in front of NASA headquarters in Houston while the staff inside helped Armstrong and Aldrin navigate their lunar descent. Tilford had with him a different kind of bubble: an inflated architectural environment dubbed *Space Egg* that was slightly larger than the Lunar Module Eagle (fig. 15.1). Made of black plastic and draped with cream parachute fabric, it billowed in the wind as an attached fan helped it maintain its shape with a constant influx of air. Tilford was a member of Ant Farm, a San Francisco–based architectural collective formed in 1968. Ant Farm members had made a temporary home in Houston to teach students at the University of Texas their ideas about nomadic plastic architecture just as the Apollo missions were on the brink of the first moon landing.

The global coordination required for the space effort offered new aesthetic priorities for avant-garde architects who were frustrated with the rigidity of Modernism. Many 1960s architects embraced flexible, often nomadic architectural forms made of readily deployable materials. Like numerous other architects and artists throughout the 1960s and 1970s, Ant Farm was deeply influenced by the logics, aesthetics, and materials of the Space Age. In the period 1965 to 1972, structures like the *Space Egg* were prevalent in architecture and art across the United States and Europe: immersive, environmen-

Figure 15.1. Charley Tilford with *Space Egg* inflatable, July 20, 1969, outside Johnson Space Center, Houston, Texas. Photo by Chip Lord, courtesy University of California, Berkeley Art Museum and Pacific Film Archive.

tal sculptural forms made of inflated plastics and textiles appropriated from space industries. This chapter explores how art and architecture in the 1960s transformed in the context of a culture that became more aware of our unique place in the universe through space travel. Specifically, the chapter examines how a preoccupation with breathable air made its way into aesthetic forms that included air-supported structures, artistic balloon flights, and interactive sculptural bubbles. It centers on three protagonists: artist-balloonist Vera Simons, artist Piero Manzoni, and the architectural collective Ant Farm. Their stories came together in Houston, through exhibitions at the Contemporary Arts Museum Houston (CAMH). That museum is the fourth character in the study. Situated in the city that hosted NASA's Manned Spacecraft Center, a key node in the nervous system of the race to the moon, the museum's location, unique institutional structure, and curatorial choices reflect the influence of the space industry. This chapter elucidates both the history of CAMH in its Space City context and the influence of the space race on cultural production with inflatable materials.

The Nested Houston Arts Scene

CAMH was conceived in 1948 as one of the first institutions in the United States dedicated to contemporary art. Several factors suggest why it was well suited from its inception to engage with inflatable art informed by the space race in Houston, one of NASA's most important hubs. They include its founding commitment to exposing the relationships between art and everyday life, an interest in kinetic art, and the museum's network of colleagues and collaborators.

CAMH aimed to put the role of art in the modern design of everyday life on display. One of the museum's first presentations was a memorial exhibition for the influential Hungarian artist László Moholy-Nagy, who had died in 1946. Moholy-Nagy began his career in 1923 as a teacher at the Bauhaus, a German school in Dessau that was instrumental in formulating the tenets of modern design in the 1920s and 1930s. The Bauhaus emphasized function and practical forms that were meant to link art in all media with daily living. In 1937, Moholy-Nagy brought his ideas to Chicago as director of the New Bauhaus school, the Institute of Design, where he continued to explore how scientific and technological discoveries inform the artistic imagination. Moholy-Nagy experimented with perception, kinetic forms, and air. For example, he used the continuous, forceful supply of an air compressor to keep an object such as a chisel afloat.[1] This was a precursor to later experiments with air through inflatable structures. The moving components of his kinetic sculptures used scientific knowledge of mechanics for artistic ends. Controlled air flow, such as compressed air, regularly served as a structural element in his sculptural forms.[2]

Kinetic art grew to be a subset of the twentieth-century modern art movement in the west, though it was better known in Europe than in the United States. Kinetic art deployed scientific phenomena in aestheticized forms, producing moving sculptures. The movement began before World War I as artists searched for new forms to link art and everyday experience and to make sense of life in a rapidly changing world. Several artists experimented with air in the prewar years. In 1914, Pablo Picasso and Marcel Duchamp began discussing the possibilities of air as an aesthetic component of kinetic sculpture. Five years later, Duchamp inflated a small glass globe with 50 cubic centimeters of air, one of the earliest inflatable artworks.[3] Kinetic sculptures continued to evolve and develop in response to interwar and then postwar technologies and conditions. They were an intuitive reflection on the instability of reality. The moments of surprise they produced did more than represent or describe an artist's experience in the world; they captured and mirrored the phenomena of modern life.[4]

The Center for Advanced Visual Studies (CAVS) at the Massachusetts Institute of Technology was the base for the kinetic art movement in the United States in the 1960s and 1970s. CAVS was formed in 1967 by Hungarian-born artist György Kepes, who had collaborated with Moholy-Nagy beginning in the 1930s in Berlin and then had taught at the New Bauhaus in Chicago.[5] Is it possible that there was another such base in Houston that had been slowly brewing since the museum's inception in the late 1940s but has been largely overlooked? Furthermore, is there an unexamined relationship between the two institutions that explains CAMH's support of later kinetic art? The museum's genealogy offers some clues.

Houston-born painter Robert Preusser, who had studied under Moholy-Nagy in Chicago in the early 1940s and then returned to Houston, cofounded CAMH in 1948. Nine years later, he left Houston to join MIT as professor of visual design in the Department of Architecture. He later became the director of education for CAVS. The careers of Vera Simons and Piero Manzoni suggest additional links between CAVS and CAMH. Simons became a visiting artist at CAVS in 1969, the year after a major solo exhibition at CAMH. This was thanks to the fact that James Harithas, the curator of her exhibit, introduced her to György Kepes, then CAVS director.[6] Simons consistently framed her inflatable art practice in terms of kinetic art. Piero Manzoni died very young in 1963, but during his lifetime he was a collaborator in Group Zero, a collective formed in Düsseldorf, Germany, by Otto Piene in the late 1950s that quickly grew to include artists from across Western Europe. Piene, in turn, was the first fellow of CAVS in 1968 and later succeeded Kepes as its director for several decades. He was a renowned kinetic artist who was especially known for Sky Art, a series of experiments with suspended inflated plastic forms, and experiments with light and fire. The CAMH was committed to exhibiting "the interrelationship of the various arts and the role of the arts in the modern design for living"[7] with a focus on integrating the arts, sciences, and modern technology industries, a focus that makes clear its alliance with the Bauhaus. But the museum's institutional history also reveals its affinity for the kinetic art movement through its network with CAVS.

As an institution, CAMH understood its position in a networked web of local and national connections. The museum embodied a new kind of institution that was nomadic and flexible. This way of working evolved out of necessity but became a creed and an important feature: the institution was dynamic and highly responsive to its environment. From its early days, the museum relied on allies for physical and creative capital. Until the founders of CAMH, who organized as the Contemporary Arts Association, constructed the museum's own building in 1950, exhibits were hosted by the larger Museum of Fine Arts in Houston. Although the eventual CAMH building was small, it was designed to meet the security and insurance requirements of the Museum of Modern Art in New York. This enabled the young museum to borrow materials from the renowned, well-established New York institution. It also meant that CAMH did not have the burden of housing its own permanent collection. The connections of board members ensured that the Museum of Modern Art was willing to make such loans. In 1966, the campaign for a new CAMH building began. During the years of construction, the museum maintained central offices but produced numerous exhibits in various institutions across Houston. Director Sebastian Adler viewed the networked format as "an exchange of energy between 'a main space station (the future museum) and its satellites,' using the city itself as the real museum."[8] Adler's metaphor of the space station suggests the influence of the NASA activity in Houston at the time and the incorporation of that agency's global networks of production into the logics of the art institution, even into manners of speech.

CAMH showed the work of artists who were highly experimental, many of whom would leave significant marks on the art world. The exhibition offerings did not distinguish between art and architectural media: in summaries of exhibitions or guest lectures in the museum archives, which are held at Rice University's Woodson Research Center, artists such as Alexander Calder and Max Ernst are regularly listed alongside their contemporary in architecture, Philip Johnson, while mentions of later artists John Cage and Robert Rauschenberg appear with architect R. Buckminster Fuller. This lack of distinction between disciplines is especially salient when we consider that with inflatable structures the professionals in the two fields began to use the same media, for example plastics such as polyethylene and Mylar, both of which were popularized by the airborne structures of NASA.[9]

In the 1960s, American and European architects and artists adopted the materials and aesthetics of the space race to create immersive air-filled environments especially attuned to Earth. However, there were significant differences in the operating logics of space structures and the earth-bound forms they informed. While the pursuit of spaceflight since the mid-nineteenth century had emphasized the production of artificial, fully controlled atmospheres that kept the environment *out*, inflatable architectures invited the outside *in*. These forms continually registered and mediated a relationship between circulating air and the plastic membrane that together formed a non-rigid structure and the human body that occupied the resulting space. It is as if, through inflatable forms responsive to environmental forces, architects and artists anticipated that NASA's successful mission to the moon would recommit humans to life on Earth.[10] The inflatable experimentation that artists and architects did were characterized by a fundamental inquiry into our human relationship with earthly air and with each other.

CAMH Exhibitions

During the late 1960s and early 1970s, exhibitions in the United States and Europe showcased experiments with air. Many made explicit their inspiration from and material echoing of the space race. From about 1966 through 1972, art institutions from small artist-run galleries to established names like the Walker Art Center in Minneapolis captured and fomented this "inflatable moment."[11] The CAMH exhibitions examined here illustrate that the institution was a conscious participant in the international inflatable moment. The museum's proximity to the control center of the American space program makes it an especially interesting case study of the links between space and artistic practices.

The journey of Vera Simons, a pioneering balloonist-turned-artist, reveals how CAMH captured the influence of the space race in art. In the 1968 catalog accompanying Simons's solo exhibit at CAMH, her largest show to date, curator James Harithas wrote that Simons was an artist who had the scientific knowledge to deploy space age technology for space age forms.[12]

Simons was indeed unique in her scientific understanding of such forms. In the 1950s, Vera (then Vera Winzen) founded and ran Winzen Research Inc. in Minneapolis with her first husband, Otto Winzen. The company pioneered the use of polyethylene high-altitude balloons, which were lighter, thinner, and more resistant to radiation than earlier fabric balloons. This meant that they could achieve greater heights. The US Navy and the US Air Force contracted Winzen Research to produce balloons for efforts such as Project Manhigh, which aimed to determine the human limits of space flight in the years immediately preceding the formation of NASA. These flights sent three men into the stratosphere in 1955–1958. They were attached to Winzen plastic balloons, whose manufacture Vera Winzen closely supervised. One of the aeronauts was David G. Simons, who would become Vera's second husband. David Simons had been involved in space research since the 1940s, when he had worked on the first American animal research flights aboard V-2 rockets. He later participated in balloon research for the US Air Force.[13] In 1957, during the second Manhigh mission, he broke altitude records and stayed aloft for over thirty hours. That balloon flight tested whether humans could survive at the edge of Earth's atmosphere. During her tenure with Winzen, Vera Winzen received four patents that included improvements to the balloon's skin, sealing mechanism, structure, and capsule in order to protect human passengers and cargo.[14] She was deeply involved in balloon design and manufacturing and in managing the factory workers during the delicate production process. In 1957, she obtained a gas balloon pilot's license and became a respected balloonist, a female pioneer.[15] After NASA was formed in 1958, the contribution of ballooning research to space exploration such as through the Manhigh project was overlooked, at times discarded. Books such as Craig Ryan's *The Pre-Astronauts* (1995) have attempted to recover the story of twentieth-century aeronautics as a foundation for NASA's 1960s pursuits. Simons's later artwork reflected critically on the competitive aspects of space exploration, which continued beyond the transition from the balloon to the rocket.

After she divorced Otto Winzen in 1958, Simons sold off her two-thirds interest in the company and resumed her art education, which her marriage had forestalled. Her father, a photographer, had early on sparked her interest in astronomy and meteorology, and she gradually combined her art practice, her ballooning experience, and her passion for exploring the sky. After the divorce, she completed an art degree at the Corcoran School of Art in Washington, DC, and moved to Houston in 1960. By the time of Vera Simons's 1968 exhibit at the CAMH, David Simons was on the board of directors of the institution. Further archival research is needed to understand the board's influence and role in determining the exhibition topics and whether David Simons's presence affected links between the institution and Houston's Space City activities at the Manned Spacecraft Center.

What is known is that, unlike many of her male counterparts in 1950s high-altitude ballooning, for Simons the height or endurance records of flights in enclosed space capsules ferried by balloon were less important than a desire to connect the minds and eyes of people on Earth with the atmosphere via the balloon and its open gondola: "I want to do something more romantic, more people-oriented. I like the connection with people on the ground vs. the balloon, establishing this dialogue cross-country. . . . It is so romantic, so exciting, and so elevating—emotionally and physically."[16] Her artwork, which included site-specific inflatable installations and sculptures as well as drawing, painting, and photography, explored the elements of fire, earth, and water and their embeddedness in time.[17]

Vera Simons's exhibit at CAMH in 1968, *Aerial Sculpture*, featured her inflatable sculptures (fig. 15.2). The silver laminated plastic balloons that hung from the ceiling responded to the movement of the audience through the tri-

Figure 15.2. Installation at the exhibit *Vera Simons: Aerial Sculpture*, Contemporary Arts Museum, Houston, Texas, April 16–May 31, 1968. Courtesy of Center for Advanced Visual Studies special collection, Massachusetts Institute of Technology, Department of Distinctive Collections, Cambridge, Massachusetts.

angular steel framework of the CAMH gallery. These were echoes of Andy Warhol's 1966 installation *Clouds* at the Leo Castelli Gallery in New York. *Clouds* was a series of Mylar pillow-shaped balloons filled with helium and air that floated about the gallery space. Warhol was inspired by NASA's use of Mylar and had worked with Bell Laboratories engineer Billy Klüver to produce the balloons.[18] Because of her intimate knowledge of ballooning, Simons could not only produce her own shapes but could also experiment with their forms. Her exhibit included six hanging geometric sculptures, several in multiples. For example, one piece consisted of a stack of five cylindrical floating modules. The exhibit also included balloons in the shape of a round disk, a cube, a barrel, seven horizontal hexagonal modules, and one vertical hexagon. They varied in size from roughly three to six feet in at least one dimension. The forms were not free floating but were attached with minimal string to the ceiling, swaying in response to changing air currents.[19]

Harithas, the curator of the exhibition, was at the time assistant director and curator of the gallery of Simons's alma mater, the Corcoran Gallery in

Washington, DC. In 1974, he would become the director of CAMH. In the exhibition catalog, he wrote:

> Many contemporary artists have attempted to apply space-age technology to the creation of works of art but often without the scientific knowledge necessary to create the relevant forms. Vera Simons is an exception. . . . Vera Simons' experience in making balloons and the years devoted to painting have led inevitably to the conception of a lighter-than-air art form. . . . What the aerial sculpture in this exhibition does, is make an aesthetic environment out of a simple interior space. The forms constantly shift with the changing air currents. As a result the space around them is constantly being reorganized; unlike traditional sculpture, this sculpture excludes any fixed points of reference. The shifting forms also have the important effect of making one acutely aware of the physical properties of space. Vera Simons has laid the groundwork for this sophisticated and timely art form.[20]

Marcel Duchamp focused on the sculptural object that resulted from his air experiments. Similarly, kinetic artists tended to produce interactive forms. However, although Simons insisted on contextualizing her projects as kinetic art,[21] she did not focus solely on the individual sculptures she produced. Instead, her goal was to transform familiar spaces like the gallery into aesthetic and responsive environments via the air structures she created taken together. In describing similar sculptures for a later exhibit, she wrote: "Tethering the sculpture and mounting on swivels permit it to be responsive to any air movements. It is kinetic without being programmed movement. The modules also seem to have an affinity for each other which adds to their interaction."[22] This approach, which balanced systems of input with chance output, reveals a sphere of concern that was similar to that of contemporaneous architectural experiments with air, as will be explored in the following sections.

The year after the Houston exhibit, in 1969, Simons joined CAVS as a fellow. There, artists such as the German-born Otto Piene, the founder of Group Zero, were also exploring kinetic air art forms and the relationship between art and technology through experiments with the elements. Long after her tenure as a fellow ended, Simons continued to attend Sky Art conferences organized by Piene that gathered artists from around the globe who pursued experiments with air.

Beginning in the late 1960s, Simons's balloon flights were enmeshed with her artistic research and involved collaboration with scientists to investigate ecology, aesthetics, and the elements. For example, during a record-setting five-day trans-American flight from Oregon to Ohio in 1979, the Da Vinci

Transamerica, Simons and an accompanying scientist tracked pollutants to find that they migrated at unexpectedly great distances. Simons flew the balloon at low altitude so that she could also continually record sound, shoot video and photographs, and drop leaflets, postcards, seedlings, and other mementos such as a colored smoke cloud from the balloon to people on the ground to "establish a ground to air connection."[23] Unlike the goals of other balloonists, who were focused on altitude or speed, Simons's goal was a way to link people across great stretches of the United States. Also on board was an NBC cameraman, who parachuted daily tapes along the way that aired on the *Today Show*. Simons hoped to "get people to look at the sky . . . the idea was that here was a vehicle moving across the sky with the wind."[24] In this desire, Simons reflects a sentiment that was observed from the earliest balloon flights and carried into the Apollo moon missions of the late 1960s: that is, by looking up and away, humankind could reflect on their existence on the planet anew. In the *Earthrise* photograph taken by astronauts aboard Apollo 8 in 1968, the same year as Simons's CAMH exhibit, the Earth viewed from space appeared as an unprecedentedly marvelous site, an exception in the vast cosmos that made it uniquely, preciously tailored to human existence.[25]

The CAMH director during the time of Simons's 1968 solo exhibition was Sebastian Adler, who four years later invited her to participate in *10*, an experimental group exhibition in 1972. Adler anticipated that future art would have a very different direction from past art. *10* delineated the contours of that direction: its young, emerging artists displayed "a heightened sensitivity to the urban environment and the human condition; the use of nontraditional media in conjunction with collaborative effort with persons versed in other disciplines; and an attempt to reach the widest possible audience and to extend the exhibition of work beyond the walls of the museum."[26] Many of the featured artists were concerned with ecology and the elements. Vera Simons, for example, used her understanding of plastics in collaboration with the expertise of storm researchers to produce a wave generator sculpture on the museum grounds. Like several other projects in this exhibition, the endeavor never quite functioned the way it was envisioned: the tendency of artworks to malfunction in the exhibit cost Adler his tenure at CAMH. However, the deeply exploratory demonstration was a sign and reflection of changing times in art, fueled by larger changes in culture and society. The exhibit was an important steppingstone for a number of the featured artists, such as Newton Harrison, William Wegman, and Michael Snow, all of whom became well-recognized figures in art.

Before his departure, Adler brought a Piero Manzoni exhibition from the Sonnabend Gallery in New York to CAMH. Although Manzoni had died in

1963, early in the space race, his work was highly relevant to the art of the early 1970s. It built off Duchamp's aesthetic explorations with air and kinetic art's experimental and conceptual approaches to art and natural phenomena such as gravity. Manzoni used his own body as not only the subject but the object of art. That extended to breath: the CAMH exhibit included his *Artist's Breath* piece from 1960, one of a series of balloons tied to a piece of string on which were two lead seals attached to a wooden base.[27] As the balloon deflated, its rubber skin imprinted onto the wood, capturing the futility of permanence. Several sculptures from another series linked to breath were featured in the exhibit. For example, in *Corpo d'Aria*, Manzoni offered a tripod, a deflated balloon, and a mouthpiece. During his lifetime, the artist could be asked to inflate the balloon for a small fee.

In 1960, Manzoni created *Placentarium*, a model for "air architecture" to house Otto Piene's *Light Ballets* of 1959, kinetic artworks that projected lights through rotating stenciled cylinders. Manzoni's unrealized *Placentarium* structure was a plastic sphere eight meters in diameter supported by atmospheric pressure inside. The piece was meant to seamlessly blend acoustic, tactile, and visual sensation.[28] Manzoni is today associated with both kinetic art and the Italian Arte Povera movement, although this movement was not articulated until after his death. While Simons's CAMH exhibition had been about the space-transforming capacity of inflatable forms, Manzoni's was about the exchange of air between artist and audience via the medium of the balloon. His work gave form to the human body's reliance on air, which preoccupied aeronautical engineers during the 1960s, as it limited the body's suitedness for space. Such limitations were important factors in the years preceding the formation of NASA.[29] Manzoni did not address space travel explicitly in his work, but his engagement with air and his interest in gravity are evidence of a preoccupation with the challenges that immanent extraplanetary travel would present for the human body.

Finally, the last protagonist to explore in the narrative of CAMH and the space race context is the architectural collective Ant Farm. After Sebastian Adler's departure from the museum in December 1972, the institution was without a director for over a year until James Harithas took the helm in 1974. During the year between directors, acting administrator Margaret Prince and the CAMH staff and board maintained the museum's mission of offering contemporary art in Houston. Ant Farm was invited during this time to produce the exhibit *20/20 Vision: Ant Farm* as an example of the institution's support of anticipatory art.[30] By that point, the collective's relationship to Houston and to the museum was already several years old.

Ant Farm's members were a generation or two younger than Simons and Manzoni and had been trained in architecture rather than art. Their air forms were large enough for a body or multiple bodies to inhabit. The movement of the body inside an inflatable form affected the movement of the plastic membrane, which in turn transformed the body's experience within the space. The interest in air was not new in architecture, just as it was not new in art. The first inflated buildings initially emerged in the second decade of the twentieth century, around the same time as the first kinetic artworks. English engineer Frederick W. Lanchester, who was inspired by nineteenth-century balloon and airship construction, developed the concept of an air-supported field hospital in 1917.[31] Although his ideas were never realized, they inspired engineering-architectural works in subsequent decades. Time and again, air structures presented practical solutions to the need for lightweight, easily deployable shelters that could both support a large floor plate free of columns and disappear quickly. One such structure was engineer Walter Bird's 1948 radome, a spherical, air-supported enclosure used to protect sensitive radar equipment from weather and wind without interrupting the signal. Another well-known example is the 1960 air-supported exhibition pavilion for the United States Atomic Energy Commission that Bird designed with architect Victor A. Lundy. This 22,000-square-foot structure made of vinyl-coated nylon took just a few days to erect and tear down, making it possible for the US Atoms for Peace program to promote nuclear energy use across the globe for over ten years.[32]

By the mid-1960s, young architects were increasingly preoccupied by depictions of life in a bubble as imagined in space colonies and by a growing awareness of environmental issues. Critical of mainstream modernist architectural practices and in search of new physical and social forms for postwar life, these architects found in air structures a typology ripe for experimentation with form and social engagement. Earthly air as an art form was not something to observe from an enclosed form, like a spacesuit. The experimental architectural inflatables, like the earlier air-supported military structures, invited people to enter inside the balloon. Unlike their defense predecessors, however, these new inflatables were not meant to efficiently support practical functions. Instead, they were highly responsive registers of the smallest activities of the environment, and their objective, if they had one, was to dance with the surrounding forces. As viewers lifted or unzipped the entrance flaps of an inflated bubble, the form would deflate, the walls billowing to accommodate the escaping air. A fan outside a structure that was connected via a tube continually blew in new air, so a bubble would fill out again once the entry was closed. The overall effect was of a lively, breathing, creaturely dome that responded to its environment.

The architects engaged in this domain were especially attuned to advances in ecological ideas, which were in turn informed by cybernetics and systems theory.[33] Cybernetic systems, which Norbert Wiener had conceptualized for his work to calculate the trajectories of anti-missile guns during World War II, emphasized continuous adjustment based on information received through feedback loops. Gradually, the difference between open and closed systems was articulated: while closed systems relied on enclosure and prioritized total control and predictability, open systems emphasized indeterminacy and the process of change. Open systems posited human actors not as managers positioned outside a system and thus afforded a perfect overview but as participants inside the system who contribute to its unpredictability and have only a partial view of the whole.[34] Architects and artists took from the open system a celebration of entropy, indeterminacy, and chance and began to emphasize the processual development of a work rather than the finished product. Landscape architect Lawrence Halprin, who Ant Farm member Curtis Schreier worked with for several years before joining Ant Farm,[35] and artist John Cage were both proponents of chance operations in their art inspired by systems theory. Cage explored his ideas of systems through a fruitful relationship with the architect R. Buckminster Fuller starting in 1948. Fuller's research and designs were in turn deeply influential for the inflatable practices of Ant Farm.[36]

Recent architecture school graduates Chip Lord and Doug Michels formed Ant Farm in San Francisco in late 1968. They were soon joined by another newly graduated architect, Curtis Schreier. Throughout its tenure, the collective welcomed multiple transient members, but this trio remained the most prominent. Their decade-long practice began just before the Apollo 8 mission, which transmitted the now-famous *Earthrise* photograph. Ant Farm's early projects self-consciously borrowed NASA iconography and materials and soon they were heavily involved with inflatables. The perceived shift from American car culture toward communications media underscored for them a rapid exchange of ideas independent of place. This in turn facilitated nomadism, the idea that one could simply pick up and go elsewhere so long as they remained connected to their world via telecommunications.[37] Michels and Lord had been car enthusiasts since young adulthood. In the 1960s, they were deeply marked by the shift away from the American landscape and the romanticism of the road trip toward simultaneous data access via the television set; by the possibility of traveling without leaving home. During the years of the Apollo missions (1968–1972), they were especially attuned to the temporal and spatial unsettling brought about by the flood of expanding information and its juxtaposition with the enduring pace of lived life, which did not change all that visibly. In a statement written during their time in Houston, they observed: "Global-conscious twenty-first century man opened a tin can yesterday with a machine designed in the 19th century sat down and watched a live broadcast from the moon. America, in an allegorical time warp using pre electronic, print age techniques welcomes returning space heroes [with a] ticker tape parade down Fifth Avenue."[38] Their practice during these years was primarily focused on inflatables.

Ant Farm members both embraced technology and were unnerved by its implications for the future, reflecting an impulse art historian Pamela Lee described as "chronophobia."[39] As a result of these contesting attitudes toward technological progress, a radical presentism prevails in their work, an attempt to make sense of the changes happening around them, including NASA's progress toward the moon. Because they were part of a historic moment in time, they were unable to fully grasp its implications. Instead, they experimented with architectural forms that corresponded to the properties of new materials such as plastic sheeting and parachute fabric and explored new ways of relating through these forms. While earlier modernist art and architecture strove to perfect control over then-new materials (such as concrete) to construct their vision of a future world, Ant Farm's inflatables embodied the anticipated disorientation of living in that world. Architectural historian Felicity Scott frames the connection between the space race and countercultural, often drug-induced, architectural ideas like Ant Farm's as seeking to produce equally extraordinary "launchings" into space and time and attempting to articulate emerging subjectivities in the cybernetic era.[40] How could architecture relate to contemporary social and economic forces, and techniques of power beyond perpetuating a techno-euphoria?

The concerns of Ant Farm members were shared by many of the artists and architects who experimented with inflated forms in the late 1960s and early 1970s. The characteristics of mobility and a continuous engagement with the present resonated across both disciplines during these years. Air structures gave perceptible shape to the notion of constant becoming. Instead of shaping form, they continually remade space that responded to its environment, including people and the elements, a sign of "unlimited elasticity in the present-day world."[41] That is, they were not concerned with producing fixed, finished form; they produced forms appropriate for a reflexive, ever-changing, and unpredictable present.

Ant Farm's first engagement in Houston was in 1969, when they were employed to teach in the College of Architecture at the University of Houston. Ant Farm members Doug Michels and Chip Lord organized two workshops:

Astrodaze in the spring semester, and Time Slice over six weeks in the summer. Astrodaze was an overnight experimental trip to Freeport Beach in Texas with a group of students and three guest speakers from East Coast Ivy League institutions. Doug Michels's 60-foot nylon surplus cargo parachute, the *Dreamcloud*, was a primary, especially photogenic protagonist: a shelter, a projection screen, and a climbable inflatable structure (fig. 15.3).

In describing the parachute, Felicity Scott wrote: "Dreamcloud was exemplary of a type of kinetic architecture that could be fueled by both natural and technological systems. It was a mobile 'response environment.'"[42] It is notable that although Scott used the term kinetic, she, like other scholars, did not consider the link between inflatable architecture and kinetic art.

Time Slice was a series of gatherings that involved costumes, performative movement through Houston that engaged multiple senses and, importantly, the construction of "disposable environments."[43] The first month, or phase one, involved the gathering and production of "enviropaks," or materials to prepare for an excursion of "media nomads" to Padre Island in Texas that occurred over two days in early July. During this excursion, the group flew parachutes large enough to sail with, tested various recording media, and performed ritualistic dances. Nomadic gear that students were assigned to develop or contribute to the excursion included geodesic domes, inflatable structures, sleeping bags, and communications devices: tape recorders, cameras, projectors, and sound equipment. They framed the entirety of Time Slice as theater or life art and invited notable and like-minded architects as speakers throughout its duration.[44]

For Ant Farm, consciousness of a shifting relationship to time and information was inseparable from a growing consciousness of ecological issues that were influenced by images of the whole Earth captured from space. An autobiographical Ant Farm data sheet from these years states: "[Ant Farm's] immediate concerns are for expanding environmental awareness and ecological consciousness through interaction with institutions, natural and created groups, families, and individuals and all available forms of media. The catalyst for interaction is often inflatable structures, but the basic information exchange is at the essence of our commitment to alternate lifestyles."[45] Alternate to what?, we might ask. A few years after Ant Farm's visit in Houston, the publication *Limits to Growth* made explicit that the Earth's capacity to support human life was in fact limited and conditional on our behavior.[46] Groups such as Ant Farm sensed that the approaching lunar landing would only exacerbate just how earthbound we truly are. Their call for alternatives was thus a call for ecological sustainability coupled with an urgent sense of a need for social transformation in order to continue collectively inhabiting the planet.

Figure 15.3. Ben Holmes in *Dream Cloud* parachute, June 1969, Freeport Beach, Texas. Photo by Ant Farm, courtesy University of California Berkeley Art Museum and Pacific Film Archive.

The Time Slice archival documents are replete with iconography that collapses multiple temporal scales: memorabilia from Project Apollo and the anticipated moon landing, which occurred just as the workshop wrapped up; Texas mythologies, including cowboys and the nineteenth-century battle at the Alamo; maps of contemporary Houston annotated for walking assignments; and late eighteenth-century etchings of hot-air balloons collaged with photographs of romantic twentieth-century ballooning excursions. This heterogeneity gives form to Pamela Lee's argument that creative practices at this moment recognized the shift from chronological time to simultaneity in a period that undermined notions of determinism.[47] For Ant Farm, this left only the present as a site of intervention. And, at the crossroads of leaving the planet and realizing its tethers, the present demanded heightened awareness.

In September 1969, just before Ant Farm left Texas, CAMH board member Marilyn Lubetkin invited the collective to make a performative room, or environment, for a CAMH fundraiser staged at the Alley Theater. They called it *Space Cowboy Meets Plastic Businessman*, underscoring the collapse of timescales brought about by modern technology and the space race. The event featured collages; slide shows; an inflatable environment; rock music by Steve Miller, who wrote "Space Cowboy" to symbolize Houston; and many American flags. Six Ant Farm members participated as technicians and performers, and it was the group's first use of a video camera.[48]

This began a relationship with CAMH that would involve a time capsule event in 1972 and the production of their first exhibition, *20:20 Vision*, in 1973–1974. Research for the exhibition was funded by the National Endowment for the Arts. That funding enabled Doug Michels to explore the archives of automobile manufacturers such as Ford Motors.[49] The exhibition was organized around key moments of envisioning the future: the 1939 New York World's Fair; 1959, which Ant Farm identified as the apex of American car design; the Orwellian 1984; and 2020, a distant future for speculation. A vehicle represented each era; for the 2020 vision, NASA loaned the group a lunar rover prototype.[50] Ant Farm projects and drawings filled out the gallery, addressing the central question "Can man control technology's domination of nature?"[51] However, by the time of this exhibit in 1974, the world was facing an oil crisis and many of Ant Farm's visions of the future appeared bleak to critics, especially car-dependent mobility and the idea of post-Earth worlds. According to several assessments of this exhibition, the show was received as a testament to a way of life, an envisioned future that was no longer possible; a sort of future-past.[52] The Apollo era was over, as were ambitions of space travel. In Western Europe and the United States, inflatable forms had similarly become less prominent in experimental architectural and artistic practices by the early 1970s.

Conclusion

The networked effort required by the space program extended strategically all around the globe, although NASA controlled it closely to facilitate an explicit goal. In contrast, artists' and architects' embrace of inflatable forms in reaction to the space race had no articulated goal. The forms instead exposed a search for an aesthetic and phenomenological response to the disruption that the impending moon landing was anticipated to have on humans' perception of themselves as a planetary society. In the late 1960s, air structures in art and architecture facilitated explorations of the changing conditions of spacetime. Houston was an important site of networked nodes, both in the ways NASA operated the management of logistics around the globe and in the ways that CAMH was comfortable functioning as a distributed organization across several physical locations with a nationwide network of support and communication. Both were open systems with many feedback loops, and the museum's exhibitions, whether consciously or not, fostered reflection on both the formal and the existential implications of NASA's successful moon landing. It appears that the museum served less as a command center and more as a gravitational force, the orbit of which facilitated conversation and contributed to the sudden and expansive output of inflatable projects in the late 1960s and early 1970s. In the case of Ant Farm, the group's cultural explorations were deeply enhanced by the willingness of an established educational institution, the University of Houston, to support their experimental pedagogies.

NASA coordinated its own art program during the Apollo years, inviting selected artists to visit its facilities and attend space launches. The goal of the NASA Art Program was to translate or interpret for the public the other-worldliness of humans' departure from Earth through representation, primarily through painting and drawing.[53] The inflatable forms did not attempt to *represent* this otherworldliness but instead embodied the sense of disorientation. They created on Earth experiences of the post-lunar world, a world in which humans have traveled to another celestial body.

Despite the weak link between kinetic art and architectural inflatable forms in existing scholarship, the interdisciplinarity of CAMH allows us to see the outline of a bridge between the two fields. Interest in air as an artistic medium moved through the twentieth century from free-floating kinetic sculptures to immersive architectural environments that recreated an other-worldly experience on the familiar planet and reflected the major technological transformations—and preoccupations—of their time. While New York or London may have been more important artistic centers during the 1960s and early 1970s, the dialectical Houston context—its mythological and timeless Wild West stereotype and the centrality of its role in the futuristic Space Age—uniquely captured art's reflexivity as prompted by the space program. This segment of Houston's cultural scene and the artists drawn to it were deeply attuned to the activities of NASA, activities that prompted reflection on space, time, and the changing delineations of the future.

Notes

1 László Moholy-Nagy, *Vision in Motion* (Chicago: Paul Theobald, 1947).
2 Jorge Glusberg, "air art. one. The Dialectic between Container and Content," *Art and Artists* 3, no. 10 (1969): 42–46.
3 Glusberg, "air art"; Eventstructure Research Group, "air art. two. Concepts for an Operational Art," *Art and Artists* 3, no. 10 (1969): 47–49.
4 Jean Clay, "Painting—a Thing of the Past," *Studio International* (July/August 1967): 12–17.
5 "History," Art, Culture, Technology MIT, n.d., http://act.mit.edu/about/history/.
6 György Kepes to James Harithas, January 16, 1969, box 1, folder Harithas, Vera Simons Papers, National Air and Space Museum, Smithsonian Institution, Washington, DC (hereafter Vera Simons Papers).

7   Board letter to Contemporary Arts Association Membership, October 9, 1948, quoted in Contemporary Arts Museum, *In Our Time: Houston's Contemporary Arts Museum, 1948–1982* (Houston, TX: Contemporary Arts Museum, 1982), 8.

8   This quote and the summary of the museum's history are taken from Contemporary Arts Museum, *In Our Time*, 40. The Sadler quotes originally appeared in "The City Is the Real Museum," *Houston Chronicle*, February 13, 1970.

9   For the use of Mylar for the first two balloon satellites in NASA's Project Echo, which launched in 1960, see C. L. Staugaitis and L. Kobren, "Mechanical And Physical Properties of the Echo II Metal-Polymer Laminate," NASA Technical Note, Goddard Space Flight Center, Greenbelt, Maryland, 1966, https://apps.dtic.mil/sti/pdfs/ADA307366 .pdf. The aesthetics of the enormous balloons were reflected in numerous artworks.

10 This renewed commitment has been examined through studies in literature (e.g., Ronald Weber, Seeing *Earth: Literary Responses to Space Exploration* [Athens: Ohio University Press, 1985]) and history (e.g., Benjamin Lazier, "Earthrise; or, The Globalization of the World Picture," *American Historical Review* 116, no. 3 [2011]: 602–630). Part of this response was driven by recognition of Earth's ecological limits.

11 *The Inflatable Moment* is the name of a 1998 exhibition by Marc Dessauce presented by The Architectural League of New York that reflected on the 1968 moment through the work of French architectural collective Utopie. In the late 1960s, important exhibits included those organized by the Museum of Contemporary Craft in New York—e.g., *Body Covering* (1968) and *Plastic as Plastic* (1968–1969, which promoted the notion of plastic as a new artistic medium)—and several congresses on pneumatic structures by the International Association of Spatial Structures and the University of Stuttgart starting in 1967. *Air Art*, an exhibition that traveled across the United States in 1968, curated by Willoughby Sharp, and *Earth, Air, Fire, and Water: Elements of Art* at the Boston Museum of Fine Art in 1971 further explored air structures in kinetic art.

12 See the catalog: Vera Simons, *Aerial Sculpture: April 15 through May 31, 1968* (Houston, TX, Contemporary Arts Association, 1968).

13 Craig Ryan, *The Pre-Astronauts: Manned Ballooning on the Threshold of Space* (Annapolis, MD: Naval Institute Press, 1995), 21–22.

14 See the results of a Google patents search for Vera H. Winzen at https://patents.google .com/?inventor=Vera+H+Winzen.

15 The biographical information on Simons is from Ryan, *The Pre-Astronauts*; and from the Vera Simons Collection, Center for Advanced Visual Studies special collection, Department of Distinctive Collections, MIT Libraries, Cambridge, MA.

16 Simons quoted in Ryan, *The Pre-Astronauts*, 274.

17 Richard Demarco, untitled text in the Simons files in the Center for Advanced Visual Studies archive, January 1986. This document is likely linked to her project *Aerial Crown*, which she mounted on the roof of the National Gallery of Scotland for the Edinburgh International Festival, with which Demarco was involved.

18 Willoughby Sharp, *Air Art* (New York: Kineticism Press, 1968).

19 Simons, *Aerial Sculpture*.

20 Simons, *Aerial Sculpture*.

21 Simons quoted in David Saltman, "Kinetic: Vera Simons & Her Sculpture in Scotland," *Washington Post*, August 4, 1986, B7. Saltman suggests that this insistence was driven

in part by the difficulty of being recognized in her new career: "[The artwork *Da-Vinci TransAmerica*], she says, was wrongly interpreted by just about everyone as a feat of ballooning, instead of as kinetic art. She's not happy about it, wishing that people would think of her as an artist and not a balloonist." Simons's personal correspondence in the National Air and Space Museum archives suggests that this difficulty with recognition was in part due to her status as a woman. I suspect that the frustration was additionally due to Simons's privileging of artistic rather than competitive goals in her projects, even if she could not help but break records in the process. She did not want to be recognized for the records as much as for her artistic capacities.

22 Vera Simons to Virginia Gunter, June 23, 1969, box 1, folder Kepes MIT, Vera Simons Papers.

23 Questionnaire filled out by Simons, 1992, Vera Simons archive at Center for Advanced Visual Studies at MIT.

24 Simons in David Saltman, "Kinetic: Vera Simons & Her Sculpture in Scotland," *Washington Post*, August 4, 1986, B7.

25 Hans Blumenberg, *The Genesis of the Copernican World* (Cambridge, MA: MIT Press, 1986), as summarized in Benjamin Lazier, "Earthrise; or, The Globalization of the World Picture," *American Historical Review* 116, no. 3 (2011): 602–630.

26 Contemporary Arts Museum, *In Our Time*, 42.

27 Contemporary Arts Museum, "Works included in the Piero Manzoni Exhibition," box 26, folder 31, Contemporary Arts Museum of Houston records, Woodson Research Center, Rice University, Houston, TX.

28 Piero Manzoni, *Manzoni* (Milan: Electa, 2007), 242. See also Sharp, *Air-Art*, 8.

29 For example, attempts to alter the human mechanism via internal chemical interventions and external protective suits meant to accommodate the body during spaceflight led Manfred E. Clynes and Nathan S. Kline to famously arrive at the term cyborg to define this extended human-machine system in 1960. Manfred E. Clynes and Nathan S. Kline, "Cyborgs and Space," *Astronautics*, September 26–27, 1960, 74–76. See also Nicholas De Monchaux, *Spacesuit: Fashioning Apollo* (Cambridge, MA: MIT Press, 2011), esp. chapter 6, "Cyborg."

30 Contemporary Arts Museum, *In Our Time*.

31 Whitney Moon, "Environmental Wind-Baggery," E-Flux, August 2018, https://www.e-flux.com/architecture/structural-instability/208703/environmental-wind-baggery/.

32 Moon, "Environmental Wind-Baggery."

33 Margot Lystra, "McHarg's Entropy, Halprin's Chance: Representations of Cybernetic Change in 1960s Landscape Architecture," *Studies in the History of Gardens & Designed Landscapes* 34, no. 1 (2014): 71–84.

34 Cary Wolfe, "Lose the Building," in *What Is Posthumanism?* (Minneapolis: University of Minnesota Press, 2009), 203–238.

35 Curtis Schreier, interview with author, November 2019.

36 Eva Díaz, *The Experimenters: Chance and Design at Black Mountain College* (Chicago: University of Chicago Press, 2015). See also Lystra, "McHarg's Entropy, Halprin's Chance."

37 Ant Farm and Chip Lord, *Automerica: A Trip Down U.S. Highways from World War II to the Future* (New York: E. P. Dutton & Co., 1976). Lord credits the writings of media theorist Marshall McLuhan in articulating these observed changes.

38 Ant Farm, *No. 853 Allegorical Time Warp*, Ant Farm Archives, Berkeley Art Museum and Pacific Film Archive. A similar point, related to this quote, appears in Felicity D. Scott, *Living Archive 7: Ant Farm: Allegorical Time Warp: The Media Fallout of July 21, 1969* (Barcelona: Actar, 2008).

39 Pamela M. Lee, *Chronophobia: On Time in the Art of the 1960's* (Cambridge, MA: MIT Press, 2004), xix–xxii.

40 Scott, *Living Archive 7*, 22.

41 Sharp, *Air Art*, quoted in Glusberg, "air art," 42.

42 Scott, *Living Archive 7*, 41.

43 As mentioned in Ant Farm, "Time Slice Official Notice: Score First Day June 3," folder Time Slice, Architecture & Art Special Collections, William R. Jenkins Architecture, Design and Art Library, University of Houston, Houston, TX.

44 Speakers included architects Cedric Price, Forrest Wilson, and Dennis Crompton (from the architectural collective Archigram). Ant Farm, untitled document, folder Time Slice, Architecture & Art Special Collections, William R. Jenkins Architecture, Design and Art Library, University of Houston, Houston, TX.

45 Ant Farm Data, 1970, folder Truckstop Network Proposals, Art Collection, Berkeley Art Museum and Pacific Film Archive.

46 Patrick McCray, *The Visioneers: How a Group of Elite Scientists Pursued Space Colonies, Nanotechnologies, and a Limitless Future* (Princeton, NJ: Princeton University Press, 2013).

47 Lee, *Chronophobia*.

48 Ant Farm, "Some Notes on Ant Farm and Performance," n.d., box 12, folder 1, Doug Michels Architectural Papers, Special Collections, University of Houston Libraries, Houston, TX (hereafter Doug Michels Architectural Papers). See also Constance Lewallen and Steve Seid, *Ant Farm, 1968–1978.* (Berkeley: University of California Press, Berkeley Art Museum, and Pacific Film Archive, 2004), 45.

49 For Doug Michels's research at Ford, see 20/20 Vision booklet, box 12, folder 14, Doug Michels Architectural Papers. For the NEA's funding of Ant Farm, see D. L. Prince to Director of Public Relations, Texas A&M University, October 3, 1973, CAMH # 87.39. box 28, folder 10, "20/20 Vision (extra copies correspondence)," Contemporary Art Museum of Houston Records, Woodson Research Center, Rice University, Houston, TX.

50 Lewallen and Seid, *Ant Farm*, 65

51 Thomas H. Garver, "Can Man Control Technology's Domination of Nature?" box 12, folder 14, 20/20 Vision Catalog, Doug Michels Architectural Papers.

52 20/20 Vision press clippings, box 12, folder 14, 20/20 Vision Catalog, Doug Michels Architectural Papers.

53 Anne Collins Goodyear, "The Relationship of Art to Science and Technology in the United States,1957–1971: Five Case Studies" (PhD diss., University of Texas, Austin, 2002).

Conclusion

STEPHEN P. WARING

In the 1980s, the Center for the Study of Southern Culture at the University of Mississippi organized *The Encyclopedia of Southern Culture*, a sprawling 1,634-page compendium of scholarly essays of the American South in fact and symbol. Published in 1989, its diverse contributors included historians, novelists, social scientists, theologians, lawyers, and journalists. The editors clustered pieces such as "Azaleas" and "Moon Pies" under broad topics such as Art and Architecture, Black Life, History and Manners, Music, and Violence. "NASA" had six entries, none of which focused on nationally known and important projects or administrators. Instead, references to the agency appeared in entries on a southern painter, technical education, Georgia Institute of Technology, the Piney Woods, the military and the economy, and the city of Houston. These references testified not only to the diversity of the encyclopedia but also to the manifold relationships of NASA to southern life and culture. Such copious and complex relationships of agency and region create openings for further investigation.

To assist this possible work, this essay will explore three overlapping conceptual classifications. Since the proposed research would explore regional topics, these categories follow approaches to the history of the South rather than approaches to NASA studies. Admittedly the classifications overlap, and obviously these classifications do not intend to end conversations but to foster more. The first classification is the South as a geographic region where national and international events occur. The second is where southern culture and NASA organizational culture interact or are said to interact. The last category is how NASA fits into the diversity of the region as one aspect of the many Souths.

Some topics consist of events that occur in the South as a geographic region. They are similar to, even part of, events that are national or even international in character. For these types of events, no claim is made that the South has exceptional cultural or political patterns. Studying aspects of NASA's impact on the South creates opportunities to examine wider patterns. For example, from the middle of the twentieth century, the federal government made a big footprint on the nation. The government's strong presence in the South has been particularly transformative. This makes the region a test case for diverse impacts of different types of spending. The Tennessee Valley Authority, the military, and NASA affected the region's landscape, demography, education, and social-economic development. How were those impacts similar and different? What were the impacts of a NASA installation compared to a military one? How were research and development programs different from infrastructure programs? Along the same lines, when NASA constructed new facilities, did it pioneer or merely reflect new trends? When it built new office buildings for professional workers, did it typically choose architectural designs in the International Style and Corporate Modernism? Was it a trendsetter in bringing architectural change?

Another fruitful direction for research is the study of the intersections of NASA culture and regional culture. Here the research task is to investigate the mentalités and social patterns in and outside the agency. The aim is to move historical work beyond aerospace and scientific projects, events, and trends; instead, the goal is to add the description and analysis of the ways people in particular times and places thought about, interacted with, and classified the world around them. Here the studies of NASA and civil rights provide direction and have shown how ideas about race, class, and gender have affected education, employment, and politics in agency centers and surrounding communities. Many more opportunities exist. Historians have long studied conceptions of honor and white masculinity in the South. How might NASA's professional culture among managers, engineers and astronauts replicated those traditional ideas—and conflicted with them? Stereotypes of the South are common in American society and the installations associated with human space flight are in the region. How have such stereotypes about regional backwardness or grift affected media and political conversations, and decisions, about human space flight?

A similar way of exploring the interaction of NASA and the southern region is to use a concept of "many Souths." Accepting that the South as a region is as diverse as most, seen especially in its rural and urban divides with traditional evangelical Christianity and modern universities and professions. Historians would look for ways the agency's organizational culture relates to regional ones. NASA has many southern employees and contractors who walk in the worlds of both science and religion. How have they balanced their work and private beliefs? Another topic for study would be how NASA's institutional mandate for expertise and science has interacted with the South's religious identities and corporate libertarianism. A way to investigate this contrast would be the issue of global warming and the human role in climate change. NASA has studied human-induced climate change and presented it as a scientific fact. In contrast, a dominant conservative culture of white southerners has emphasized benevolent divine harmony in nature, which precludes human ability to change God's creation. Many of the region's members of Congress have resisted the implications of climate science and even questioned whether climate change exists. Ironically, many of these same congressional officeholders supported NASA funding. All Americans, of course, and not just southerners, experienced the same misleading messages that created doubt in climate science. However, several southern states hosted NASA installations.

In conclusion, the study of NASA is open to many approaches. Regional studies are only one approach, and they too can lead in many directions. Nor should regional studies of NASA be restricted to discussions of the South. Historians may also find it fruitful to study the agency's relationship to the Northeast and the West. Future studies might also consider international and transnational contexts. Hopefully this volume will stimulate new research across new boundaries.

CONTRIBUTORS

Drew Adan is an archivist at the M. Louis Salmon Library of the University of Alabama in Huntsville. Before coming to UAH, Drew worked for ten years in the Yale University library system at the Lillian Goldman Law Library and the Beinecke Rare Book & Manuscript Library. Drew holds an MLIS and is currently pursuing a master's degree in history.

Katarzyna Balug is a historian of modern and contemporary architecture and assistant professor at Florida International University. Her research explores the feedback loops between techno-political change, subjectivity, and architectural experiments. Balug's current book project examines inflatable structures produced in the context of the 1960s environmental movement and the space race. Through curatorial work and public art practice, she endeavors to cultivate public space as a site of critical civic engagement. Balug's writing been published in, among others, *Geoforum*, *Critical Sociology* (co-authored) and *New Geographies 11: Extraterrestrial*. She holds a PhD in architecture from Harvard University.

Douglas Brinkley is the Katherine Tsanoff Brown Chair in Humanities and professor of history at Rice University, the CNN presidential historian, and a contributing editor at *Vanity Fair*. He works in many capacities in the world of public history, including on boards and with museums, colleges, and historical societies. His book *Cronkite* (2012) won the Sperber Prize and *The Great Deluge: Hurricane Katrina, New Orleans and the Mississippi Gulf Coast* (2006) received the Robert F. Kennedy Book Award. *The Nixon Tapes* (2015–2016), which he co-edited with Luke Nichter, won the Arthur S. Link–Warren F. Kuehl Prize.

Max Campbell is reference archivist for the Winthrop Group who is currently working as a contractor at the IBM Corporate Archives in Poughkeepsie, New York. Prior to working at the Winthrop Group and IBM, Max was a museum technician at Independence National Historic Park and spent two summers interning for Allan Needell in the Space History Department of the National Air and Space Museum. Max holds BA and MA degrees in history from Purdue University, where he worked for three years in the Barron Hilton Flight and Space Exploration Archives.

Andrew J. Dunar is professor of history emeritus at the University of Alabama in Huntsville. He is coauthor with Stephen Waring of *Power to Explore: A History of NASA's Marshall Space Flight Center, 1960–1990*, which won the American Institute of Aeronautics and Astronautics History Manuscript Award in 2001. He has served as editor of the *Oral History Review* (1999–2006) and is coauthor with Dennis McBride of *Building Hoover Dam: An Oral History of the Great Depression* (1993).

Kari Edwards is a PhD candidate in history at the University of Mississippi. She holds a BA in Religious Studies from the University of Tennessee at Chattanooga and an MA in Southern Studies from the University of Mississippi. Her primary area of research is American religion in the twentieth century, focusing on Christian fundamentalism and conflicts between religion and science. She is currently working on her dissertation, which explores the religious history of the Space Race in the United States.

Arslan Jumaniyazov is a graduate from Purdue University. He has a PhD in American studies with a specialization in history. His research interests include the study of US history from a global and transnational perspectives. Lately, Jumaniyazov has been focusing on comparative studies of center-driven development policies the United States and the Soviet Union pursued during the Cold War.

Rachael Kirschenmann is a graduate of the University of Maryland, where she earned an MA in history and an MLIS. She currently works as a digital asset management contractor at the National Air and Space Museum. Her professional interests as an archivist include preserving collections related to the history of science and technology, in particular the records of communities that interact with scientific innovation.

Roger D. Launius is principal of Launius Historical Services. From 2002 to 2017 he worked as a senior official at the National Air and Space Museum in Washington, DC. From 1990 to 2002 he served as chief historian of the National Aeronautics and Space Administration. Launius is the author, most recently, of *The Smithsonian History of Space Exploration: From the Ancient World to the Extraterrestrial Future* (2018); *Apollo's Legacy: The Space Race in Perspective* (2019); and *Reaching for the Moon: A Short History of Space Race* (2019).

Emily A. Margolis is curator of contemporary spaceflight at the Smithsonian's National Air and Space Museum, where she is also responsible for the Mercury and Gemini collections. She previously served as curator of American women's history, a joint appointment with the Smithsonian Astrophysical Observatory.

Margolis earned a PhD in the history of science and technology from Johns Hopkins University in 2019. She is writing a history of space tourism from the 1950s through 2021.

Jeffrey S. Nesbit is an architect, urbanist, and assistant professor at Temple University. His research focuses on urban theory, infrastructural urbanization, and "technical lands." He recently completed a doctor of design degree. Nesbit is the author of several journal articles and book chapters. He is the co-editor (with Charles Waldheim) of *Technical Lands: A Critical Primer* (2022), the co-editor (with Guy Trangoš) of *New Geographies 11: Extraterrestrial* (2020), and the editor of *Nature of Enclosure* (2022). He is currently writing a monograph on the American spaceport complex at the intersection of architecture, infrastructure, and aerospace history.

Brian C. Odom is chief historian at the National Aeronautics and Space Administration, where leads the program responsible for capturing, preserving, and disseminating the agency's history. Odom is co-editor (with Stephen Waring) of *NASA and the Long Civil Rights Movement* (2019), which was awarded the American Astronautical Society's Eugene M. Emme Astronautical Literature Award. Before his current position, he was the historian at NASA's Marshall Space Flight Center.

Jennifer Ross-Nazzal is the National Aeronautics and Space Administration's Human Spaceflight historian. Ross-Nazzal is the author of numerous articles about the space program and the astronauts and of *Winning the West for Women: The Life of Suffragist Emma Smith DeVoe* (2011). Her latest work, *Making Space for Women: Stories from Trailblazing Women of NASA's Johnson Space Center*, focuses on the history of the Johnson Space Center through the experiences of its female employees.

Stuart Simms is a doctoral student at Auburn University. His research focuses on the nexus of power between the federal government and local communities during the Cold War transformation of the US South with a focus on environmental approaches and local voices. His hope is to generate a more holistic examination of sites of dispossession throughout the US South.

Caroline T. Swope is historic preservation specialist for the city of Decatur, Alabama. Trained as an architectural historian, she has worked in cultural resource management for more than thirty years. She has consulted on a variety of preservation projects for private, governmental, and nonprofit organizations, including Native American sites, barns, bridges, churches, schools, and private residences. Her work has included federal tax credits and HABS/HAER documentation for

Historic Structure Reports and National Register nominations. She was a major contributor to the Society of Architectural Historians' Archipedia on Alabama. She holds an MS in historic preservation from Ball State University and a doctorate in art and architectural history from the University of Washington, Seattle.

Stephen P. Waring is professor of history and department chair at the University of Alabama in Huntsville. Waring has degrees from Doane College in Crete, Nebraska, and the University of Iowa. His research specialty is management and technology in contemporary business and government. His first book, *Taylorism Transformed: Scientific Management Theory since 1945* (2016) examined such topics as worker participation and managerial use of mathematics. His book with Andrew Dunar, *Power to Explore: A History of NASA's Marshall Space Flight Center, 1960–1990*, won the American Institute of Aeronautics and Astronautics History Manuscript Award in 2001.

INDEX

Printed in the United States
by Baker & Taylor Publisher Services